Public Administration and Information Technology

Volume 27

Series editor

Christopher G. Reddick, San Antonio, TX, USA

More information about this series at http://www.springer.com/series/10796

Rowena Cullen · Graham Hassall
Editors

Achieving Sustainable E-Government in Pacific Island States

Springer

Editors
Rowena Cullen
School of Government
Victoria University of Wellington
Wellington
New Zealand

Graham Hassall
School of Government
Victoria University of Wellington
Wellington
New Zealand

Public Administration and Information Technology
ISBN 978-3-319-50970-9 ISBN 978-3-319-50972-3 (eBook)
DOI 10.1007/978-3-319-50972-3

Library of Congress Control Number: 2016960724

This Springer imprint is published by the registered company Springer Nature Switzerland AG
The registered company address is: Gewerbestrasse 11, 6330 Cham, Switzerland

Foreword

I am delighted to be invited to write a Foreword to this very timely volume which is about a subject of considerable importance to the small island states of the Pacific. E-government, the application of information and communications technologies in the public sector in Pacific Island countries, has the potential to make a major contribution to their developing economies and to link our small countries to the global knowledge economy. As the Pacific Regional Digital Strategy of 2010 noted, information and communications technologies (ICTs) are 'universally acknowledged as powerful tools for development essential to social development and economic growth.' They are also 'critical to the development of good governance [and] vital for sustainable development.'

In the Pacific, we face many challenges including lack of economy of scale, lack of resources, tyranny of distance, lack of capacity and expertise, poor connectivity, expensive and centralized services, to name a few. This book demonstrates that ICT is a key enabling and transformational tool essential to addressing these challenges. Due to the lack of resources, there are competing interests when allocating these. Do we use resources to build roads, climate proof schools or promote e-government? For this reason, the knowledge and insights presented in the volume are important to us as Pacific leaders. The book gives us an overview of what has been achieved to date, and reminds us of the important role of leadership and good public policy in bringing technological change into our countries, in seeing policy through to implementation, and ensuring that such change is sustainable, and leads to good governance. So it is valuable to read of some of the exciting initiatives that have been taking place in both our regional organizations, and in individual countries and the real gains that have come from them. It is even more valuable to have them all brought together in one place, and to be able to get such a comprehensive overview of e-government in Pacific Island countries and the benefits it brings. In fact this book localizes e-government, making it more e-government for SIDs. In addition, the book offers some valuable analysis of what works and doesn't work, and what makes for sustainable change, applying relevant theoretical approaches to help us see the bigger picture.

Most Pacific countries consist of widely dispersed islands that are poorly served by airline and shipping services. The poor transportation services are further compounded by lack of connectivity in these outer islands. The connectivity options which are discussed in the book, especially the provision of services using mobile phones, afford the most promising solution to mitigate the lack of accessible services in these remote communities. Government services that can be provided on a mobile phone, mobile and online banking promoting financial inclusion that can enable government officials and customers to manage their financial affairs online, online tutorials that can supplement the lack of teachers, e-health systems that can improve health services, and early warning systems that can provide life saving information; are all part of a range of e-government services that can be provided to the comfort of your beautiful Fale in your remote island.

But these services can only be provided through better and more affordable connectivity, both internationally and domestically. Since 2010 we have made great progress. We are building infrastructure and capacity, and learning how to make the most of these new technologies for growth and prosperity. Despite the small size of Pacific countries, collaborations with development partners, reductions in technology costs and increased demand for capacity have made it possible to improve levels of connectivity through the use of fibre optic cables and satellites to reduce the isolation of remote communities. Who would have believed that countries with about 100,000 people could justify connecting up using these technologies. But more than 70% of Pacific SIDs are either connected or in the process of connecting up via fibre optic cables. With increased affordability, communication services are more accessible and are making the lives of our people better and providing access to global knowledge and global partnerships. E-government applications, such as e-procurement, e-business registration, and information portals, are starting to transform our governments, making them more efficient, relevant and accountable to citizens. The use of new communications technologies such as the world wide web, mobile technology and social media is enabling Pacific governments to reach more citizens and helping them become more informed and aware about the roles and responsibilities of government and citizens.

At the same time I am very proud of the achievements of our regional organizations in leading e-government in the region—the Pacific Islands Forum, the University of the South Pacific, the establishment of Japan Pacific ICT Centre at USP and development of new ICT programmes, and the Pacific Community (SPC). The role of the Forum in setting regional policy and assisting member countries to develop good cross-cutting policy and practice, and the role of the SPC with its focus on scientific and technical expertise, its regional databases and support for national e-government initiatives are rightfully acknowledged by the authors. Indeed, the fact that Pacific Leaders in 2015 identified ICT as a regional priority is a testimony to the regional organizations' recognition and advocacy of the key role of ICT in development. From my personal experience in ICT in both Tonga and at the SPC, I have learned what it takes to ensure that projects succeed and are sustainable. E-government is a critical issue for us and we need to exploit the opportunities it presents but be mindful that we do not have the resources to become a testing ground for ideas or ambitious agendas. We need national champions to drive its implementation but we still need to build carefully, reflecting on what works within our own cultures and meets our particular needs, selecting carefully from the opportunities that present themselves, building capacity and learning from our own as well as others' experience.

The editors, Graham Hassall and Rowena Cullen, have done an excellent job in ensuring that so many perspectives on e-government have been covered, from global and regional policy, to infrastructural and regulatory issues, to key government sectors that have been transformed by ICTs. The book's emphasis on having the right legislative and policy framework and the importance of educating the lawmakers and policy makers is crucial as Pacific leaders can't, or shouldn't, regulate what they don't understand. The authors have integrated their extensive knowledge and experience of governance, their insight and knowledge of the region, to ensure that issues are considered within the Pacific context. The book also identifies many of the factors that that are important for successful and sustainable e-government, and the importance of local ownership. There is no silver bullet to a successful e-government programme. Each Pacific country should identify where it is in developing e-government, focus on its priorities, then develop its own e-government roadmap that can include legislative review to enable e-government services, establishing an independent regulator, ICT capacity building, defining data architecture, improving international and local connectivity, using ICT for climate change and disaster risk reduction, or implementing e-procurement.

I hope that politicians, policy makers, administrators, technical advisors and development partners will all read this book; I am sure all will learn something of value.

Let's celebrate these successes while we take note of the challenges we face as small isolated island nations, and the magnitude of the task that lies ahead. We are ready for it, and we embrace it—we have come a long way, but ahead lies an even more exciting e-Pacific Island country journey.

Malo 'aupito

Hon. Siaosi 'Ofakivahafolau Sovaleni
Deputy Prime Minister of Tonga and Minister for Meteorological
Services, Energy, Information, Disaster Risk Management
Environment, Climate Change, and Communication (MEIDEC)

Acknowledgements

We would like to express our sincere gratitude to the many individuals and organizations who have contributed to this research in some way.

We are grateful for research funding support from the Victoria University of Wellington University Research Fund, the Victoria University School of Government, and Digicel Papua New Guinea. Thanks also are due to our research assistants, Poto Aiafi and Bertrand Guyander who worked on the project at various times, to Megan Clark of the Philson Medical Library at the University of Auckland, and especially to our editor Vic Lipski, whose careful work and expertise have contributed in great measure to the quality and consistency of the text.

We would like to express our gratitude to all our colleagues at Victoria University and elsewhere in New Zealand, Australia and the Pacific for reviewing chapters: Arthur Jorari, Ioana Chan Mow, Keith Davidson, Emele Duituturaga, Vince Galvin, Gerald Haberkorn, Lenore Hamilton, Maureen Hilyard, Karl Lofgren, Frank March, Jessica Noske Turner, Geoff Rashbrooke, Simone Sala, Ellen Strickland, Salanieta Tamanikaiwaimaro, Janet Toland, Amanda H. A. Watson, and Ruby Va'a.

We would also like to thank our co-authors who have brought their considerable experience and knowledge of the Pacific Islands to this project, and Siaosi Sovaleni for his Foreword.

Finally, we would like to thank the large number of people who gave generously of their time and insights in our many visits to the Pacific Islands, and who understood the importance of getting this story told. Warm Pacific greetings and thanks to:

Charles Ah Poe, Tuiamalo Ah Sam, Taniela Aiafi, Ronnie Aiolupotea, Cheryl Ala, N. Sadiq Ali, Iluminado Aloaina, Ronald Amigo, Winifred Kula Amini, Anthea Arnhambat, Howard Aru, Paul Asitewa, Audrey Aumua, Josephine Baig, Paul Barker, Jacquie Berrell, Ron Box, David Braun, Derek Brian, Manu Bobola, Alifereti Bulivou, Brett Collett, Marion Crawshaw, Douglas Creevey, Donnie De Freitas, Phoebe Dengate Thrush, Simon Donald, Nial Downey, Bredina Drollet, Lynlee Earles, Sam Era, Mana Etches, Richard Falvey, Peter Forau, Derek French,

Tau Geno-Hoire, Hon Nandi Glassie, Mary Harris, Kelly Haines-Sutherland, Lenore Hamilton, Hugo Hebala, Douveri Henao, Bernie Hill, Tepua Hunter, Margaret Inifiri, Elizabeth Iro, Kari James, Stuart Jones, Arthur Jorari, Lawrence Kalino, Julius Kane, Kevin Hwasihao, Kamrooz Khademazad, Narith Khamla, Cheryl King, Peniana Lalabalavu, Jeffrey Tila Langata, Tomas Lamanauskas, Tupaimatuna Iulai Lavea, Aiono Fanaafi Le Tagaloa, Julienne Leka-Maliaki, Brian Lenga, Brian Louey-Gung, Marke Lowen, Joe McCarter, Dan McGarry, Darren McLean, Sammie Maeasi, Moffett Maenao, David Main, Papalii Malietau Malietoa, Anju Mangal, Anna Marikawa, Ellasalah Matatier, Roana Mataitini, Kyle Matheson, Robert Matheson, Leasiolagi Malama Meleisea, Dennis Meone, Sina Meredith, Samson Metofa, Jackson Miake, Ane Moananu, Adolfo Montenegro, Emmanuel Narokobi, Khamla Narith, Louise Nasak, Ann Naupa, Richard Neves, Jimmy Nipo, John Nirua, Rodel Nodora, Kora Nou, Ezekiel Nukuro, Roy Obed, Une O'Ome, Hinauri Petana, Charles and Barbara Pierce, Ivan Pomelau, Sascha Piggott, Aaron Pitqaue, Ian Rakafia, Georgina Roberts, Gregory Rofeta, Seyha Ros, Anthony Saaga, Paul Sai'i, Manu Samoa, Fred Samuel, Nicole Sarkis, Sir Bruce and Lady Keithie Saunders, Emma Scadeng, Cheryl Scott, John Secker, Gary Seddon, David Sheppard, Andrew Simpson, Christian Slaven, Rara Soro, Siaosi Sovaleni, Terence Sua, Peter Swain, Epeli Tagi, Taggy Tangimetua, Laeimau Oketevi Tanuvasa-Savea, Len Tarivonda, George Tasso, Ambong Thompson, Ian Thomson, Gibson Tito, Simon Tiwok, Sylvain Todman, Llewellyn Toulman, Ronald Tomausi, Sir Andrew Trawen, William Tuivaga, Beau Tydd, Lasse Vestergaard, Josiah Vira, Ione Viji, Save Vocea, Korai Vurobaravu, Johnson Wabaiat, Jeremy Watson, Thomas Webster, Shadrach Welegtabit, Ashley Wickham, Charles Yala, and Henry Yamo.

Wellington, October 2016

Rowena Cullen
Graham Hassall

Contents

Editors and Contributors

About the Editors

Dr. Rowena Cullen is Emeritus Professor of Information Management in the School of Government at Victoria University of Wellington, New Zealand, and a former Associate Dean for Research in the Victoria Business School. She has been working in the field of e-government for more than 20 years and on e-government in small island developing states for the past 12 years. She has published over 30 articles and book chapters on e-government and e-health and collaborated with the New Zealand State Services Commission on several e-government projects, including the effective use of government web sites for communication with citizens, government's use of online consultation, and citizens' trust in government concerning their personal information.

She is the author of Health Information on the Internet (2005) and was joint editor of Comparative Perspectives on E-Government (2006). She also taught the first university course on e-government in New Zealand, offered as part of the Master of Information Management degree at Victoria in 2006.

Dr. Graham Hassall is Associate Professor in the School of Government at Victoria University of Wellington, New Zealand. He has previously taught at the University of Melbourne, Landegg Academy, the University of the South Pacific, and the University of Papua New Guinea. His research focus has been government in the Pacific Islands, including Pacific regionalism, in the context of globalization. He is currently academic director of PNG's Foreign Service Training Programme and is on the advisory boards of the United Nations Association of New Zealand, the New Zealand Centre for Global Studies, and the editorial boards of the Commonwealth Journal of Local Governance, Comparative Law Journal of the Pacific, and the Australasian Parliamentary Review.

Contributors

Dr. Transform Aqorau is currently Technical and Legal Adviser/Support to the PNA Office, and an Adjunct Visiting Fellow in the School of Government, Development and International Affairs at the University of the South Pacific. He is also a Visiting Fellow at the Australian National Centre for Oceans, Resources and Security at the University of Wollongong, Australia. Prior to his current role he was Deputy Director-General and Legal Counsel of the Pacific Islands Forum Fisheries Agency (FFA), Legal Adviser and Acting Deputy Secretary of the Ministry of Foreign Affairs in Solomon Islands, Legal adviser to the Pacific Islands Forum Secretariat (PIFS), and CEO of the PNA Office. Transform Aqorau is a Ph.D. graduate of the University of Wollongong, Australia, and is author of a number of articles in international journals, and conference papers on fisheries law and governance, and governance and development.

Ioana Chan Mow is currently Associate Professor in Computing and Computer Education at the National University of Samoa. She is currently the Samoa Country Focal Point for the Commonwealth of Learning. She has been the project coordinator for NUS for the NIU Edulink project and the PACENETPlus project. Her research interests are in computer science education, ICT4D and ICT policy frameworks in Samoa, areas in which she has published in both journals and conference proceedings.

Len Cook was Government Statistician of New Zealand from 1992 to 2000, having worked in Statistics New Zealand in a variety of roles since 1971. From 2000 to 2005 he was National Statistician of the United Kingdom and Registrar-General of England and Wales. Len has been Vice-President of the International Statistics Institute (2005-2009), and also was Vice-President of the International Association of Official Statistics. Len Cook worked with Dr. Masasso Paunga to develop a 10-year strategy for official statistics in the Pacific Region for SPC in 2010, and has been the Friend of Chair of the Pacific Statistics Steering Committee since 2013.

Len is an Honorary Life Member of the NZ Statistical Association, Past President of the NZ Institute of Public Administration, and is currently the Chair of the board of Superu, a New Zealand crown agency operating as a "what works" centre. Len's prime interests are in the areas of population change and public policy, public administration, official statistics and the place of science in policy.

Gisa Fuatai Purcell is Regional Advisor for the South Pacific for the Commonwealth Telecommunications Organisation. Gisa was formerly the Head of the ITU's Division on LDCs, SIDS, and LLDCs; Climate Change Adaption and Disaster Risk Reduction; and Emergency Telecommunication and prior to that the Head of the ITU's LSE Department. Before joining the ITU, she was the Assistant Chief Executive Officer for the Ministry of Communication and Information Technology in Samoa after working as an ICT and Financial consultant in Wellington, New Zealand for many years. Gisa has a Master of Commerce and

Administration degree from Victoria University of Wellington, New Zealand with a major in telecommunication and information systems.

Dr. Sarah Logan is an honorary research fellow in the Department of Government and International Relations at the University of Sydney. She was previously the inaugural digital politics research fellow in the State, Society and Governance in Melanesia Program at the Australian National University. Her research interests include the political and social impact of the Internet on emerging democracies, and the impact of the Internet on international politics.

Brian Louey-Gung is a telecommunications technology and regulatory strategist with over 30 years' experience in both industry and government. In 2013 and 2014 he was the Technical Advisor, and later Acting Director of the Pacific ICT Regulatory Resource Centre, based at the Suva campus of the University of the South Pacific, and contributed to the work of the centre in assisting the policy makers of fourteen Pacific Island countries to improve their ICT regulatory regimes to best support their economic development. Prior to that, he held senior executive positions at the Australian Communications and Media Authority, Iridium South Pacific, Link Telecommunications, and Telstra. He has a Master of Science and Technology Commercialisation from the University of Adelaide, a Bachelor of Science (Honours) from the University of Melbourne, and a Graduate Diploma of Information Technology from RMIT University.

Anju Mangal is a Knowledge Management Specialist who works for the Pacific Community (SPC) as a trainer in the use of ICT, knowledge management, social media and e-learning platforms in the Pacific. She holds a Master of Arts degree in Governance, and a Bachelor of Science in Information Systems and Geography (GIS). Anju was a tutor for DiploFoundation's ICT Strategy Course, and is an active member of DiploFoundation's research and teaching faculty. She is an alumna of the Internet Governance Capacity Building Program in 2009 and was an ISOC Ambassador to the IGF in 2008 in Hyderabad, India, and a Commonwealth IGF fellow and member of the UN-Internet Governance Forum and Secretariat team in Sharm El-Sheikh. She has served as a fellow at the UN IGF Secretariat in Geneva, Switzerland, as a board member of the Pacific Islands Chapter of the Internet Society (PICISOC), and was the founder of the PICISOC Women and ICT group. Anju coordinates and develops innovative ideas on ICT and knowledge management for development and advocates on Internet governance issues in the Pacific.

Katalina Pasiale Taloka is currently the Director of Education for the Ministry of Education in Tuvalu. She is also the Country Focal Point for Tuvalu for the Commonwealth of Learning. Katalina has been an active participant in all education forums in the Pacific region and has contributed to most of the educational initiatives within Tuvalu and also within the Pacific.

Dr. Netatua Pelesikoti is Director of the Climate Change Division at the Secretariat of the Pacific Environment Programme (SPREP). She is a coastal ecologist specializing in coastal systems monitoring (Coral Reef, Sea Grass and Marine Water Quality), and was a lead author for Chapter 29 (Small Island Chapter) of the 5th Assessment Report for the Intergovernmental Panel on Climate Change (IPCC). Dr. Pelesikoti, who is from Tonga, has more than 20 years of national experience in environmental impact assessment, monitoring and policy development, and project management in various aspects of climate change.

Dr. Joseph Kim Suwamaru earned his Ph.D. for his research on aspects of mobile phone usage in socio-economic development in Papua New Guinea (PNG). Prior to completing his Ph.D. he was the Executive Director of the Engineering Department within the former ICT regulator in Papua New Guinea, PANGTEL. He also served as Vice Chairman of the Asia Pacific Telecommunity Study Groups. On completing his Ph.D. Dr. Suwamaru became a senior lecturer within the Department of Information Systems at the Divine Word University in Madang, PNG. During 2016 Joseph was a member of the board of directors of a new state-owned enterprise in PNG, DATACO, which is tasked with rolling out terrestrial and undersea submarine fibre optic cables across PNG and Melanesia. He is currently working as an independent consultant.

Ian Thomson is currently working at the University of the South Pacific as a Senior Fellow in e-Learning. In this role, he jointly manages the Teachers Education Resources and e-Learning Centre and focuses on helping Pacific ministries of education develop strategies, policies, capacity building plans and trails of ICTs to help improve education outcomes.

Ian has 10 years' experience working in the region, starting at SPC as the inaugural ICT Outreach coordinator. He helped develop and review the regional ICT and ICT for Education Frameworks and has led many initiatives, including the OLPC and Rural Satellite connection projects and more recently has been involved in OER and MOOC developments in the region. He is currently working with five Pacific ministries of education and developing regional approaches that will assist many more. His engineering background and Pacific experience help him develop ICT solutions appropriate for Pacific education systems based on his research into the latest international developments.

Dr. Emma Kruse Vaai is Professor of English and Applied Linguistics in the Faculty of Education at the National University of Samoa where she has held both senior academic and management positions. She completed postgraduate studies at the University of New South Wales and Victoria University Wellington. She is also a graduate of the Kennedy School of Government Executive Programme for Leaders in Development and Management of Higher Education Institutes Programme, Galilee International Management Institute in Israel. She was a recipient of the NZ Prime Minister's award for Emerging Women Leaders from the Pacific; and was also the 2014 Ian Gordon Fellow in Linguistics at Victoria University Wellington.

Emma Kruse Vaai represents the Pacific Region as a member of the Board of Governors and Executive Committee of the Commonwealth of Learning (COL) and was the first Chair of the Virtual University of the Small States of the Commonwealth (VUSSC). She has published widely in the fields of language, politics, education and is also well known as a Pacific writer.

Peter Walton is an agricultural information specialist and a librarian by profession. Peter's work has primarily been in the Pacific Islands region, where he has worked on a range of information, communication and knowledge management activities in the agricultural sector for the past 31 years. Peter has been involved in the development of agriculture libraries, information centres, information networks and systems, scientific, technical and extension publications, training materials, products and services. He is the designer of the Melanesian Agricultural Information System (MAIS), the co-author of CTA's Facilitator's Guide to Agricultural Information and Communication Management Strategy Development, and the last elected President of the International Association of Agricultural Information Specialists (IAALD).

Dr. Amanda H. A. Watson is a Visiting Fellow with the State, Society and Governance in Melanesia Program, Coral Bell School of Asia Pacific Affairs at the Australian National University. She completed her Ph.D. at Queensland University of Technology on the responses of rural villagers to the early spread of mobile phone networks throughout Papua New Guinea. Dr. Watson conducted research on the strategic use of mobile phones in rural service delivery in her role as Mobile Communications Research Consultant with the Economic and Public Sector Program in Papua New Guinea from 2012 to 2016. Her current research focuses on strategic use of mobile phones across a range of sectors, including health, education, justice and governance. Dr. Watson's research has appeared in a range of outlets, including the following journals: Mobile Media and Communication, Journal of Personalized Medicine, Australasian Psychiatry, Pacific Journalism Review, Media Asia, and Australian Journalism Review. For more details, please visit http://www.ahawatson.com.

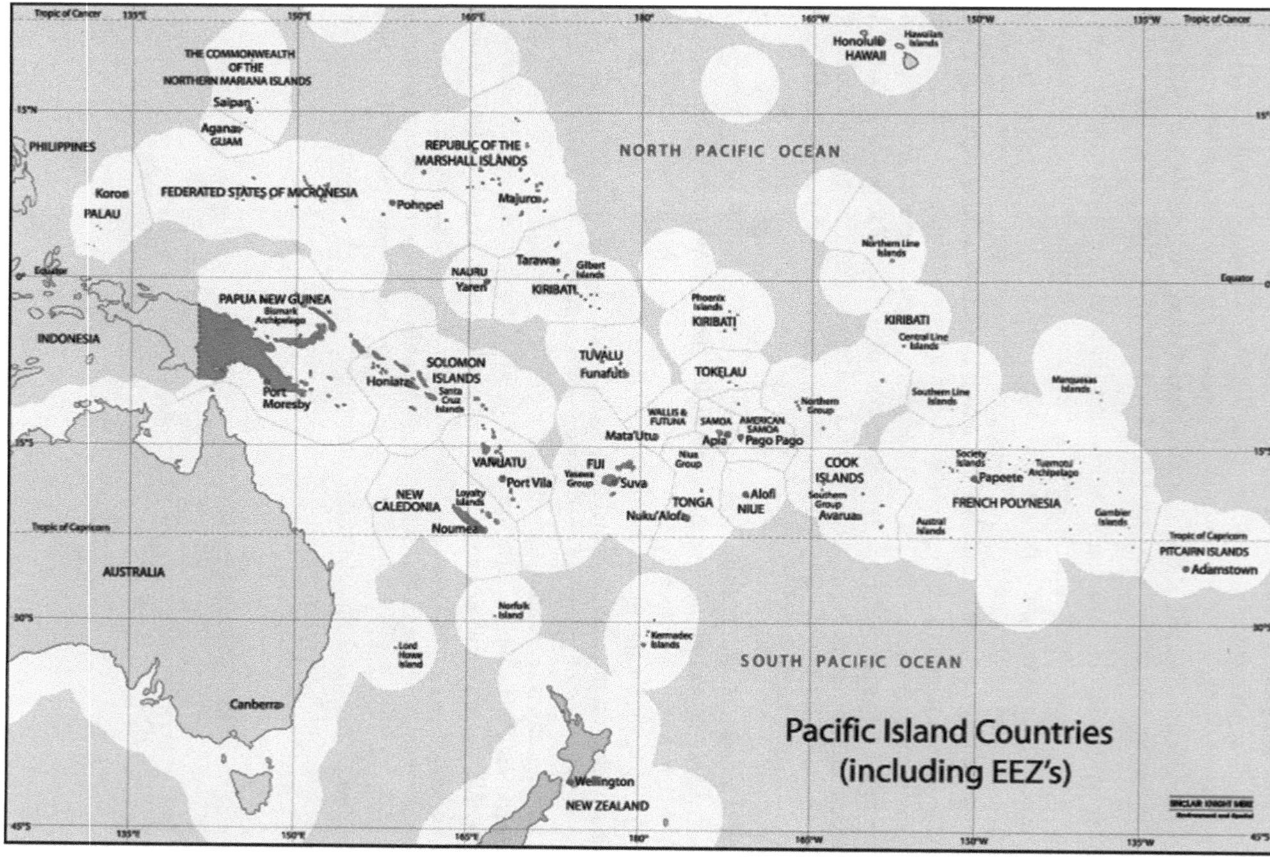

Pacific Island Countries
(including EEZ's)
NORTH PACIFIC OCEAN
SOUTH PACIFIC OCEAN
THE COMMONWEALTH OF THE NORTHERN MARIANA ISLANDS
Saipan
Agana
GUAM
PHILIPPINES
Koror
PALAU
FEDERATED STATES OF MICRONESIA
Pohnpei
REPUBLIC OF THE MARSHALL ISLANDS
Majuro
Tarawa
Gilbert Islands
NAURU
Yaren
KIRIBATI
INDONESIA
PAPUA NEW GUINEA
Bismark Archipelago
Port Moresby
SOLOMON ISLANDS
Honiara
Santa Cruz Islands
TUVALU
Funafuti
Phoenix Islands
KIRIBATI
TOKELAU
Honolulu
HAWAII
Hawaiian Islands
Northern Line Islands
KIRIBATI
Central Line Islands
Southern Line Islands
Marquesas Islands
Northern Group
WALLIS & FUTUNA
SAMOA
AMERICAN SAMOA
Mata'Utu
Apia
Pago Pago
Niua Group
VANUATU
Port Vila
FIJI
Yasawa Group
Suva
NEW CALEDONIA
Loyalty Islands
Noumea
TONGA
Nuku'Alofa
Alofi
NIUE
COOK ISLANDS
Southern Group
Avarua
Society Islands
Papeete
Tuamotu Archipelago
FRENCH POLYNESIA
Austral Islands
Gambier Islands
Tropic of Capricorn
PITCAIRN ISLANDS
Adamstown
AUSTRALIA
Canberra
Norfolk Island
Lord Howe Island
Kermadec Islands
Wellington
NEW ZEALAND
Tropic of Cancer
Equator
135°E
150°E
165°E
180°
165°W
150°W
135°W
15°N
0°
15°S
30°S
45°S

Acronyms

ADB	Asian Development Bank
CC	Climate change
CCA	Climate change adaptation
COP	Conference of the Parties
CROP	Council of Regional Organizations of the Pacific
DM	Disaster management
DM	Disaster risk management
DRR	Disaster risk reduction
ESCAP	Economic and Social Commission for Asia and the Pacific (ch.1)
FAIDP	Framework for Action on ICT for Development in the Pacific
FAO	Food and Agriculture Organization
FEMM	Forum Economic Ministers Meeting
FFA	Forum Fisheries Association
GDP	Gross domestic product
ICT	Information and Communication Technologies
ICT4D	Information and Communication Technologies for Development
ISP	Institutional Strengthening Project
ITU	International Telecommunication Union
M4D	Mobile phones in Development
MDGs	Millennium development goals
MfR	Ministry for Revenue (Samoa)
NDMO	National Disaster Managers Office
OECD	Organization for Co-operation and Development
PCCR	Pacific Climate Change Roundtable
PCRAI	Pacific Catastrophe Risk Assessment and Financing Initiative
PFTAC	Pacific Financial Technical Assistance Centre
PIANGO	Pacific Islands Association of Non-Governmental Organisations
PIC	Pacific Island country
PICISOC	Pacific Internet Society
PICTs	Pacific Island countries and territories

PIFACC	Pacific Islands Framework for Action on Climate Change
PIFS	Pacific Islands Forum Secretariat
PITA	Pacific Islands Telecommunications Association
PNG	Papua New Guinea
SDGs	Sustainable development goals
SIDs	Small Island Developing States
SOPAC	South Pacific Applied Geoscience Commission, In 2011 it was replaced by the Geoscience Division of the SPC
SPBEA	South Pacific Board of Educational Assessment. Formerly a CROP agency, later integrated into SPC
SPC	Secretariat of the Pacific Community (SPC)
SPREP	Secretariat of the Pacific Regional Environment Programme
SPTO	South Pacific Tourism Organization
SRDP	Strategy for Disaster and Climate Resilient Development in the Pacific
UNDP	United Nations Development Programme
UNCTAD	United Nations Conference on Trade and Development
UN DESA	United Nations Department of Economic and Social Affairs
UNESCO	Education and Scientific and Cultural Organization
UNFCCC	United Nations Framework Convention on Climate Change
UNISDR	United Nations Office for Disaster Risk Reduction
UNPAN	United Nations Public Administration Network
USP	University of the South Pacific
WHO	World Health Organization
WSIS	World Summit for the Information Society

List of Figures

List of Tables

List of Boxes

Part I
E-Government in Pacific Island States

Chapter 1
E-Government in Pacific Island Countries

Rowena Cullen and Graham Hassall

Abstract This chapter explains the purpose of the book, which is to examine the state of e-government in the Pacific Islands, and how information and communication technology is changing the practice of government in the region. It outlines the structure and scope of the book, which covers the Pacific Island countries and territories which are members of the Pacific Community, and provides some background for later chapters by describing the dominant cultural groups (Melanesian, Micronesian and Polynesian) and the role of traditional forms of governance in the Pacific. It also describes the characteristics of Pacific small island developing states and how their geographical and economic environment impacts on their development. The role of telecommunications and the development potential of affordable and reliable telecommunications is also outlined. The concept of e-government, how it is appropriately defined in small island developing states and how Pacific Island countries can benefit from e-government is discussed. Their limited resources and dependence on development partners and international investment are also discussed as factors affecting the adoption of e-government. Finally, some conceptual frameworks that have been found useful in exploring e-government in the small island developing states of the Pacific are outlined: Bekkers and Homberg's 'information ecology' approach; Heeks' ICT4 2.0 Manifesto; and elements of public policy that can usefully be applied. Factors in the ongoing sustainability of e-government projects, including the role of leadership, are identified along with questions and issues that will be raised in the chapters included in the volume and addressed further in the conclusion.

1.1 Introduction

This book is concerned with the ways information and communication technology (ICT) is changing the practice of government in the Pacific Island countries and territories. These countries are seeking, to varying degrees, to use ICTs to create

R. Cullen (✉) · G. Hassall
Victoria University of Wellington, Wellington, New Zealand
e-mail: Rowena.cullen@vuw.ac.nz

R. Cullen and G. Hassall (eds.), *Achieving Sustainable E-Government in Pacific Island States*, Public Administration and Information Technology 27,
DOI 10.1007/978-3-319-50972-3_1

more efficient processes in government and deliver information and services to citizens and the business community online—an endeavor usually referred to as e-government. In the global development discourse Pacific Island countries (PICs) are also identified as small island developing states (SIDs), a group of countries that share similar environments and development problems. Although all countries are confronting the challenges of adopting e-government, one of the principal tenets of this book is that implementing e-government is especially challenging for these small states which have the same roles and responsibilities as larger states, but far fewer resources and capacity.

The book has two parts. In Part I, this chapter explains the geographic, demographic, economic and political circumstances that make the Pacific region unique; introduces key ideas about e-government; discusses the use of ICTs in government in the developing country context; describes the challenges that SIDs face in implementing ICTs in government; and provides some conceptual frameworks that are helpful when analyzing the state of e-government in the region, such as Heeks' conceptualization of appropriate technology use. Chapters 2 and 3 examine global policies for e-government in SIDs; their impact and effectiveness in the Pacific Islands context; how regional institutions and policies contribute to the use of ICTs by Pacific governments; and the importance of national ICT and e-government policies. Part I concludes with Chaps. 4 and 5. Chapter 4 looks at telecommunications in the region as a part of the essential infrastructure of e-government, the role of competitive telecommunications markets and of independent regulation bodies in them as factors in e-government success; Chap. 5 explores the consequent expansion of mobile technology across the region, and the potential for mobile technology (or m-government) as a subset of e-government.

Part II looks at how ICTs are used by government, starting with central financial and administrative agencies, the institutions of democracy (such as parliamentary and electoral processes and the justice sector), the role of ICTs in official statistics and in other key areas of government activity such as agriculture, forestry and fisheries, climate change and disaster management, health, and education. Each chapter asks pertinent questions: What benefits are ICTs bringing? What challenges do countries face in implementing them? How well are e-government initiatives aligned with local contexts and local needs? How sustainable are they? Who is driving these initiatives and how are they funded? We ask these questions in order to identify factors critical to the successful implementation of e-government in Pacific Island states or that lead to failure. Chapter 13 explores the role of civil society in promoting e-government in SIDs, and the last chapter adds some final reflections on these questions as well as some comment on the value of the conceptual frameworks applied, the factors that have been most influential in promoting effective e-government, and further defines the dimensions that are critical to e-government sustainability.

1.2 The Small Island Developing States (SIDs) of Oceania

Twenty-two Pacific islands and territories (PICTs) are members of the Pacific Community (see the SPC (Secretariat for the Pacific Community) website)[1] and are considered to be within the scope of this volume. They are: American Samoa, Cook Islands, Federated States of Micronesia, Fiji, French Polynesia, Guam, Kiribati, Marshall Islands, Nauru, New Caledonia, Niue, Northern Mariana Islands, Palau, Papua New Guinea, Pitcairn Islands, Samoa, Solomon Islands, Tokelau, Tonga, Tuvalu, Vanuatu, and Wallis and Futuna (See map, Fig. 1.1). Not all are independent states: Cook Islands and Niue are self-governing countries 'in free association with New Zealand', which means that they control all government functions except defense and foreign affairs, while American Samoa, Guam, Wallis and Futuna, French Polynesia, New Caledonia, Pitcairn Islands, and Tokelau are non-self-governing overseas territories of the United States, France, the United Kingdom and New Zealand respectively.

The twelve independent states in this group (Federated States of Micronesia, Fiji, Kiribati, Marshall Islands, Nauru, Palau, Papua New Guinea, Samoa, Solomon Islands, Tonga, Tuvalu and Vanuatu) face the challenges of not only administering their domestic affairs on tight budgets, but representing their own interests on an increasingly crowded global stage. However, irrespective of their size or degree of sovereignty, PICs face similar challenges and frequently cooperate in shared solutions. Although not all of these countries are considered in equal detail in this book, examples and case studies are drawn from many of them, and most of the research and initiatives described in this volume, include lessons that can be of benefit to them all.

Although they are also Pacific countries, to include New Zealand and Australia in this study would suggest that solutions which work in developed, highly connected, predominantly urban societies are appropriate for SIDs, some of which fall into the United Nations' category of Least Developed Countries which inhabit a completely different environment and which face a very different set of challenges. To argue against such assumptions is one of the main themes of this volume. As this chapter and many later chapters show, the resources, environment and political context of these countries are very different from their neighbors and development partners, Australia and New Zealand, as they are from the rest of the industrialized world. Different challenges face them, different solutions apply. What works elsewhere is not necessarily appropriate for PICs. This volume is focused on identifying what does work, and how ICTs and e-government can bring promised benefits to government in the Pacific Island states.

---

[1]The former Secretariat of the Pacific Community was recently renamed the Pacific Community (SPC), a form used throughout the volume. Australia, New Zealand, France, and the United States of America, are four of the 'founding members' of the Pacific Community, although as industrialized countries, development partners and colonial powers in the region are not actively served by the SPC.

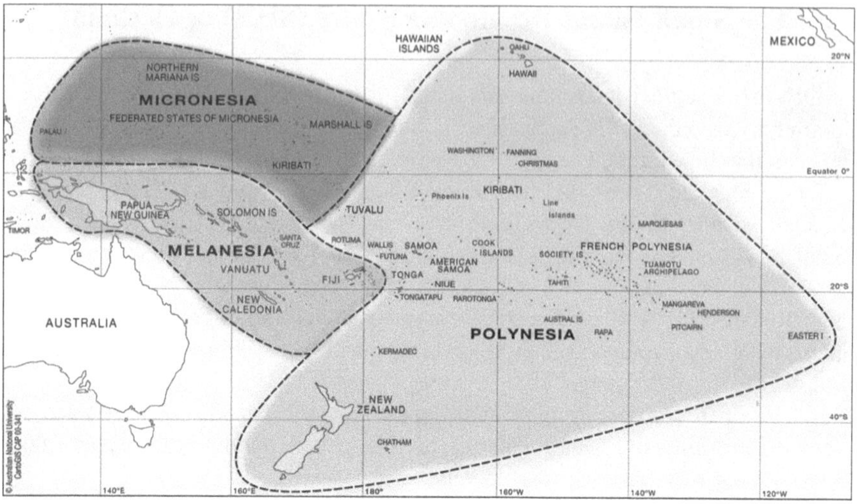

Fig. 1.1 The Islands of Micronesia, Melanesia and Polynesia. Used with the permission of CartoGIS, Australian National University

1.2.1 Cultural Groupings Among Pacific Island States

Each country and territory listed above traditionally belongs to one of three cultural regions: Melanesia, Micronesia, and Polynesia. Melanesians make up 89% of the indigenous populations of the region, Polynesians account for about 6%, and Micronesians for 5%.[2] While the origin of the Micronesians[3] is uncertain, they are believed to have settled in the region some 5000 years ago from the Philippines and Indonesia. They are regarded as ethnically and culturally separate from the Polynesian and Melanesian populations who originated from the indigenous Austronesian people of Taiwan, and who are believed to have migrated eastward across the Pacific between 5000 and 2000 years ago (and to New Zealand from east Polynesia 800–900 years ago), using highly developed navigational skills (Crocombe 2001). Elements of traditional culture remain significant in contemporary Pacific societies and systems of government, although these differ from country to country. Polynesian societies with hereditary systems of chiefly authority, for instance, are inherently more hierarchical than Melanesian societies in which chiefly offices exist but are filled competitively rather than by birth (Sahlins 1963).

All of these countries have considerable internal cultural and linguistic diversity. Papua New Guinea, for instance, has over 850 indigenous languages, and many of its villages, situated either in the lowland rain forests or the populous highlands, are

---

[2]http://prism.spc.int/images/downloads/Pacific_Population_Poster_StatisticsFINAL.pdf.

[3]The inhabitants of islands lying between the Philippines and Hawai'i, which includes Guam, the Federated States of Micronesia, the Gilbert Islands, the Marshall Islands and Palau.

as isolated by rugged terrain as the Pacific atolls are by distance across the sea. Vanuatu has the highest number of languages per capita in the world (Crowley 2000). A form of Creole or pidgin is the lingua franca in countries that have such linguistic diversity, but literacy levels in pidgin, English or French are low, and the local language often has no written form. Education is a challenge, especially in those countries made up of many small islands and those with a dispersed largely rural population, because of the difficulty of training and retaining teachers and the lack of educational materials.

Although inter-governmental organizations exist at regional levels, the diversity of the Pacific states and peoples has up to this point in time generated what has been described as a 'soft multilateralism' (Strickland 2015), examined in more detail in Chap. 3, based more on shared identity than on firm power-sharing arrangements (Graham 2007).

1.2.2 The Vulnerabilities of SIDs and their Impact on Development

Small island developing states have certain vulnerabilities, some related to economic 'shocks' and natural hazards and others as a consequence of their geography; these vulnerabilities are often articulated by the states themselves as well as by development partners in order to clarify priorities for national, regional and global development agendas. Most Pacific SIDs are archipelagos of isolated islands and atolls[4] situated in tropical regions with limited land suitable for cultivation of crops. Exceptions include Papua New Guinea (which comprises 183 islands and atolls in addition to half of the island of New Guinea, has a land mass almost as large as California and considerable unexploited deposits of minerals), and Samoa (which has two main islands, surrounded by reefs and a few smaller islands). (See Table 1.1)

The fourteen PICs that gained independence in the second half of the twentieth century still suffer the after-effects of colonial rule (such as the territorial boundaries determined by Germany and Great Britain, and neglect of social, economic and political development suffered at the hands of the colonizers). Their economies are limited by remoteness from major markets and by their small internal markets and limited resources (although their marine resources should, if well managed, compensate for some of these disadvantages, a point discussed in Chap. 9). Governance in Pacific Island societies is shifting from rural, village-based traditional life-styles and governance structures based on chiefly authority to urbanized, mostly elected, town, provincial or regional authorities and national parliaments. A number of countries, in addition to having national Parliaments, retain some form of chiefly

[4]Atolls, which are ring-shaped coral reefs surrounding a lagoon, have little soil and no minerals.

Table 1.1 Characteristics of Pacific Island countries

Country/Territory	General characteristics				
	Last population census	Population count at last census	Land area (km^2)	Density (persons/ km2)	Urban population (%)
Melanesia					
Fiji	2007	837,271	18,333	47	51
New Caledonia	2009	245,580	18,576	14	67
Papua New Guinea (PNG)	2011	7,059,653	462,840	16	13
Solomon Islands	2009	515,870	28,000	22	20
Vanuatu	2009	234,023	12,281	22	24
Micronesia					
Federated States of Micronesia (FSM)	2010	102,843	701	147	22
Guam	2010	159,358	541	323	94
Kiribati	2010	103,058	811	134	54
Marshall Islands	2011	53,158	181	299	74
Nauru	2011	10,084	21	499	100
Northern Mariana Islands (CNMI)	2010	53,883	457	122	90
Palau	2012	17,445	444	40	77
Polynesia					
American Samoa	2010	55,519	199	284	50
Cook Islands	2011	14,974	237	64	74
French Polynesia	2012	268,270	3521	74	51
Niue	2011	1611	259	6	na
Pitcairn Islands	2012	57	47	na	na
Samoa	2011	187,820	2934	64	20
Tokelau	2011	1205	12	98	0
Tonga	2011	103,252	749	138	23
Tuvalu	2011	10,564	26	420	47
Wallis and Futuna	2008	13,445	142	85	0

Source http://prism.spc.int/images/downloads/Pacific_Population_Poster_StatisticsFINAL.pdf

system exercising power over traditional or 'custom-based' land tenure systems; local chiefly roles may also involve dispensing justice and settling civil disputes.

More than half of all Pacific Islanders continue to live in villages, observe customary law and land ownership and continue to rely on subsistence agriculture in addition to some cropping, fishing or forestry for export markets. Tensions between the traditional and introduced systems of governance often impact on the effectiveness of both.

In addition to these factors Pacific SIDs are prone to natural disasters: cyclones, volcanic activity, earthquakes and tsunamis may cause severe damage that can take several years to repair. Climate change, which is intensifying in the 21st century, is leading to more violent tropical cyclones as well rising sea levels, loss of coastal arable land and increased salinity of inland water supplies, affecting small islands disproportionately to the extent that these consequences can be considered more potential causes of economic ‘shock’ (Haskins 2012). As Lino Briguglio, Director of the Islands and Small States Institute at the University of Malta and Chair of the Board of Trustees of the Small States Network for Economic Development,[5] has observed, “small size, insularity, remoteness, and proneness to natural disasters … render the economies of these states very vulnerable to forces outside their control—a condition which sometimes threatens their economic viability” (Briguglio 1995, p. 1615). Moreover, as he also notes, vulnerabilities other than size and environmental fragility that affect SIDs include dependence on foreign sources of finance and demographic changes. In the Pacific, such demographic changes would include an altered profile (e.g. either dramatically reduced or increased fertility, internal and international migration), departure of educated youth for better jobs in Australia and New Zealand or elsewhere and dependence on remittances from migrant Pacific Island workers undertaking seasonal work as agricultural laborers or in extractive industries in Australia or New Zealand (UNFPA 2014; Haberkorn 2008).

The vulnerabilities of island developing states were raised during a meeting of the United Nations Conference on Trade and Development, UNCTAD III, in 1972. Concern about their insularity and remoteness was subsequently documented in a manifesto prepared for an expert-group meeting on island developing states held in Malta in 1988 (UNCTAD 1988), and these concerns continue to be reiterated by the United Nations and UNCTAD. In 2014, the International Year of Small Island Developing States, the third UN Conference on SIDs held in Samoa focused on sustainable development, especially in the context of the commencement of the Global Development Agenda. This “post-2015 development agenda” otherwise known as the Sustainable Development Goals (SDGs) comprises 17 goals to be reached by 2030 (United Nations 2014a). Adopted by the 193 UN member states at the General Assembly in September 2015 to supersede the Millennium Development Goals of 2000–2015, the SDGs address such political and socio-economic issues as inequality, poverty, hunger, gender equality, industrialization, sustainable development, full employment, human rights, quality education, climate change and sustainable energy for all (United Nations 2015). They are discussed in more detail in Chaps. 2 and 7 and in other chapters when they are relevant to issues being considered.

1.2.3 ICT in Pacific SIDs

The basic conditions of life in most Pacific Island countries—a warm humid environment that is damaging to the equipment needed to support ICT, low levels of

[5]Funded by the World Bank.

English language, computer literacy and technical training, economies based on the export of one or two commodities and on subsistence agriculture—militate against wide and effective use of ICT. But a growing community of professionals educated overseas, foreign technical advisors and a burgeoning tourism industry have created demand for access to the Internet in national and provincial capitals, and to a large extent this has been met with services available for business and domestic use, Wi-Fi in hotels and cafés, and 3G and 4G mobile telephone networks in most towns. (The rapid growth of mobile technology and its impact on e-government is discussed in Chap. 5.) Businesses, government agencies and urban residents now expect to be able to use standard office ICTs and have access to the Internet for a range of purposes, supported by a variety of local computer companies, Internet providers, and telecommunications providers (see Chap. 4).

But Internet access remains costly, and reliable bandwidth and Internet service is patchy. Retaining skilled staff, maintaining equipment, negotiating access to networks and building capacity is an on-going challenge for the telecommunications industry and for Pacific governments. Nevertheless, information and communication technologies are having considerable impact in reducing the effects of some of the vulnerabilities outlined above and are making a major contribution to local economies. This impact was noted in a recent report commissioned by the Pacific Region Infrastructure Facility (Minges and Stork 2015), which observes that liberalization of the telecommunications industry, increased competition and increased global bandwidth has had both a direct impact on Pacific economies through growth in the industry itself, as well as indirect impacts through deployment of government networks (although less so on citizen and business-oriented services), and mobile services such as text alerts for severe weather, ferry schedules, voting locations and the status of pension fund accounts. In addition, important sectors such as tourism and remittances, which contribute significantly to GDP, are making use of ICTs. One specific impact of tourism is the number of tourists purchasing SIM cards for their phones on arrival or using locally based roaming services (p.2).

As well as helping build the internal economy, ICTs are also becoming an essential tool for participation in the global economy. An UNCTAD report (2004) commented: "globalization offers small island developing states (SIDs) valuable economic opportunities in the same way as it does with other countries." But the report also notes that, "because of their intrinsic disadvantages, most SIDs will be unable to seize these opportunities unless certain special measures to compensate their disadvantages are granted to them by their development partners." These measures, initially conceived as trade preferences, lower tariffs, non-tariff barriers and soft loans, some of which have proved difficult to sustain in the face of global trade partnerships (United Nations 2013, p. 6), are now focused in an increasingly globalized and interdependent world on assisting SIDs to actively participate in the knowledge economy and to increase connectivity (p. 9).

In the Pacific region connectivity currently means international and national access to the Internet through satellite, fiber-optic cables and terrestrial wireless (Wi-Fi/Wimax) systems. In recent years the debate between the comparative benefits of cable and satellite have become fiercer and more urgent. The World Bank's Pacific Regional Connectivity program has assisted installation of underwater fiber-optic cables linking Tonga with Fiji, (World Bank 2011), and other cables linking Vanuatu and Samoa. New companies (BlueSky and Hawaiki) have recently proposed new cables across the Pacific linking Australia and New Zealand with the continental United States, connecting several smaller Pacific countries on the way.[6] By contrast, the SIDs Action Development Plan no 2751 "Addressing Connectivity for the Sustainable Development of SIDs", developed after the Third International SIDs Conference (entitled Island Voices: Global Choices), is an alliance of the UN and several commercial partners (Intelsat, Inmarsat, Kacific Broadband Satellites, and the International Telecommunications Union). Plan no. 2751 is dedicated to bringing "low cost, reliable, diverse satellite communications capacity for the socio-economic development of the Pacific Islands region utilizing un-used satellite capacity" (United Nations 2014b). These issues are discussed in more detail in the following chapters.

In addition to the crucial need to provide affordable and reliable network configurations for economic development, and their essential role in emergency communication for disasters (see Chap. 10), several proposed initiatives are outlined in the SIDs briefing papers and subsequent SIDs Action Platforms, including the ICT4SIDS Partnership (2016). These projects are intended to take advantage of unused satellite capacity and focus on bundled (and affordable) software packages for community enterprises. They include content management software, security software, ICT applications for telemedicine and healthcare services, online and distance learning, and agriculture support systems, as well as tourism, and the marketing of local products. The projects will also include capacity development through training and outreach programs.[7] Numerous existing projects using similar applications are discussed throughout this book; examples include tablets used for data collection and transmission, biometric personal identity systems, SMS messaging to warn vulnerable populations when cyclones threaten and GIS systems to record essential data to deal with emergencies. These and many other examples show how ICT, largely put in place as part of, or through government programs, can help overcome the problems that dispersed remote rural populations face, the challenges of geography and remoteness from global commerce. These are innovative small-scale e-government applications, far-removed from common definitions of e-government projects, but they are equally transformational in their context.

The SIDs Action Platform places considerable emphasis on regional cooperation, knowledge sharing and regional partnerships. For example, the Pacific National Sustainable Development Strategy (NSDS) Regional Support Partnership SIDs Action Plan no. 2328 (United Nations 2014c) is an initiative of the Pacific Island Forum leadership that focuses on aligning policies and activities of all Pacific

[6]Construction of the Hawaiki cable (hawaikicable.co.nz) commenced in 2016.

[7]http://www.sids2014.org/partnerships/

countries and territories, development partners, the private sector and NGOs. These efforts will be dependent on ICTs for communication and access to knowledge bases. In addition, the role of ICTs and new developments in technology are noted as new opportunities for regional action, the sharing of best practice, a way of diversifying the economy, building capacity and providing opportunities that will help Pacific SIDs counteract some of the vulnerabilities noted above. The link between global and regional policies and the role of ICTs and e-government in achieving the new sustainable development goals is further discussed in Chaps. 2 and 3, and in Part II of the volume.

1.3 E-Government in Pacific SIDs

All the factors outlined above have an impact on the application of ICTs in the operations of government in Pacific SIDs, and on the speed with which they are able to adopt e-government policies and practice. But what form does e-government take in this environment? While many commentators acknowledge the particular circumstances and vulnerabilities of small island developing countries in the Pacific, it is easy to assume that models of e-government appropriate to more industrialized countries also apply in this context or that measures and ways of evaluating e-government readiness, adoption and success applicable elsewhere are equally relevant here. But this is simply not the case. The factors that create the vulnerabilities of SIDs and hinder the adoption of ICTs are the same factors that make it necessary to reconsider what e-government means in the Pacific Islands region, and what are reasonable indicators of success.

This also applies to the concept of sustainability, which is often heralded as one of the most important success criteria for e-government in the development context —a context where failure and waste of exceedingly scare resources (whether financial, human or political capital) are all too common. One of the core concepts that drove the SDGs was that development goals need to shift from a perspective based on economic growth to one that was informed by ecological economics, a framework in which the focus "must shift from merely growth to 'development' in the sense of improvement in sustainable human well-being, recognizing that growth has significant negative by-products" (Le Blanc 2012, p. 16). Similarly, the role of government/governance (see Sect. 1.3.1 below) must change from the neo-liberal model of minimal government intervention and market-driven services to one where "government should play a central role, including new functions as referee, facilitator, and broker in a new suite of common-asset institutions" (p. 16). In terms of these constructs, we propose that a concept of sustainable e-government for Pacific SIDs would depend on a set of criteria that address the constraints that PICS face. These would be:

- adequate resources to continue innovation in the future including staffing commitments and budgetary allocations for hardware and software (from either development partners or the national government);
- attention to capacity building during implementation of a project and ongoing commitment to staff development in annual budgets;
- a critical mass of developers and users who find sufficient value in a project, and changes in business process, outputs and outcomes that its continuation is assured (this does not preclude investment in more efficient or updated software);
- strong local leadership and ownership; and
- a clear contribution to national or local well-being.

1.3.1 Defining E-Government

Taking these criteria for sustainability into account, we have chosen to adopt definitions of e-government that fit this concept. E-government for Development Information Exchange website (EGov4Dev) defines e-government as "… the use of information and communication technologies (ICTs) to improve the activities of public sector organizations" (eGovernment for Development 2015a). The website, coordinated by the University of Manchester's Institute for Development Policy and Management, authored by Richard Heeks, notes that some people "… restrict e-government to Internet-enabled applications only, or only to interactions between government and outside groups" (Heeks 2008a). On the EGov4D website all digital ICTs and all public-sector activities are included. The site includes three main activities in its definition:

- improving government processes, or eAdministration;
- connecting citizens: eCitizens and eServices; and
- building external interactions: eSociety, which includes "*working better with business*" and "*building partnerships*".

This provides a useful framework for the purposes of this book, which focuses more on the first activity because that is the reality of the situation in the Pacific, although as we note throughout the book (especially about mobile government in Chap. 5), the transformational value of mobile technologies in regions where technology is expensive, and where electricity supply is irregular or non-existent, the rapid uptake of mobile technology in developing countries[8] opens up significant opportunities for Pacific Island governments to deliver services to citizens and businesses through mobile technologies. (In many texts on e-government, the three activities above are sometimes referred to as G2G (government to government

[8]ITU-ICT Facts and Figures 2015—online at https://www.itu.int/en/ITU-D/Statistics/Documents/facts/ICTFactsFigures2015.pdf.

activities), G2C (government to citizen) and G2B (government to business). While the authors of some chapters have occasionally used these terms, we have not used these terms as a framework, preferring the terminology of the EGov4Dev e-government for development framework as more suited to the situation in the Pacific islands.)

Also relevant is a recent definition of the World Bank, a major donor in the ICT for Development (ICT4D) sector, that reflects the values it seeks to promote through its sponsorship of technology and e-government initiatives:

> e-government refers to the use by government agencies of information technologies (such as Wide Area Networks, the Internet, and mobile computing) that have the ability to transform relations with citizens, businesses, and other arms of government. These technologies can serve a variety of different ends: better delivery of government services to citizens, improved interactions with business and industry, citizen empowerment through access to information, or more efficient government management. The resulting benefits can be less corruption, increased transparency, greater convenience, revenue growth, and/or cost reductions (World Bank 2015).

These terms are not the same as those currently used in the industrialized world which is moving toward the use of the term 'digital government' as more relevant in the 21st century. Nor is it the same as 'e-governance', which we define as a broader concept referring to the application and relevance of ICTs in relation to the commonly adopted concepts of governance (Bannister and Connolly 2012), although the term is used by UNESCO and by some authors, especially in the Indian sub-continent, to mean something similar to what is understood as e-government in the definitions above.

1.4 Harnessing the Benefits of E-Government for Development in the Pacific

The benefits of e-government for developing countries are not dissimilar to those identified for more developed states, taking into account the nature of government in these different environments and their political and social contexts. Most literature concerning the transformational nature of e-government focuses on developed western nations or industrialized developing nations. The concept of transformation is applied in two key ways: it includes transformation of the processes within government, and the transformation of external processes, which can be identified with either government or governance (Cullen 2010). From an information systems perspective, ICT is seen as both a facilitator and driver of organizational change, leading to efficiency, better and more responsive (or citizen-centric) services, and often aligned with a new business process model (also known as business process reengineering, or BPR.) At the same time, the public administration literature has tended to focus on the capacity of ICT to change the relationship between government and citizens by changing governance processes and transforming democracy through the use of ICTs for e-consultation, e-participation and crowd-sourcing. Both models are credited with increasing transparency and reducing corruption.

This is emphasized in the Foreword to the United Nations 2014 survey E-government for the Future We Want[9] by Wu Hongbo, Secretary-General for the 2014 Conference on Small Island Developing States, who says:

> At the United Nations Conference on Sustainable Development held in Rio de Janeiro in June 2012, a global consensus was reached that to achieve our sustainable development goals we need institutions at all levels that are effective, transparent, accountable and democratic…. e-government holds tremendous potential to improve the way that governments deliver public services and enhance broad stakeholder involvement in public service.

Section 1.4.1 of the report notes that the unique geographical challenges faced by SIDs, that is, their small size, isolation, and geographic dispersion, which results in the very high cost of providing goods, services and infrastructure, including telecommunications, create real problems for governments of SIDs in coordinating and delivering services; it notes that Pacific SIDs rate very poorly in its e-government Development Index, with only Fiji ranking 85th in the top 100 countries.

> The challenges faced by SIDs have intensified with the natural disasters associated with climate change and the lasting negative impacts of the global financial crisis. This highlights the structural nature of the constraints of the SIDs and the lack of effective national and international response mechanisms. Several of these difficulties are amplified by the SIDs' limited access to modern technologies. (UN DESA 2014, p. 37)

However, the report also emphasizes the value of e-government to SIDs, not only in disaster management in locations where populations are widely dispersed, but in other ways as well.

> e-government is of special importance in SIDs also in relation to citizen engagement and improving the livelihood of people. With e-participation, citizens in even the most remote and far-scattered islands can be connected to their government and be consulted in decision-making processes. In commerce and improving the livelihood of people, ICT enabled tools such as mobile apps for fishers play an important role in reducing poverty. (UN DESA 2014, p. 38)

Even so, many Pacific SIDs have not experienced (or have not been able to measure) the increased efficiency, productivity, accountability or economic growth that e-government promises, nor the greater citizen involvement that e-governance potentially offers. Indeed, the inherent challenges facing Pacific Island countries make it difficult to capitalize on these benefits, and this is at the heart of the question of how Pacific Island countries can benefit from e-government. Moreover, securing the benefits of e-government is also dependent on good governance, even while it helps to ensure it.

[9]The 2016 UN DESA e-government Survey is discussed below.

Fijian economist, public servant, educator and passionate advocate of good governance, Savenaca Siwatibau, was one Pacific leader who quickly identified the potential benefits of e-government for PICs: (Siwatibau 2009, p. 13). In a 2001 paper "EGovernance, Governments and The University of the South Pacific", Siwatibau noted benefits of early adoption of Internet-based information exchange to the public, the business community and the public sector. These benefits included better communication between government and citizens, better communication between and within government agencies allowing for more collaboration and better planning, efficiencies and transparency in procurement (government purchasing), and improved service to government employees. Siwatibau identified five characteristics of governments that sought to embrace the potential offered by e-government:

- a commitment to invest in IT;
- adoption of appropriate policies;
- collaboration with other organizations;
- re-engineering of business processes; and
- employment of people with the right skills and attitudes.

Siwatibau's insights regarding successful deployment of new technologies are important, for they deal as much with policy leadership, public sector management and cultivation of capacity, as they do with the technologies available. These are also concerns raised in this volume.

1.5 Government in Pacific SIDs

The small size and economies of most PICs result in limited resources for government, making economies of scale hard to achieve when administrations must reach out to a population spread over many islands. Limited natural resources are stretched when a small and isolated independent state must fulfill all the roles now required of a nation, including raising revenue; employing and remunerating staff; managing finances and the economy; securing and controlling borders; policing territorial waters; managing airports; building transport, water, waste and communications infrastructure; administering the electoral system, the legislature, justice and police; licensing vehicles, drivers, professions and businesses; and providing essential health and education services to its population. In addition, these states must manage their international affairs, participate in regional and global intragovernment organizations, treaties and agreements, and protect their people when the natural hazards of the region result in disasters. Their small economies also mean that government is the biggest employer in most PICs and government expenditure plays a significant part in driving the economy. Government itself is therefore 'small' in such jurisdictions (though not in the neo-liberal sense), confining its activity through necessity to essential roles; the ability to go beyond this to provide modern welfare systems and other protections for citizens is limited. However, while the economies of most PICs remain affected by their reliance on remote rural agricultural subsistence, new emerging industries such as fisheries, pearl culture, forestry, tourism and the extractive industries require engagement with government on an unprecedented scale. In addition, the growth of ICT use by residents and businesses is bringing economic growth, and with it a demand for online services from government; it also impels higher rates of literacy and an increase in ICT skills as the industry develops.

To varying degrees, all Pacific Island governments depend on funding from donors and the aid agencies of wealthier countries. The major regional players are Australia (which provides 46% of regional aid), New Zealand, the United States, France, Japan, and some EU institutions.[10] More targeted aid agencies, usually units of inter-governmental organizations such as the UN (through e.g., the UNDP, UNCTAD, UN Women, FAO, UNFPA), the IMF (which has a Pacific branch PIFTAC), the World Bank (WB), and the Asian Development Bank (ADB) support projects and technical advisors throughout the region, many of which involve e-government initiatives. Aid agencies have their own agendas, and lack of awareness in Pacific countries of the potential of ICTs and how they can help governments achieve their goals can lead to development agendas and priorities for e-government projects being set by their development partners (Boase 2009; Budden 2005; Cullen and Hassall 2013). Several examples in this book describe projects that have been successfully integrated into a country's own development policies and plans; but some projects have languished because they have not been sufficiently grounded in local needs, have ignored contextual factors, or simply have ended when the project's funding ends. These failures mitigate against sustainability of e-government in PICs; they also lead to waste of effort and resources, often characterized as financial, political, opportunity and 'future' costs (the last creating barriers to future development investment (eGovernment for Development 2015a, b)). That is also a frequently-raised concern in the volume.

Moreover, aid dependency itself is a growing concern (Pryke 2013a, b). In the more economically stable Pacific Island states, especially those with abundant natural resources that can be exploited, aid is not as high a proportion of the Gross National Income (GNI) as might be expected. Table 1.2 shows World Bank data reporting net Official Development Aid (ODA) received as a percentage of Gross National Income for Pacific Island countries. Net official development assistance consists of loan disbursements made on concessional terms (net of repayments of principal) and grants by official agencies of the members of the OECD's Development Assistance Committee (DAC), by multilateral institutions (such as the World Bank and the UN), and by non-DAC countries to promote economic development and welfare in countries and territories in the DAC list of recipients. Such assistance includes loans with a grant element of at least 25% (calculated at a rate of discount of 10%).

[10]OECD (2016). Development Aid at a Glance. Statistics by Region. 6. Oceania. Paris, Organization for Economic Cooperation and Development.

Table 1.2 Official development aid (ODA) received as a percentage of gross national income for Pacific Island countries

Country	Percentage of GNI (%)	Trend
Federated States of Micronesia	41.8	Increasing
Fiji	2.2	Declining
Kiribati	23.2	Declining
Marshall Islands	41.5	Increasing
Palau	16	Increasing
Papua New Guinea	4.4	Declining
Samoa	15.5	Stable
Solomon Islands	27.4	Declining (from 49.7% in 2011)
Tonga	18.1	Slight increase
Tuvalu	48.3	Increasing, following a recent sharp decrease from 67.2%
Vanuatu	11.4	Declining sharply[a]

[a]This may have changed in 2015 following the severe impact of cyclone Pam
Source World Bank 2016

1.6 E-Government Readiness in Pacific SIDs

The United Nations uses various metrics to assess the e-government achievements and readiness of its member states (that is, the extent to which they meet a pre-determined set of criteria that purports to measure their potential for successful e-government adoption). Table 1.3 shows the 2016 results of the two-yearly country rankings published in the UN E-government Survey: e-government In Support of Sustainable Development (UN DESA 2016). As the Executive Summary of the 2016 Survey notes, the tools of e-government are being applied to the 2030 Agenda for Sustainable Development Goals, and toward a world free of poverty, hunger, disease and want. E-government can contribute significantly and in practical ways to realizing these goals.

> Through advanced electronic and mobile services, e-government aims at improving the relationship between people and their government. It aims to make public services delivery more effective, accessible and responsive to people's needs. It also aims at increasing participation in decision making and making public institutions more transparent and accountable. (UN DESA 2016, p. 1)

New trends include facilitating integrated policies and services, establishing more open data in order to promote effective, accountable and transparent institutions and advancing e-participation to increase possibilities for participatory decision-making and service delivery.

Table 1.3 E-government development in the Oceania region

Rank	Country	EGDI	Online service component	Telecom infrastructure component	Human capital component
2	Australia	0.9143	0.9783	0.7646	1.0000
96	Fiji	0.4989	0.4130	0.3326	0.7509
145	Kiribati	0.3122	0.2101	0.0665	0.6599
156	Marshall Islands	0.2695	0.0290	0.0849	0.6947
146	Federated States of Micronesia	0.3103	0.1449	0.1197	0.6663
152	Nauru	0.2868	0.0942	0.2448	0.5214
8	New Zealand	0.8653	0.9420	0.7136	0.9402
111	Palau	0.4546	0.1087	0.3684	0.8867
179	Papua New Guinea	0.1882	0.1667	0.0739	0.3240
121	Samoa	0.4019	0.3406	0.1576	0.7076
164	Solomon Islands	0.2406	0.1667	0.1150	0.4402
105	Tonga	0.4700	0.3696	0.2302	0.8102
151	Tuvalu	0.2950	0.0217	0.1981	0.6651
149	Vanuatu	0.3078	0.1667	0.1684	0.5884

Source drawn from the United Nations E-government Survey 2016 (UN DESA 2016)

The rankings are based on the E-government Development Index (EGDI), a composite score based on measures for online services (e.g. government web sites, portals and functions available), telecommunication infrastructure (which includes fixed telephone lines, mobile subscribers and Internet users as a percentage of population), data obtained from the ITU and human capacity (adult literacy, combined primary, tertiary, and secondary education enrolment per head of population, expected years of schooling and mean years of schooling). The focus in the measures has moved over time from e-government readiness, based on infrastructure and human resource endowments, to "a country's capacity to engage in the information society, without which e-government development efforts are of limited immediate use" (UN DESA 2016, p. 2).

The EGDI by which countries are ranked is therefore impacted by many of the factors outlined in this chapter, including the disadvantages PICs suffer due to distance, small size, vulnerabilities, and their small economies. The fact that the PICs are ranked in the lower half of a global EGDI survey is not a great concern, since small states cannot be expected to compete on the same level as larger and better endowed ones. What such exercises do provide, on the other hand, is feedback on the extent to which a country is improving on its own previous score. The results reported in 2016 indicate that only three PICs (Papua New Guinea, Solomon

Islands, and Vanuatu) have advanced in global ranking since the 2014 report and Palau is noted for the significant progress made since the 2003 survey (when its EGDI was 0.009), ranked last at 173. However, the decline in ranking for the remaining PICs was due more to a decline in the human capital component rather than a decline in online services or telecommunications infrastructure and partly to changes in the way the three component scores are measured.

However, many highly valuable e-government initiatives, many of which are related to the new and emerging industries noted above or which mitigate the threats to SIDs through their heightened exposure to natural disasters, suggest that some of the real benefits of ICTs in government in PICs may often rely on technologies and applications that are not easily measured by global assessment systems. Not all countries have the same requirements; not all will use ICTs to resolve the same problems. Metrics such as those used in the UN E-Government Survey often measure what can be universally measured not what necessarily should be measured.

But there is another dimension that makes e-government in Pacific Island states harder to achieve and to measure, and this is the social and political dimension, driven by the meld of western democratic systems and traditional governance structures that characterize most PICs, as well as the "social structural complexities including languages and the relative immaturity of the political orders in western terms" (Budden 2005, p. 4). The introduction of democratic processes and international standards of justice and policing, alongside traditional forms of governance, therefore present challenges to PICs, which in turn affect the adoption of e-government in these areas. As Budden commented when reporting on the lack of progress of a number of UNDP projects:

> The reasons for this relative lack of progress are many: indecisive leadership and frequent changes of direction not to mention government, overall slow economic development, absence of infrastructure to ensure ubiquitous access and justify development, and donor driven development are among them…. The relatively low needs for government administrative services e.g. tax and licensing, in village settings probably impacts on demand as does the traditional decision making. (Budden 2005, p. 6)

Many of these factors still apply. Governments in many PICs have limited tenure; coalition governments comprising members from large numbers of parties are unstable (see Chap. 7) and corruption remains an ongoing issue despite several Pacific governments signing up to the UN Convention against Corruption, and participating in the UN Pacific Regional Anti-corruption Project (UNPRAC 2012) and ongoing workshops. Public servants are still frequently appointed by patronage, and lack security of tenure and a shortage of skills militate against developing a strong ICT staff who can support e-government initiatives. All these are e-government readiness factors in the Pacific as they are in most other SIDs. But equally, as we hope this volume will show, Pacific communities have readily adopted e-government solutions that fit local needs, and in many areas they have indeed been transformational.

1.7 International Investment in E-Government in PICs

International intervention, regional development partnerships and assistance, and ODA are critical parts of many e-government initiatives in the region. But they are by no means evenly distributed across all the functions of government. It is clear from the evidence in Chap. 6 that there is strong international interest in maintaining states in order to avoid the emergence of failed states, or centers for international criminal activity that could damage other states in the region. Thus, international intervention tends to be proactive, well-funded and focused on systems and capacity building. Some countries included in our investigations report significant benefits from institutional strengthening programs in the financial sector (Cullen and Hassall 2016a–d). State of the art information systems are routinely installed and supported by development partners in these core aspects of government operations. But this leads to an interesting phenomenon, as the various chapters in this book will show, and that is, that the real gains from the use of ICTs in government in the Pacific countries included in this study have mostly been in these core aspects of government, i.e. the management of the economy, collection of revenue, management of cross-border trade, allocation of government funds, management of the bureaucracy—all supporting a money-based economy.[11] They are usually well managed, employ relatively sophisticated ICT applications, and are well supported internationally by global agencies which have an interest in seeing effective ICT systems in place to keep the country functioning and participating in the global economy. (Also included in this core is a group of government activities well supported by ICT—disaster preparedness and management, where regional collaboration and support includes the use of sophisticated technology.)

Outside this inner core are other key tasks of the modern state—an impartial police and criminal justice system to maintain law and order, and a parliamentary system which reflects the values of democracy and global citizenship. The use of ICT in these areas is patchier. Although some ICT systems are used across this group of activities, the application of ICT in these second-tier activities is inconsistent, more ad hoc, and dependent on specific aid grants or the interests of development partners. In general, ICT is not systemic across an agency or sector the way it is in the core group of agencies. An outer shell represents areas of government where traditional practice is far more prevalent; these sectors are further from the centers of government and, with some notable exceptions,[12] their use of ICTs has not been recognized until recently as contributing to the economy, or essential to the well-being of the people and the protection of their rights. Areas such as education, agriculture and health are in this third shell, along with preservation of the indigenous culture and the records of government. This is represented in Fig. 1.2.

[11]It must also be acknowledged that in some PICs the unmeasured, indigenous rural subsistence economy may be greater than the money-based economy.

[12]See for example the Fisheries Information Management System outlined in Chap. 9.

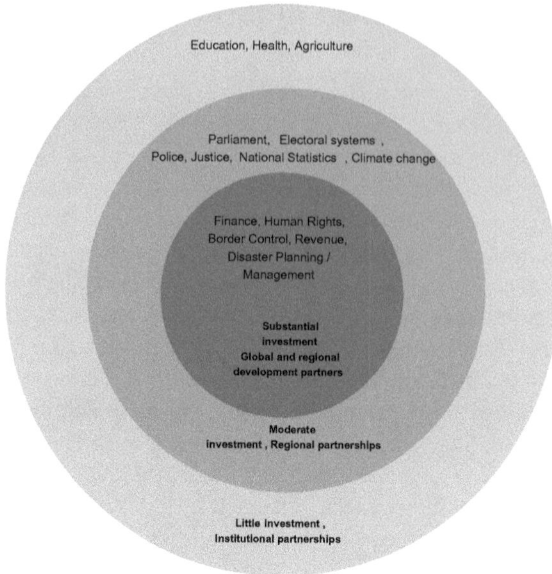

Fig. 1.2 ICT and international investment in government in PICs. *Source* Authors

1.8 Conceptual Frameworks and Models of E-Government Transformation that Inform the Approaches Taken in the Volume

***It is clear from many observations above that the adoption of e-government in Pacific SIDs does not follow the models that are put forward for industrialized nations and cannot be discussed using the same frameworks. Whether an information systems approach or a public policy approach is taken the political context, environment, capacity and resources are so different that assumptions commonly made in the e-government literature must be challenged. In this section we put forward some alternative conceptual frameworks drawn from the information systems perspective that we believe are more useful in exploring the adoption of e-government in PICs.

1.8.1 The Information Ecology Approach

In addition to the economic, developmental and physical impediments to the adoption of e-government noted above, some authors have suggested that there are other more deep-seated factors involved, and that adequate theory for exploring these is lacking:

> theory on how the government sector in particular in LDCs [Least Developed Countries] should operate appears to be almost entirely lacking… One problem is that [established theories on ICT adoption] have been developed and tested primarily in the context of developed or Western countries and are possibly not relevant in the context of many developing countries and LDC's. (Imran and Gregor 2005, p. 32)

This has been described by Heeks as the 'if it works for us it will work for you' mentality of IT multinationals, consultants and aid agencies (Heeks and Bhatnagar 1999, p. 70). Instead, as Budden argues, "we should… be wary of burdening ICTs with expectations as their potential cannot be realized in isolation from the social and cultural environment in which they exist" (Budden 2005, p. 4).

There is evidence, some of it presented in this volume, e.g. in Chap. 6, that these assumptions are being modified and that in the past decade the Pacific countries which are the subject of this volume are taking more ownership of their ICT and e-government policies and adoption, and that their development partners are showing greater sensitivity toward, and trying to work within, individual countries' national objectives and regional policies. This is an essential step for the sustainability of e-government projects and for the transformation that e-government has the potential to bring in Pacific SIDs.

The assumptions and approaches that inform this book are influenced by the work of Bekkers and Homberg, whose application of the 'information ecology' framework for e-government (Bekkers and Homberg 2005) derives from Davenport and Prusak (1997). This way of looking at information later emerges as the concept of an 'ecosystem' in the ICT for Development framework (ICT4D), in Heek's ICT4D 2.0 Manifesto, and in emerging principles and debates within the e-government for development community (Egov4Dev). Applying an 'information ecology' approach, Bekkers and Homberg (2005) argue, enables us to investigate the "evolving interactions and relations between a diversity of actors, their practices, values and [use of] technology within their specific and local environment." But as they also note:

> … the interactions between these stakeholders do not occur in a vacuum. They are embedded in a specific cultural, political, intellectual and economic environment in which specific 'rules' guide the behavior and interactions between these stakeholders as well as the meaning that will be attached to the use of ICTs. (Bekkers and Homberg 2005, p. 9)

This understanding provides a more appropriate framework for the cultural and governance systems of the Pacific communities we described above. It encompasses a radically different approach from the techno-centric model that has driven e-government implementation and research to date; it focuses on the local, and asks: 'Whose interests? Who benefits? Who are the actors here? What motivates them? What are the models of governance?' We believe this approach will allow us to take into account the information cultures of the Pacific Island states we are investigating and, with better understanding of the actual situation on the ground, to develop a Pacific model of e-government that has a 'Pacific' concept of sustainability at its heart (Cullen and Hassall 2013).

1.8.2 Heeks' ICT4D 2.0 Manifesto and the Sustainability of E-Government Projects

The second conceptual framework we have applied reflects and acknowledges the influence of Richard Heeks and the work of the University of Manchester's Institute for Development Policy and Management, the Centre for Development Informatics, and the eGovernment for Development Information Exchange. Heeks' ICT4D 2.0 Heeks (2009), grounded in Computer Studies, Information Systems and Development Studies, brings together many of the factors that influence both how ICTs are applied in government in developing countries and the many relationships that impact on the sustainability of both projects and policies. The Manifesto argues that however well motivated, the original ICT4D movement (now styled ICT4D 1.0) was grounded in a techno-centric approach, focused on ICT innovations, an "invention-down" approach bringing new technologies into development contexts. It argues a decade later that the new ICT4D model should focus on a bottom-up approach, shifting the balance from technology invention to technology use. This would mean:

- less emphasis on what might be used (the Internet and PCs), and more emphasis on what is actually used (mobiles, radio and television);
- less emphasis on fundamental technical innovation and more emphasis on application and business model innovation; and
- less emphasis on piloting and sustaining new applications, and more emphasis on assessing and scaling existing applications. (Heeks 2009, p. 5)

As a consequence, Heeks argues, new innovation and implementation models are needed which will require new categories of funders, new actors involving the private sector, and public/private partnerships, (which raises questions about sustainable markets), new approaches and techniques. This new approach is summed up in the following diagram (Fig. 1.3).

The implications of this model and strategies for ICT *in* development, rather than *for* development, means not just integrating ICT into development goals (as it may be argued that the new SDGs do) but focusing on the value ICT brings to a community. The ICT4D 2.0 model thus makes ICT the platform for economic and social development, through the following phases.

- Readiness: do we have the policies and infrastructure to make ICT availability possible?
- Availability: how can we roll out ICTs to the poor to help them become users?
- Uptake: in what ways can we implement and apply ICT to make it useful?
- Impact: how can we use ICTs to make the greatest development impact? (Heeks 2009, p. 28)

Many examples of e-government initiatives in this book will show how some of these questions have been answered.

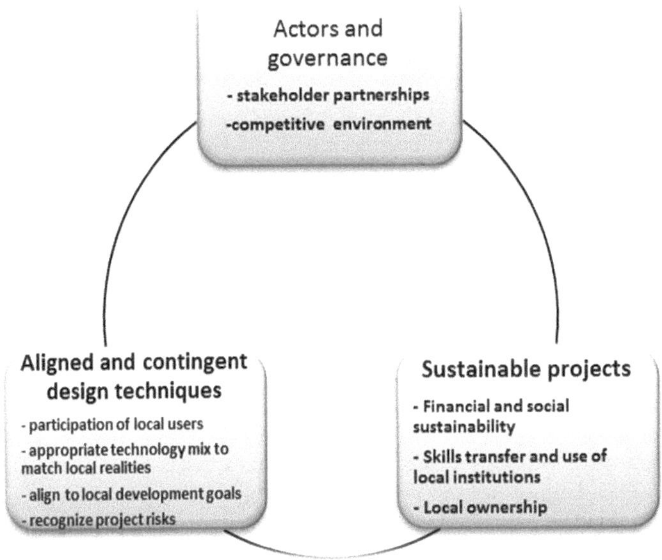

Fig. 1.3 Adapted from Heeks' Good Practice for ICT4D 2.0 Implementation (Heeks 2009, p. 20)

In addition, we propose to explore some of these ideas (although not formally evaluating individual projects because our focus covers 22 countries and territories) by applying some of the factors determining e-government failure and success that Heeks sets out on the eGovernment for Development website. This includes the ITPOSMO model first developed in Heeks and Bhatnagar (1999) and used in Heeks' widely used textbook (2006). This is an information systems perspective that focuses on the 'design-reality gap' based on the following dimensions: Information; Technology; Processes; Objectives and values; Staffing and skills; Management systems and structures; and Other resources. Within this model further investigation can be made into the Drivers/Enablers (such as Strong external pressure and Internal political desire), and the Constraints (Poor change management, Lack of overall vision and strategy, Dominance of self-interest) (Heeks 2008b). These factors are part of what Heeks describes as an ICT4D 2.0 ecosystem, defined as,

> the combination of organizations (government, private sector, NGO/community, etc); institutions (policies, culture, etc), technologies (standards, infrastructure, architecture, applications, etc), and other resources (money, skills, etc) which allows ICTs to have a transformational effect. (Heeks 2014).

Understanding the ecosystem of ICT innovations (or their 'information ecology' as outlined above) is critical to their success. These ideas illuminate many of the issues that this volume addresses and bring a range of diverse themes together in a way that meaningfully reflects the ecosystem of e-government in the Pacific Island states.

1.9 The Public Policy Dimension

We also include a public policy perspective in the volume. The policy frameworks outlined here describe some aspects of the policy dimension necessary for effective e-government planning and implementation. The first of these is the "policy cycle" framework (Jann and Wegrich 2007), based on Lasswell's historic work (Lasswell 1951), which is often used in academic public policy programs as a readily comprehended model that is applicable to several empirical fields. This begins with agenda setting, which requires identification of the critical issues to be addressed and identifies the participating actors; policy formulation, which involves articulation of policy settings, legal frameworks and identification of funding implications; implementation, at which stage vision and planning are tested in the realities of deployment; and evaluation, a phase sometimes given too little attention, but which if undertaken properly can yield lessons for subsequent planning and implementation. The policy cycle model is often criticized as being naïvely idealistic, rarely applying in the 'real' world, because it represents a legalistic top-down rationalist approach to policy, focusing on process not content, which does not reflect the way policy is made and implemented (Nakamura 1987).

Real-world policy processes are, of course, far messier than simplified models suggest; they are often iterative rather than unidirectional, and invariably shaped by constraints such as time, resources and politics—key elements in the ecosystem of e-government in the Pacific. A slightly modified form of the policy cycle was used in the *Australian Policy Handbook* (Bridgman and Davis 1998) "to help new public servants make sense of the policy task", although the authors acknowledge it is "just a heuristic, an ideal type", a 'tool' that helps policy makers understand the policy process and the nature of policy dialogue in the complex environment of government (Bridgman and Davis 2003, p. 100). The *Australian Policy Handbook* is now in its 5th edition (2015) and undoubtedly influential in the Australian public service. There is some evidence that this approach to policy has some influence in regional, national and agency level policy-making in the Pacific in which Australian aid workers and advisors have played a part, although in the pragmatically-driven resource-constrained environment of government in the Pacific islands it is sometimes difficult to identify the role of policy at all, or any evidence of the role of policymakers in e-government implementations.

However, the debate over the validity and relevance of the policy cycle certainly raises critical questions about how policy is made and implemented in the SIDs context. When most e-government projects are driven by development partners, do they set about developing policy in the local context, through dialogue, consultation and coordination, as Bridgman and Davis (1998, 2003) advocate? And what is the impact of partners bringing their own perspective on how policy should be made and implemented? On the one hand, the imposition of policies from other countries and contexts without due regard to local circumstances, can result in rapid project failure as Budden (2005), Boase (2009) and (Cullen and Hassall 2013) have noted. On the other hand, reluctance to learn from comparative experience can similarly lead to poor results (eGovernment for Development). When development partners advocate for policies and projects that have been implemented elsewhere, what power do Pacific governments have to insist on having a part in the policy process? And what, if this does occur, do they bring to it? These are crucial questions. To avoid potential project failure (in particular, of the kind that Heeks (2008a) identifies) appropriate consultation about policy in which local actors communicate effectively among themselves as well as with external actors is essential. All institutions involved in policy development must share an understanding of the policy environment, the government's priorities and the range of solutions within reach. The public policy perspective argues that such policy discourse should be part of any government's normal practice, but in the context of developing countries it necessarily involves a broader development dialogue involving regional collaborations, as well as development partners, professional bodies, user groups, and service providers.

The examples shared in this book suggest that projects are often scoped in short timeframes, without such consultation, implemented by consultants with very little prior knowledge of local context, who attempt to apply a generic solution. This could be described as a form of 'policy transfer' (Nedley 1999, cited by Stone 2001) and although the term is not widely used in the development field, it can often be seen in the development of policy templates. The UN's Economic and Social Commission for Asia and the Pacific (ESCAP) Asian and Pacific Training Centre's Policies and Strategies Guides (APCICT 2013) for example, and the World Bank's e-government web site offer advice on "strategy, policy, regulatory and legal aspects, institutional frameworks, enterprise architecture… training and change management" (World Bank 2015). This includes a set of e-government resources, and all are promoted with the implication that generic policies can be 'cut' or 'tailored' to meet domestic conditions. But even when expert advice is available policy must still be developed in collaboration with multiple stakeholder groups, through information sharing and consultations that in the Pacific need to be conducted 'face-to-face' in the Pacific way. Individual policies must be developed in the context of each country's development plan and aspirations, and with sensitivity to the needs of marginalized groups as well as those who are closer to the levers of power. To what extent is this occurring in the Pacific islands? And if it is not, how can policies and projects be sustainable? These questions must be asked.

1.10 E-Government Leadership

The 2014 UN E-government Survey (UN DESA 2014) highlights strong political will and collaborative leadership among the main enabling factors for achieving e-government goals,[13] and identifies the lack of leadership, as well as lack of

[13] Leadership, especially collaborative leadership, is highlighted in the 2016 Survey as necessary for a commitment to open data.

resources, ongoing digital divides and inadequacy of institutional change processes among the challenges facing countries seeking to achieve their e-government goals. Effective leadership in the areas of policy, legislation, regulation and operations is required not only from the public sector, from commercial operators and stakeholder groups throughout civil society, but most importantly at the political level. But the effective application of e-government applications in environments where systems and policies are not well established, and where ICT skills are scarce requires more this; it requires a very special kind of leadership.

In their review of national policy leadership in ICTs, Biggs and Polomska (2013) point out that policy leadership is required, among other things, to "…highlight the role of broadband in national development, provide an enabling environment for private investment, coordinate dialogue, and encourage work across different sectors and ministries" (p. 55). Such leadership requires a unique set of competencies, which include both technical and management skills. Technical demands include an understanding of the available and appropriate technologies, the ability to map government's current needs and future aspirations to potential solutions and the construction among stakeholders both within and beyond government of a shared 'mental model' of what is possible (including, most importantly, buy-in from political leaders). Managerial skills include an appreciation of models for financing infrastructure, an ability to incorporate Official Development Assistance (ODA) resources (such as technical assistance and financing) into domestic operations, liaising effectively with donors and the ability to proactively solve potential conflicts among stakeholders that can threaten the progress of an entire e-government project in a country or region. When small states struggle to identify leaders with such a repertoire of skills, knowledge and expertise, leadership is often sourced internationally. In such cases, it needs to be introduced in such a manner that cultivates national leadership rather than suppresses it. Ironically, such internationally-sourced leadership, which requires the ability to work effectively across cultures in addition to the other professional skills, may be as difficult to procure as the domestic leadership it stands in place of. Thus, we identify not only leadership but sustainable, transferred leadership, embedded in the local organizational culture, as one of the critical factors in e-government success in the Pacific.

1.11 Conclusion

These and many other issues are explored in the following chapters through an examination of policies, initiatives and examples which cover:

- the institutional and policy context of e-government adoption in Pacific Island countries and territories (PICs);
- global, regional and national influences on ICT policy in PICs;
- appropriate IT/e-government governance for PICs;

- the contribution of ICTs and e-government initiatives to some of the core tasks of government that developing countries must focus on; and
- some innovative solutions that provide models for effective sustainable e-government in PICs.

The final chapter in the volume will address the issues raised here and other questions identified in the ensuing chapters. Out of these we hope will come a picture of the unique nature of e-government in the Pacific islands, the critical factors contributing to its many successes, and the challenges to be overcome. When a government decides to re-engineer its systems and service delivery through the transformative power of e-government it places extensive demands on its individual and collective leadership for reform of law, policy, administration and technology. It requires a systems approach to administrative process, receptivity to system feedback, and a capacity to communicate change within the bureaucracy, with development partners, the private sector and to the public. All the while it flirts with the prospect of project failure. This is a huge demand for the small island developing states of the Pacific and their under-resourced governments. But although the focus of the book is on Pacific SIDs, the outlook is also global with lessons to be learned not only by developing but developed countries as well. For whereas these small island states are each uniquely situated geographically, culturally, technologically, they are simultaneously a part of the emerging integration of economies that is concerned not so much with national borders but with the efficient delivery of services as part of a global economy. While it is essential that they retain their unique and special character, their future development depends on their mature engagement with the global economy, driving this engagement on their own terms, through e-government.

A Note on Sources: To date there has been a limited amount of scholarly investigation into ICTs and e-government in Pacific Island countries. There is a very modest number of journal articles and theses, and a small number of conference papers by academics and others familiar with the region, plus reports, surveys and policies prepared for individual governments or international agencies (such as the UN, ITU, UNDP, World Bank etc.). Most of the available and relevant material is cited in this volume in individual chapters. Individual chapters have cited what contributes to the analysis being presented, but often critical and theoretical analysis on which to draw is lacking. This particular 'information ecology' is not only the context for the current book, but provides one of the chief motivations for writing it.

References

Asian and Pacific Training Centre for Information and Communication Technology for Development (APCICT) (2013) ICT4D Policies and Strategies. http://www.unapcict.org/ecohub/resources/browse-resources/policies-and-strategies

Bannister F, Connolly R (2012) Defining e-governance. e-Service J 8(2):3–25

Bekkers V, Homberg V (2005) The information ecology of e-government: e-government as institutional and technological innovation in public administration. IOS Press, Amsterdam
Biggs P, Polomska A (2013) The importance of national policy leadership. In: Bilbao-Osorio B, Dutta S, Lanvin B (eds) The global information technology report 2013: growth and jobs in a hyperconnected world. World Economic Forum, Geneva, pp 53–65
Boase R (2009) Final report. Outcome evaluation of the information & communications technology for development (ICT4D) project of the UNDP multi-country office (MCO) based in Samoa. Vancouver: UNDP
Bridgman P, Davis G (1998) Australian policy handbook. Allen & Unwin, Sydney
Bridgman P, Davis G (2003) What use is a policy cycle? Plenty if the aim is clear. Aust J Public Adm 62(3):98–102
Briguglio (1995) Small island developing states and their economic vulnerabilities. World Dev 23 (9):1615–1632
Budden J (2005) E-government in the Pacific Island countries: opportunities and impediments—with an emphasis on the governance of ICTs. University of the South Pacific, Suva
Crocombe R (2001) The south Pacific. University of the South Pacific, Suva
Crowley R (2000) The language situation in Vanuatu. Curr Issues Lang Plann 1(1):47–132
Cullen R (2010) Defining the transformation of government: e-governance or E-government paradigm? In Scholl HJ (ed) E-government: information, technology, and transformation. ME Sharpe, Armonk, pp 52–71
Cullen R, Hassall G (2013) An information ecology approach to sustainable E-government among small island developing states in the Pacific. HICSS 13. In: Proceedings of the 46th Hawaii international conference on systems sciences. IEEE Computer Society, Washington, DC, pp 1922–1931. http://dl.acm.org/citation.cfm?id=2471128. Accessed 3 Mar 2016
Cullen R, Hassall G (2016a) E-government in the Pacific island states: ICT policy and implementation in small island developing states. Solomon islands country report. http://pippr.victoria.ac.nz/handle/123456789/32. Accessed 2 May 2016
Cullen R, Hassall G (2016b). E-government in the Pacific island states: ICT policy and implementation in small island developing states. Vanuatu country report. http://pippr.victoria.ac.nz/handle/123456789/33. Accessed 2 May 2016
Cullen R, Hassall G (2016c). E-government in the Pacific island states: ICT policy and implementation in small island developing states. Cook islands country report. http://pippr.victoria.ac.nz/handle/123456789/30. Accessed 2 May 2016
Cullen R, Hassall G (2016d). E-government in the Pacific island states: ICT policy and implementation in small island developing states. Samoa country report. http://pippr.victoria.ac.nz/handle/123456789/31. Accessed 2 May 2016
Davenport TH, Prusak L (1997) Information ecology. Oxford University Press, Oxford
eGovernment for Development (2015a) What is eGovernment? http://www.egov4dev.org/success/definitions.shtml. Accessed 5 Nov 2015
eGovernment for Development (2015b) Evaluation. The impact of eGovernment failure. http://www.egov4dev.org/success/evaluation/impact.shtml. Accessed 5 Nov 2015
Graham K (ed) (2007) Models of regional governance for Pacific island states: Sovereignty and the future architecture of Pacific regionalism. University of Canterbury Press, Christchurch
Haberkorn G (2008) Pacific Islands' population and development: facts, fictions and follies. NZ Popul Rev 33(34):95–127
Haskins J (2012) Building resilience in small island economies: from vulnerabilities to opportunities. CTA Policy Brief 8. Wageningen: Technical Centre for Agricultural and Rural Cooperation. http://www.cta.int/en/article/2013-05-30/cta-policy-briefs.html
Heeks R (2006) Implementing and managing eGovernment: an international text. Sage Publications, London
Heeks R (2008a) The impact of E-government failure. Success and failure in E-government projects. E-government for Development. http://www.egov4dev.org/success/evaluation/impact.shtml. Accessed 5 Feb 2016

Heeks R (2008b) A factor model for eGovernment success and failure. Success and failure in E-government projects. E-government for development. http://www.egov4dev.org/success/evaluation/impact.shtml. Accessed 5 Feb 2016

Heeks R (2009) The ICT4D 2.0 manifesto: where next for ICTs and international development? Development Informatics Working paper series. Manchester: Development Informatics Group, Institute for Development Policy and Management

Heeks R (2014) A Development 2.0 research agenda. ICTs for development. 25 April 2014. https://ict4dblog.wordpress.com/2014/04/. Accessed 3 Feb 2016

Heeks R, Bhatnagar S (1999) Understanding success and failure in reform, pp 49–71. In Heeks R (ed) Reinventing government in the information age: international practice in IT-enabled reform. Routledge, London

ICT4SIDs Partnership (2016) ICT Hubs for Rapid Adoption of Samoa Pathway and UN Post 2015 Agenda. http://ict4sids.com/icthubs.html. Accessed 16 May 2016

Imran A, Gregor S (2005) Strategies for ICT use in the public sector in the least developed countries: a cross-country analysis. In Campbell B, Underwood J, Bunker D (eds) Proceedings of the 16th Australasian conference on information systems

Jann W, Wegrich K (2007) Theories of the policy cycle. In: Fischer F, Miller GJ (eds) Handbook of public policy analysis: theory, politics and methods. CRC Press, Boca Raton

Lasswell HD (1951) The policy orientation. In: Lerner D, Lasswell HD (eds) The policy sciences. Stanford University Press, Stanford

Le Blanc D (2012) Back to our common future. sustainable development in the 21st Century (SD21) project: summary for policy makers. Division for Sustainable Development, Department of Economic and Social Affairs, United Nations, New York. https://sustainabledevelopment.un.org/content/documents/UN-DESA_Back_Common_Future_En.pdf . Accessed 12 May 2016

Minges M, Stork C (2015) Economic and social impact of ICT in the Pacific. http://www.theprif.org/index.php/resources/document-library/121-prif-ict-study-report-2015. Accessed 12 May 2016

Nakamura R (1987) The textbook policy process and implementation research. Rev Policy Res 7 (1):142–154

Nedley A (1999) 'Policy transfer and the developing-country experience gap: taking a southern perspective'. (Cited by Stone 2001)

Pryke J (2013a) The Pacific's aid boom. DevPolicy Blog, September 2, 2013. ANU, Crawford School of Public Policy, Development Policy Centre, Canberra

Pryke J (2013b) Rising aid dependency in the Pacific. DevPolicy Blog, September 2, 2013. Canberra, ANU, Crawford School of Public Policy, Development Policy Centre

Sahlins MD (1963) Poor man, rich man, big-man, chief: political types in Melanesia and Polynesia. Comp Stud Soc Hist 5:285–300

Siwatibau S (2009) A voice of reason: the writings of Savenaca Siwatibau. In: Lal Narsey W (ed) [Suva, Fiji]: University of the South Pacific, 2009

Stone D (2001) Learning lessons, policy transfer and the international diffusion of policy ideas. CSGR Working Paper No. 69/01 April 2001. Warwick: Centre for the Study of Globalisation and Regionalisation. http://wrap.warwick.ac.uk/2056/1/WRAP_Stone_wp6901.pdf. Accessed 18 Apr 2016

Strickland E (2015) Navigating the undercurrents: an analysis of Pacific Island regional ICT policy. (Ph.D). University of Queensland, Brisbane

United Nations (2013) Small islands developing states (SIDS) integrated and enabling cooperation framework for the Barbados programme of action and Mauritius strategy for further implementation. Outcome of the inter-regional preparatory meeting for the third international conference on small island developing states. Bridgetown, Barbados, 26–28 Aug 2013. https://sustainabledevelopment.un.org/content/documents/5184265Barbados%20outcome.pdf. Accessed 17 Apr 2016

United Nations (2014a) Sustainable development knowledge platform. SIDS member states. https://sustainabledevelopment.uSmalln.org/topics/sids/memberstates. Accessed 30 Oct 2015

United Nations (2014b) Conference on Small Island developing states. Pacific national sustainable development strategy (NSDS) regional support partnership. SIDS Action Plan no 2328. http://www.sids2014.org/index.php?type=1006&menu=1507&page=view&nr=2328. Accessed 30 Oct 2015
United Nations (2014c) Conference on Small Island developing states. Pacific national sustainable development strategy (NSDS) regional support partnership. SIDS Action Plan no 2328. http://www.sids2014.org/index.php?type=1006&menu=1507&page=view&nr=2751. Accessed 8 Jan 2016
United Nations (2015) United Nations sustainable development summit 2015, 25–27 Sept 2015, New York. https://sustainabledevelopment.un.org/post2015/summit. Accessed 8 Jan 2016
United Nations Conference on Trade and Development. (UNCTAD) (1988) Specific problems of island developing countries, LDU Miscll7. UNCTAD, Geneva
United Nations Conference on Trade and Development (2004) Is a special treatment of small island developing states possible? New York: UNCTAD/LDC/2004/1 unctad.org/en/ldc20041_en.pdf. Accessed 8 Jan 2016
United Nations Department for Social and Economic Affairs (UN DESA) (2008) The United Nations E-government Survey 2008: from E-government to connected governance, Department of Economic and Social Affair. http://unpan1.un.org/intradoc/groups/public/documents/un/unpan028607.pdf
United Nations Department for Social and Economic Affairs (UN DESA) (2014) The United Nations E-government Survey 2014. E-government for the Future We Want. https://publicadministration.un.org/egovkb/en-us/reports/un-E-government-survey-2014.pdf
United Nations Department for Social and Economic Affairs (UN DESA) (2016) The United Nations E-government Survey 2016. E-government for Sustainable Development. Workspace. unpan.org/sites/internet/Documents/UNPAN96407.pdf. Accessed 16 Aug 2016
United Nations Population Fund (2014) Population and development profiles: Pacific Island countries. Pacific Sub-regional Office, United Nations Population Fund, Suva
United Nations Pacific Regional Anti-corruption (UNPRAC) Project (2012) Project document. New York, United Nations Office on Drugs and Crime. http://www.undp.org/content/dam/rbap/docs/Research%20&%20Publications/mdg/PC_UNDP_U NODC_Joint_Pacific_ProDoc.pdf. Accessed 17 Apr 2016
United Nations (2016) Sustainable development knowledge platform. Sustainable development goals. https://sustainabledevelopment.un.org/topics http://www.un.org/sustainabledevelopment/sustainable-development-goals/. Accessed 28 Sept 2016
Waverman L, Meschi M, Fuss M (2005) The impact of telecoms on economic growth in developing countries. Vodafone Policy Paper Ser 2(03):10–24
World Bank (2011) An information-communication revolution in the Pacific. Pacific regional connectivity program. http://www.worldbank.org/projects/P113184/Pacific-islands-regional-connectivity?lang=en. Accessed 8 Jan 2016
World Bank (2014) Environmental and social framework. Setting standards for sustainable development. First draft for consultation. http://www.worldbank.org/content/dam/Worldbank/Event/ECA/central-asia/Safeguards-Consultations-2014-Overview-Presentation-en.pdf. Accessed 28 Sept 2016
World Bank (2015) What is e-government? http://www.worldbank.org/en/topic/ict/brief/e-government. Accessed 7 January 2017
World Bank (2016) Net ODA received (% of GNI). http://data.worldbank.org/indicator/DT.ODA.ODAT.GN.ZS. Accessed 9 Jan 2016
Zhao H (2015) ICTs for sustainable development #ICT4SDG. ITUblog, 23 September, 2015. International Telecommunications Union, Geneva

Chapter 2
Global Policies and Initiatives for E-Government in Pacific SIDs

Gisa Fuatai Purcell and Graham Hassall

Abstract Because the Internet is a world-wide system of intercommunications, significant policy dialogue concerning its organization and functioning takes place at global meetings of governments, technical bodies, business, development agencies, and user groups. This "global architecture" for policy-setting covers both Internet governance and e-government. This chapter focuses on the latter, particularly on the interaction between global agencies and policy dialogue processes with SIDs (Small Island Developing States), more particularly those in the Pacific Islands. While small states, such as those in the Pacific, have not played a large role in decision-making at global level, their sharing of their experience, perspectives, and preferences has contributed to the gradual recognition of their distinct development context. Since Agenda 21, through to the recognition of SIDs as a particular category of developing states in global development assistance frameworks, the challenges of e-government have been given ever-closer consideration. At the current time building e-government capacity and capability in Pacific Island SIDs is shaped by the plan of action flowing from the third SIDs Conference in Apia in 2014 and by the 2030 Agenda for Sustainable Development, a comprehensive set of development goals agreed at the United Nations for the period 2015–2030. Developing the capability of Pacific States has included increasing the extent of their participation in global e-government dialogue processes, as partners in development rather than as recipients only.

2.1 Introduction

This chapter reviews recent global policy dialogue about e-government, with particular emphasis on how it has affected the development of e-government in SIDs. The first section examines global thinking about e-government developed by the International Telecommunication Union (ITU) and other UN bodies. This is fol-

G. F. Purcell · G. Hassall (✉)
Victoria University of Wellington, Wellington, New Zealand
e-mail: graham.hassall@vuw.ac.nz

R. Cullen and G. Hassall (eds.), *Achieving Sustainable E-Government in Pacific Island States*, Public Administration and Information Technology 27,
DOI 10.1007/978-3-319-50972-3_2

lowed by sections that describe the outcomes of the World Summit for the Information Society (WSIS) (in Geneva 2003 and in Tunis 2005); introduce additional inter-governmental organizations such as the World Bank and the Organization for Economic Cooperation and Development (OECD) which operate globally and actively in the development of e-government; examine the third international conference on SIDs and its outcome, the Samoa Pathway; appraise the impact and effectiveness of this global policy architecture in the context of the Pacific Island Countries, and link each of these initiatives to the 2030 Agenda for Sustainable Development.

Internet governance is distinct from e-government. The Working Group on Internet Governance defined the former in 2005 as "the development and application by Governments, the private sector and civil society, in their respective roles, of shared principles, norms, rules, decision-making procedures and programs that shape the evolution and use of the Internet", (Working Group on Internet Governance 2005) and the latter, e-government, which concerns the ways in which the Internet serves the purposes of delivering government. There is, of course, an overlap between these two topics.

Global policy dialogue discovers felt needs, shares knowledge, generates policy options, identifies decision-making authorities, links policies with implementation strategies and appropriate resources, and includes evaluative processes that help to refine subsequent policy engagement in the same or related areas. Whether these processes are cyclic or iterative is less important than whether they are inclusive, deliberative, and, ultimately, emancipatory. Of course, in the absence of global political institutions, the term "global policy" implies recommendations for consideration and action by national governments, rather than directives. Such recommendations by networks of inter-governmental agencies, national governments and a host of other actors, constitute the global governance of e-government at the current time. Their effectiveness depends on the quality of their discursive practices rather than on any constituted authority (Risse 2004).

The Paris Declaration on Aid Effectiveness (OECD 2005) agreed by 100 developed and developing countries added an influential values framework for global dialogue on e-government policy. Borne of OECD member countries' concerns about aid effectiveness, the principles addressing ownership, alignment, harmonization, results management, and accountability seek to change the relationship between donor and recipient countries toward "partnership" for development. This was consolidated through the Busan Partnership Agreement, on which the Global Partnership for Effective Development Co-operation was established in 2011.

An established policy for e-government at global level provides options for the implementation efforts of individual countries and their various development partners. The Internet ecosystem (or Internet universe) consists of several layers of policy makers: globally through such UN agencies as its Department of Economic and Social Affairs (UN DESA), the United Nations Education and Scientific and Cultural Organization (UNESCO), the International Telecommunications Union

(ITU), the World Health Organization (WHO), the United Nations Development Programme (UNDP), and the United Nations Office for Disaster Risk Reduction (UNISDR), and regionally through Regional Internet Registries (RIRs) such as the Asia Pacific Network Information Centre (APNIC), the Asia-Pacific Telecommunity (APT), Regional CERTs (Computer Emergency Response Teams); and individual countries' regulatory systems. Private sector actors, especially network operators, also play significant roles in the development of Internet infrastructure and policy.

The United Nations has convened a series of global meetings which have brought these actors together to make decisions on policy; and UN Agencies such as ITU have been instrumental in providing assistance in such areas as Universal Service/Access, Spectrum, Next Generation Network, among others. Some International Organizations such as Commonwealth Secretariat, OECD, EU, the International Organization for Standardization (ISO), the Internet Corporation for Assigned Names and Numbers (ICANN), the Internet Engineering Task Force (IETF) and the Number Resource Organization (NRO) have also helped shaped key areas from a global policy level. The World Bank and the private sector have focused on financing and building infrastructure, although the World Bank has also developed a significant global knowledge repository. ICANN ccNSOs (Country Code Name Supporting Organizations) have helped shape the ccTLD (Country Code Top Level Domain) policies concerning country code top level domain names, and the Internet Society and NROs have helped influenced and shape policies and aid capacity building surrounding Internet Exchange Points (IXPs) and ccTLDs. Number addressing policies have been shaped by NROs and various RIRs in conjunction with their members.

The significance of participating in the "global information society" is not lost on small island and other developing states. This is confirmed by a 2010 United Nations report which notes that since the economic crisis of 2008, e-government services have assisted governments to enhance transparency, assuage public unease, increase public trust, enhance the tractability of the usage of public stimulus funds and improve the added value of government data holding. The Sustainable Development Goals, adopted by the small island states in late 2015, provide a contemporary focus for the application of ICT solutions to challenges that are global as much as local—climate change for one, and health, education, and sustainable development are also vitally important.

Many international conventions and declarations take into account the needs of SIDs: these include the Convention on the Elimination of All Forms of Discrimination against Women (CEDAW), the Convention on the Rights of the Child (CRC), and the UN Framework Convention on Climate Change (UNFCCC). Such key conventions are applicable to the social welfare of people and therefore must be considered when developing e-government. All UN agencies are required to incorporate relevant conventions into all projects. That is, any e-government policy and action plan must consider that women and children are included, even if the government has not ratified the convention. This also applies to programs addressing natural disasters and effects of climate change.

2.2 United Nations Bodies

Although discussion of information societies and indeed "knowledge societies" pre-dates the Internet, the significance of ICTs emerged through the World Summit on the Information Society[1] (2003–2005), convened on behalf of the UN by the ITU. In the years since that pivotal event, the UN General Assembly has continued to coordinate activities within the UN system focused on the expansion of ICT capability, on the articulation of Internet governance and the roll-out of e-government. United Nations General Assembly Resolution 64/187 of 21 December 2009, for instance, concerning information and communication technologies for development,[2] envisaged ICTs as contributing to solving development challenges, fostering economic growth and competitiveness, improving access to information and knowledge, eradicating poverty and improving social inclusion. It noted the role of government in designing public policies for these outcomes and providing public services "responsive to national needs and priorities", using multi-stakeholder approaches, and in support of national development efforts. To ensure effectiveness across these dimensions, the Resolution looked to a joined-up approach coordinated by the United Nations Group on the Information Society on behalf of the UN System Chief Executives Board for Coordination, and involving the ITU, UNDAD, UNDP, and UNESCO. The UN Secretariat's Department of Economic and Social Development (UN DESA) has been at the forefront of promoting e-government as a key strategy for improving government services and the effectiveness of public policies and programs[3]. And in 2014 the United Nations University established the Portugal-based Operating Unit on Policy-Driven Electronic Governance[4].

2.2.1 *The World Summit for the Information Society*

In 2001 the United Nations General Assembly endorsed the holding of a World Summit for the Information Society (WSIS) to build political will, to take concrete steps toward establishing an "information society for all", and to identify international policies necessary to ensure that every country has access to the Internet and can maintain this access for all members of society. A High-Level Summit Organizing Committee (HLSOC) was established under the patronage of then UN Secretary-General Kofi Annan to coordinate the efforts of the United Nations family to prepare, organize and hold the WSIS. Geneva-based ITU (described in some

[1]http://www.itu.int/net/wsis/index.html.

[2]http://www.un.org/en/ga/search/view_doc.asp?symbol=A/RES/64/187.

[3]http://www.un.org/en/development/desa/news/sustainable/ict-and-e-government-for-sustainable-development-in-SIDs.html.

[4]http://egov.unu.edu/about/about-unu-egov#overview.

detail below) was the lead agency, and an Executive Secretariat (WSIS-ES) was established at ITU headquarters under the authority of the HLSOC to assist in the preparation of the Summit. Host countries Switzerland and Tunisia also established Executive Secretariats to focus on preparing their respective phases of the Summit.

2.2.1.1 WSIS I: Geneva Phase, December 2003

The first phase of the Summit, held in Geneva 10–12 December 2003, marked the culmination of many months of consultations and negotiations among member states, UN experts, the private sector, non-governmental representatives and civil society, who reviewed vast amounts of information and shared a broad spectrum of experiences in issues related to the information society. These consultations and negotiations took place before the summit itself and constituted the preparatory process (ITU 2003). WSIS I focused on developing and fostering a clear statement of political will and taking concrete steps to establish the foundations for an "Information Society for All", reflecting the different interests at stake (ITU 2003). Presentations from the Pacific region were made by Samoa's minister for Communications and Information Technology Palusalue Faapo II, Niue's Associate Minister for Post and Telecommunications John Tiakia and Fiji's Head of the Delegation Abel Caine (who also chaired the Summit's credentials committee). Other PICs present included Palau, Papua New Guinea, Tonga and the Federated States of Micronesia.

At this first phase, more than 11,000 participants from 175 countries gave political support to the Geneva Declaration of Principles and Geneva Plan of Action adopted on 12 December 2003. These principles elaborated on the role of governments and all stakeholders, explained the centrality of ICT infrastructure to connectivity, stressed the importance of access to information and knowledge, advocated capacity building, building confidence and security in the use of ICTs and explained factors essential to creating enabling environments. Furthermore, they emphasized the benefits of ICTs in all sectors, and the goal of ensuring that ICTs promote cultural diversity and identity. Not least among the statement of principles was the importance of international and regional cooperation (World Summit on the Information Society 2003a, b).

The Pacific Islands' Report to WSIS

The Pacific Islands report to WSIS (Pacific Islands Forum Secretariat 2003) summarized the status of ICTs in the Pacific, the principles and strategies underpinning the regional framework, and priorities for implementation. These plans and initiatives were expressed in three documents: Forum Communication Action Plan (Pacific Forum 2002), containing ministerial decisions about what the region will pursue to develop the ICT sector; the Pacific ICT Policy and Plan (PIIPP), which was a framework to implement ministerial decisions by coordinating regional activities and offering guidelines for national activities; and Pacific I4D Initiative, a partnership initiative developed for the World Summit on Sustainable

Development, proposing partnerships and resources for implementation (these reports and plans are considered in more detail in Chap. 3).

Declaration of Principles

The two significant outcomes of this first phase of the WSIS relevant to this study were the Declaration of Principles and the establishment of the Working Group on Internet Governance (WGIG). The Declaration of Principles produced by WSIS I, which was slightly amended during the second summit, continues to be the main declaration on global ICT policy (ITU 2013). The Plan of Action includes a section on e-government, citing the experience of some developed countries on its value in the efficient and effective delivery of public services. Paragraph 16 of the Declaration focusses on the development of ICTs in SIDs, and other UN-approved categories of developing countries. Specific recommendations for the implementation of e-government set out in the WSIS Geneva 2003 Plan of Action include:

- That all countries should aim "to connect all local and central government departments and establish websites and email addresses" (WSIS 2003);
- That governments should "develop national e-government initiatives and services, at all levels, adapted to the needs of citizens and business, to achieve a more efficient allocation of resources and public goods" (WSIS 2003); and
- that governments should "support international cooperation initiatives in the field of e-government, in order to enhance transparency, accountability and efficiency at all levels of government" (WSIS 2005).

Furthermore, the Declaration also addressed the context of small island developing states, noting at paragraph 16: "…we will pay special attention to LDCs, SIDs, LLDCs and countries with economies in transition" (World Summit on the Information Society 2003a). (When first drafted this paragraph referred only to SIDs, but during the first reading LDCs and LLDCs requested that they also be included, given that they were recognized by the UN General Assembly at the Rio conference on the Millennium Development Goals as developing countries with special needs).

The WGIG was established during the Geneva summit to study the future of Internet governance. But because the Working Group was not able to file a comprehensive result of its work, the Geneva summit requested the UN Secretary General to approve continuation of its work for consideration and appropriate action of WSIS II in Tunis in 2005. The request was approved, and WGIG was asked to:

- develop a working definition of Internet governance;
- identify the public policy issues relevant to Internet governance;
- develop a common understanding of the respective roles and responsibilities of governments, existing international organizations and other forums as well as the private sector and civil society from both developing and developed countries (World Summit on the Information Society 2003b).

The Working Group's report, considered in Tunis, proposed the creation of a Global Internet Council consisting of governments and stakeholders to assume the oversight role currently held by ICANN, creation of an International Internet Council to manage most aspects of Internet administration, and creation of three additional bodies: a Global Internet Policy Council, a World Internet Corporation for Assigned Names and Numbers, and a Global Internet Governance Forum (Working Group on Internet Governance 2005)—although only the last of these, the Internet Governance Forum, has so far been established.

2.2.1.2 WSIS II: Tunis, November 2005

The main policy challenge addressed at WSIS II, held in Tunis, Tunisia, in November 2005, was management of Internet governance by the International Corporation for Assigned Names and Numbers (ICANN). Developing countries maintain that since over half of Internet users are from developing countries, a new public body should be established to be responsible for its governance, leaving ICANN free to focus on the technical side of the Internet. The United States maintained that ICANN should continue to govern the Internet, in addition to overseeing technical issues such as the top-level domain names (ccTLDs) and many others. At the eleventh hour, it was agreed that ICANN should continue to manage Internet governance, and the Government Advisory Committee (GAC) should continue to provide input into policies relating to Internet governance. The GAC's key role is to provide advice to ICANN on issues of public policy, especially where there may be an interaction between ICANN's activities or policies and national laws or international agreements. This agreement to leave the technical side of Internet governance to countries with the technical knowledge and sufficient human resources to ensure that the Internet remains available to all, with support of the GAC on policy matters, and regulation and legislative aspects of internet governance, was one of the significant outcomes of the Tunis phase. Most importantly, WSIS established mechanisms for on-going dialogue, notably the Internet Governance Forum, established in 2006.[5]

Pacific Island governments made three presentations at the Tunis Summit: Secretary for Transportation and Communications in the Republic of the Marshall Islands, Jorelik Tibon, addressed the sixth plenary session on 17th November, and Palusalue Faapo II, Samoa's Minister for Communications and Information Technology, addressed the seventh plenary session on the next day. The Federated States of Micronesia's Secretary of the Department of Transportation, Communications and Infrastructure, Robert Yatilman, also spoke that day, in the ninth plenary session. Each of these presentations emphasized the challenges faced

[5]http://www.itu.int/net/wsis/docs2/tunis/off/7.html; http://www.itu.int/net/wsis/docs2/tunis/off/6rev1.html.

by small states in establishing adequate infrastructure and the need for partnerships to achieve their ICT aspirations—but the small number of presentations by Pacific states at both the Geneva and the Tunis summits suggests a low level of engagement with ICT policy at this period.

The Tunis Commitment and Tunis Agenda for the Information Society, adopted on 18 November 2005 (World Summit on the Information Society 2005), also promoted the use of ICTs in government as a priority and crucial target area for ICT-based development interventions as well as human resource and institutional capacity in the public sector (World Summit on the Information Society 2005). Together, WSIS summits I and II established global policy networks, regional and sectoral frameworks for implementation of ICT policies. Another key aspect of the summits was the establishment of a process to encourage and monitor implementation. The WSIS Stocktaking Platform[6] includes a database of ICT-related initiatives, periodic country reports, WSIS Forum and social media presence.

Embedded in the Tunis Agenda was a recommendation that the Summit's results be reviewed after 10 years; this series of events, referred to as WSIS+10, took place in December 2015.[7] Although there were no presentations from the Pacific Islands among the 85 country and organization presentations at the WSIS Forum's High-Level Track Outcomes and Executive Brief, held in Geneva in May 2016, WSIS Stocktaking reports have noted a number of initiatives in PICs in recent years. The Report for 2013, for instance, noted under the heading "Information and communication infrastructure: an essential foundation for the information society," included Samoa and Papua New Guinea among countries in the Asia Pacific region assisted by ITU to develop their own wireless broadband master plans, "which will eventually provide access to broadband-supported services and applications at rates that are affordable and comparable to those in developed countries" (WSIS Stocktaking Report 2013, p. 15). The Report also noted that the ITU, in cooperation with the Republic of Korea and Japan, was updating guidelines on digital broadcasting "to take account of new developments in the area of DTTB and MTV implementation and convergence, while adding sections on IPTV, satellite TVtc" to countries, including Papua New Guinea, Fiji, Vanuatu, and Micronesia (WSIS Stocktaking Report 2013, p. 16).

These initiatives were followed by ITU support for developing wireless broadband master plans, with the Republic of Korea providing technical assistance for the formulation of broadband policies/plans for the Marshall Islands and a number of Asian countries (WSIS Stocktaking Report 2014, p. 17). ITU launched a project to build the first interactive online transmission maps tracking national backbone connectivity (fiber and microwave) and other key metrics in the ICT sector for countries, including Fiji and Papua New Guinea. Vanuatu was being assisted with the transition from analogue to digital terrestrial television broadcasting and for the introduction of mobile television. Micronesia, Kiribati, Nauru, Samoa and Solomon

[6]http://www.itu.int/net4/wsis/stocktakingp/en.

[7]https://publicadministration.un.org/wsis10/GA-High-Level-Meeting.

Islands were assisted with updating guidelines on digital broadcasting to take into account new developments in the area of digital terrestrial television broadcasting (DTTB) and mobile television (MTV) implementation and convergence, while adding sections on Internet Protocol television (IPTV), satellite television and other applications. All of these activities were accompanied by capacity-building training on conformity assessment procedures and testing for different domains, such as electromagnetic compatibility (EMC) and mobile terminals (WSIS Stocktaking Report 2014, p. 17–18). In the Marshall Islands, a partnership between the Mejit (a remote region of five islands and 29 atolls with just 300 to 400 inhabitants) local government, the Marshall Islands National Telecommunication Authority (MINTA) and Japan's Ministry of Internal Affairs and Communications, funded by an Asia-Pacific Telecommunity grant, resulted in a "Far Remote Island MCT project" to provide connectivity to Mejit (WSIS Stocktaking Report 2014, p. 19–20).

2.2.2 UN Department of Economic and Social Affairs

The desire to improve e-government performance globally has stimulated the development of bench-marking, indexing and diagnostic projects by a range of global agencies that includes several within the United Nations Organization. These include UN DESA's *United Nations E-Government Survey* (seven editions from 2003 to 2014), DPADM (Department of Public Administration and Development Management) and UNPAN (UN Public Administration Network); and the ITU's ICB4PAC (already noted above).

Global rankings of UN DESA's *United Nations E-Government Survey* for 2014 (United Nations Department of Economic and Social Affairs, 2014) show Fiji to be the best performing PIC (85th), with the remainder ranged between 98 and 188 out of 190 countries surveyed. Whereas such global surveys have their place, they raise the issue as to whether developing countries, particularly those that are SIDs, can benefit from being compared with significantly larger and better resourced countries. The ITU's ICB4PAC project, in contrast, concerning "Capacity Building and ICT Policy, Regulatory and Legislative Frameworks for Pacific Island Countries" (International Telecommunications Union 2013), applies a methodology premised on the need to consider the circumstances of individual countries, and to tailor development projects to the priority areas that they have identified. It was less concerned with rankings than with ensuring that each PIC developed appropriate laws and institutions that provide National ICT policy, Licensing, Cybercrime and Universal Access and Service.[8]

E-government is one of the main initiatives in the work of UNPAN, a division of the United Nations Department of Economic and Social Affairs mandated to

[8]http://www.itu.int/en/ITU-D/Projects/ITU-EC-ACP/ICB4PAC/Documents/In-country%20support%20documents/ICB4PAC_KPI_In-country_Technical_Assistance_10Oct2013.pdf.

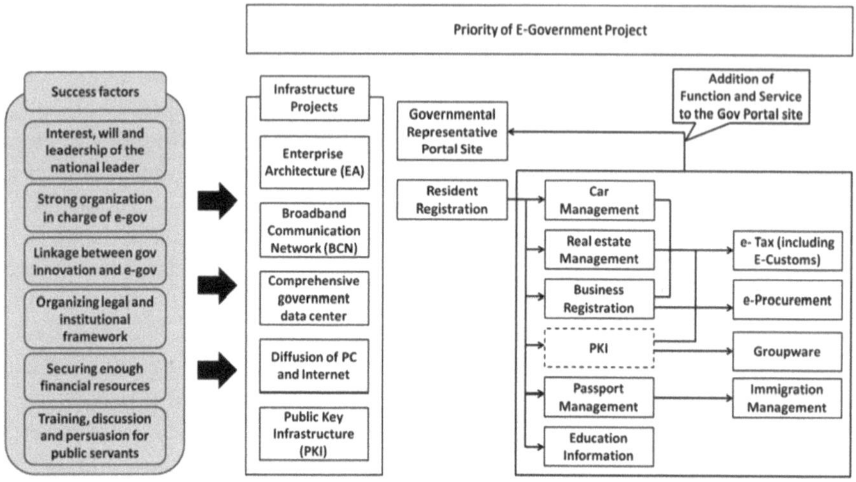

Fig. 2.1 Success factors and activity priorities for e-government in SIDs. *Source* Lee (2014). E-Government for Promoting Sustainable Development in Small Island Developing States (SIDs): Background Paper for the Special Event of the 69th Session of the United Nations General Assembly Second Committee. New York: United Nations, p. 134

provide capacity building and knowledge management to achieve effective, accountable and inclusive institutions. UNPAN provides online learning about e-government. Another of UN DESA's programs, DPADM, has a program entitled UNPACS—UN Public Administration Country studies, which maintains an e-government Development Database.[9] UN DESA's Project Office on Governance (UNPOG), established in Seoul Korea in 2005, to promote participatory, transparent and effective democratic governance in member states, has rapidly developed an active program in support of e-government that includes activities specifically addressing the needs of PICs.[10]

In addition to indexes and the challenges that they pose for comparing small states with large ones, there are UN agency studies articulating diagnostic frameworks for e-government. A background paper (Lee 2014) put to the 69th United Nations General Assembly Second Committee in 2015, for instance, laid out "success factors and activity priority flow chart" for e-government in SIDs (Fig. 2.1).

In addition to identifying such features as leadership, organizational capacity, innovation, legal environment, and financing and training, this report articulated four inter-weaving policy areas critical to the overall success in the implementation

[9]https://publicadministration.un.org/egovkb/en-us/#.V0T0amlf37Y.

[10]See "Capacity Building Training Workshop on E-Government Development in Pacific SIDs", November 9, 2015, and other activities, at http://www.unpog.org/activity/activity02.php.

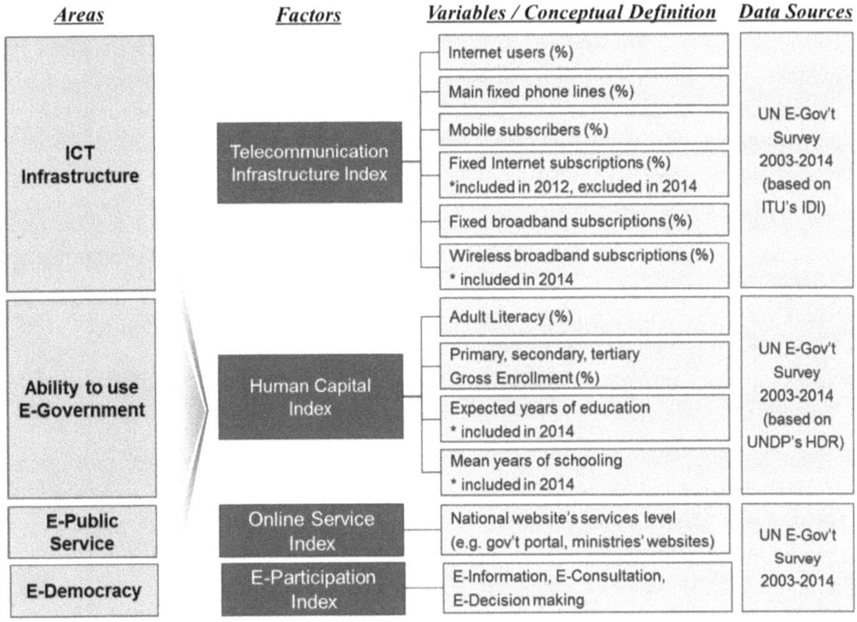

Fig. 2.2 Policy areas critical to success of an e-government strategy. *Source* Lee (2014). E-Government for Promoting Sustainable Development in Small Island Developing States (SIDs): Background Paper for the Special Event of the 69th Session of the United Nations General Assembly Second Committee. New York: United Nations, p. 16

of an e-government strategy: ICT infrastructure, ability to use e-government, e-Public Service and e-Democracy (see Fig. 2.2) (Lee 2014, p. 16).

The studies provided in the following chapters illustrate the complex interconnections between these four areas. It is often the case that those leading the development of ICT in a country are well-aware of their potential.

2.2.3 The International Telecommunications Union (ITU)

The Geneva-based International Telecommunications Union ITU, the oldest United Nations agency and one of the primary actors in the establishment of global ICT policy, holds specific responsibility for the establishment of technical standards and telecommunication rules for spectrum and frequency allocation. Its programs deal with climate change, least developed countries and small island developing states, regulatory and market environment, emergency telecommunications, cybersecurity, ICT applications, capacity building, ICT statistics and indictors, technology and network development, digital inclusion and spectrum management and broadcasting.

Under the umbrella of the WSIS Declarations, ITU is among the global agencies developing ICT capability with a focus on the Pacific Islands region. Its Development Department, in particular, focuses on policy development, regulation and legislation in all areas of ICTs; emergency telecommunication and climate change adaptation; infrastructure development and concentrated assistance to member states. This assistance includes:

- all phases of disaster management, i.e. disaster preparedness including early warning, disaster response/relief and rehabilitation of telecommunication networks;
- making a smooth transition from analogue to digital broadcasting;
- developing broadband access in urban and rural areas; and
- developing appropriate policy and regulatory framework, enhancing skills, increasing information-sharing, and strengthening regulatory cooperation.

The ITU engages with numerous global agencies and networks to promote the effective dissemination of ICTs. There is, for instance, the Partnership on Measuring ICT for Development, consisting of 14 regional and international organizations established in 2004. In 2013 the ITU collaborated with the European Community to survey the status of National ICT policies in PICs, as part of a broader study of ICT policies in Africa, Pacific and Caribbean countries (International Telecommunications Union 2013).

2.3 The World Bank

In addition to the global ICT and e-governance initiatives under UN auspices introduced above, are those of the World Bank. As a development bank that operates under considerably different governance arrangements to those of the UN, the World Bank has increasingly lent resources, knowledge and expertise to finance and otherwise capacitate the development goals set by the UN. In addition, it has developed indicators to measure the economic gains of e-government implementation (World Bank 2016).

World Bank support for development of telecommunications in the Pacific Islands dates to at least the 1960s. In Papua New Guinea, for example, it supported its first telecommunications project from 1968, a second from 1972 and a third from 1997. The first two projects focused on the expansion of phone lines and satellite, the third introduced data communications, management information systems and the organizational structure for the Post and Telecommunication Corporation (PTC).

In 2011 the World Bank initiated a five-year Pacific Regional Connectivity Program to increase bandwidth, capacity and competition in the ICT sector, and access to the Internet and telephones, including mobile services. The Program, initially designed to be implemented in five phases in five different island countries at an estimated cost of US$186.5 million (plus IDA contribution of US$88.2 million), seeks to reduce the cost and increase the availability of international bandwidth for participating countries, and thereby facilitate the development of a wide range of ICT applications to support social and economic development in the region. These projects invariably include additional development partners such as the Asian Development Bank,[11] the Pacific Region Infrastructure Facility and private sector actors.

In 2012 the WB announced a new focus on the development of ICT capability, reliant on three principles themes:

- *Transformation*: Making development more open and accountable and improving service delivery by, for instance, facilitating citizen feedback to governments and service providers;
- *Connectivity*: Scaling up affordable access to broadband, including for women, disabled citizens, disadvantaged communities and people living in remote and rural areas;
- *Innovation*: Developing competitive IT-based service industries and fostering ICT innovation across the economy with a focus on job creation, especially for women and youth.[12]

The emphasis on capital investment and economic development and growth, Strickland suggests, has emphasized the expansion of market access and neoliberalism, rather than the mere provision of infrastructure and access alone (Strickland 2015, p. 105). In the case of Solomon Islands and Papua New Guinea, for instance, the bmobile expansion project of 2011 provided US$9 million to support the expansion of bmobile's ICT infrastructure following the cessation of Telecom PNG's monopoly position.[13] Samoa's Telecommunications and Postal Sector Reform Project (WB075739 0), in operation from 2002 until early 2011, included IDA financing to undertake Competition and Sector Reform designed to improve the regulatory framework (particularly in the areas of interconnection, convergence and downstream effects of privatization) and to privatize SamoaTel.

Box 2.1 World Bank Pacific Regional Connectivity Program

Phase 1

Tonga-Fiji Submarine Cable Project (2011–2016) provided a cable, strengthened the telecommunications regulatory framework, and increased competition—to develop rural services and infrastructure, and infrastructure services for private sector development.

[11] ADB engagement with e-government in the Pacific Islands is addressed in Chap. 3.

[12] http://www.worldbank.org/en/news/press-release/2012/07/25/world-bank-group-announces-new-focus-on-using-ict-for-greater-development-impact.

[13] http://www.adb.org/projects/44937-014/main#tabs-0-2.

Phase 2

Solomon Islands Broadband for Development Project (2012–2017) to build a submarine cable connecting with an existing cable between Guam and Sydney to expand access to markets and services for Solomon Islands' geographically isolated areas.

Phase 3

Samoa Connectivity Project (2016–2020) to build a Submarine Cable System between Samoa and Fiji and landing stations; to reduce the cost and increase the availability of Internet services, support the private sector and rural areas, and build regulation and competition policy capacity.

Phase 4

The Vanuatu Telecommunications and ICT Technical Assistance Project to strengthen oversight institutions for telecommunications and ICT in Vanuatu.
Kiribati/Project (2012–2017) to provide ICT Policy and Legal Support, and ICT Regulatory Support, to enable "transition to a market-driven telecommunications sector and facilitating improved connectivity for the Outer Islands."

Phase 5

Northern Pacific Connectivity, approved in 2014, to reduce costs and increase availability of ICT services for Palau and the Federated States of Micronesia by constructing a seabed cables between (1) Palau, Yap, and Guam; (2) Chuuk and Pohnpei; and (3) a link to Kosrae.

Sources: World Bank Website; (Beschorner et al. 2015) (Beschorner 2015)

2.4 Other Global Initiatives on E-Government

A key component of the architecture of global policy for e-government is the ability of multiple government and non-government organizations to share and integrate information across their traditional organizational boundaries. E-government interoperability represents a set of multidimensional, complementary and dynamic capabilities needed among these networks of organizations in order to achieve successful information sharing. This complexity challenges researchers and practitioners to develop multiple and diverse interoperability capabilities. Researchers and practitioners alike are investing in efforts to build the necessary understanding to create new capability for coordinated action (Jansen et al. 2010). A few such global initiatives are the Internet Governance Forum, the Open Government Partnership, the OECD's digital government and the Inter-Parliamentary Union's (IPU) e-Parliament initiatives, each of which will be introduced here in brief.

As noted above, the Internet Governance Forum was established in 2006 as a multi-stakeholder forum at which public policy issues related to key elements of Internet governance could be addressed. These issues include sustainability, security, stability and the desirability of having all governments making a commitment to ensuring that these objectives are met. The IGF serves, in addition, as a platform for addressing the gaps in ICT infrastructure referred to as the "digital divide".

Increasing the transparency of government information is being promoted through the Open Government Data Project (OGD[14]). OGD assessment focuses on sectoral datasets for education, health, finance, social welfare, labor and environment. The project is also concerned about the quality of online data, and has identified eight principles for OGD: complete, primary, timely, accessible, machine-processable, non-discriminatory, non-proprietary and license-free. Of all the PICs, only Papua New Guinea has as yet associated with the Open Government Data project.[15] (PNG attended a meeting in 2014, at which it signaled its intention to join, but is yet to meet all the technical requirements for full membership.)

As described in more detail in Chap. 6, the Inter-Parliamentary Union, when calling on national parliaments to attend the WSIS in Geneva, urged them to make full use of the ICTs to enhance, among other objectives, the "effectiveness, efficiency and transparency of their activities"; to establish "… the necessary structures to follow up and monitor annually the development of e-government" in their countries; to establish "… national and regional policies that incorporate the development of information and communication infrastructures and existing resources". (Inter-Parliamentary Union 2003) The IPU has subsequently issued a *World e-Parliament Report* biennially since 2008 (Inter-Parliamentary Union 2016) and hosted a biennial World e-Parliament Conference since 2007.

Observatories for the assessment in e-government progress are also in place, such as Waseda University's Institute of e-government, which produces an annual ranking survey based on nine indicators and 32 sub-indicators, with the principal nine being Network Preparedness/Digital Infrastructure; Management Optimization; Online Services/Applications; National Portal/Home page; Government Chief Information Officer; e-government promotion; E-Participation/ Digital Inclusion, Open Government Data; and Cyber Security. Fiji is the lone Pacific Island country in this survey and in 2015 was ranked 18th of the 18 Asia Pacific countries included (Obi 2015).

---

[14]http://www.opengovpartnership.org.

[15]http://www.opengovpartnership.org/countries.

2.5 Small Island Developing States (SIDs)

As early as the United Nations Conference on Environment and Development (UNCED), held in Brazil in June 1992, SIDs have been recognized as a group of countries facing specific social, economic and environmental vulnerabilities. This recognition was made specifically in the context of Agenda 21, Chap. 17 G (United Nations 1992).[16] The SIDs agenda has evolved since the 1992 UN Conference on Environment and Development and Agenda 21 through the SIDs conferences, the first in Barbados in 1994 (Barbados Program of Action), the second in Mauritius in 2005 (United Nations. Secretary General 2005) and the third in Samoa in 2014 (The Samoa Pathway).

The Mauritius Strategy conference agreed on further implementing the Barbados Plan of Action, referred in Chap. XX to the goal of enhancing ICT development, including connectivity and access to affordable hardware and software. While there was no specific mention of e-government, the focus on ICT development and connectivity provided some stimulus in the direction of e-government.

2.5.1 SIDs 3

The third SIDs conference, held in Samoa 2–5 September 2014, brought more direct scrutiny to levels of support for e-government in PICs (20 of the 52 SIDs are PICs). SIDs III included numerous pre-events and side events, including some directly related to ICTs. The ITU, for instance, held a pre-event on "Addressing Connectivity in SIDs" on 29 August, and a multi-stakeholder dialogue pre-event on Disaster Risk Reduction and Climate Change Adaptation for SIDs on 30 August. The most significant of these side events, which addressed the theme "Bridging Barriers of Broadband for SIDs through genuine and durable partnerships" under the moderation of UNDP Administrator Helen Clark, concluded with the signing of an agreement between the ITU and 11 Pacific SIDs, and between the ITU and development agencies to develop a project for satellite connectivity in remote islands and rural areas. Expanding connectivity in remote areas via very small aperture satellites improves access to emergency telecommunications and to Internet access for education, health, climate change, agriculture, environment and other essential information. Should disaster strike, it will allow the satellites to send photos of damage and advise relevant agencies about their priority needs. In addition to an "outcomes document" (Samoa Pathway) the SIDs III Conference established 300 partnerships addressing such priority areas as Sustainable

---

[16]The United Nations also recognizes the 38 UN-Member States belonging to the Alliance of Small Island States (AOSIS), an ad hoc negotiating body established by SIDs at the United Nations. AOSIS also includes other island entities that are non-UN Member States or are not self-governing or non-independent territories that are members of UN regional commissions.

Economic Development, Climate Change & Disaster Risk management, Social Development in SIDs, Health and NCDs, Youth and Women, Sustainable Energy, Oceans, Seas and Biodiversity, Water and Sanitation and Food Security and Waste Management.

2.5.1.1 The Samoa Pathway

Although there are just three references to ICTs in the "Samoa Pathway",[17] it is nonetheless the first SIDs outcome statement to refer to the development of ICT Infrastructure. Paragraph 23 notes that one of the challenges of developing ICTs in SIDs is having the ability to sustain infrastructure networks with limited human and institutional capacity. Paragraph 27 (b) refers directly to enhancing the enabling environment at national and regional levels relating to infrastructure. Paragraph 27 (g) also urges the international community to continue to promote and enhance the use of ICTs for development and economic sustainability. In paragraph 37 the conference urged developed countries to increase technology, finance and capacity-building support to increase mitigation of challenges facing SIDs. Paragraph 49 urges the international community to continue to provide adequate support, including in the areas of capacity-building and technology transfer. More specifically, paragraph 109 (h) relates to establishing national and regional ICT platforms and information dissemination hubs to facilitate information exchange and cooperation. For the first time since the Barbados Plan of Action, a SIDs conference outcome document has included a specific section on technology. This new section, at paragraph 110, recognizes the importance of SIDs having access to appropriate reliable, affordable, modern and environmentally sound technologies in order to achieve their sustainable development objectives.

The conference also agreed in this outcome document that science, technology and innovation are essential enablers and drivers for sustainable development. The conference reaffirmed, furthermore, in paragraph 111 its commitment to help SIDs increase connectivity and the use of ICT through improved infrastructure, training and national legislation, as well as public and private sector involvement. Yet another section emphasizes the importance of data and statistics, which are essential to the measurement and evaluation of development outcomes.

These four paragraphs, 23, 27, 110 and 111, provide SIDs with the mandate to focus on developing appropriate, reliable, affordable and secure ICT infrastructure which will allow the development of e-government. They provide Pacific SIDs with an assurance that the international community will be ready when they seek financial and technical assistance for the development of ICT infrastructure, policy, regulation and regulatory frameworks. The speed at which the aspirations of the

[17]SAMOA is an acronym for SIDs Accelerated Modalities Of Action: http://www.sids2014.org/samoapathway.

Samoa Pathway are met depends in large part on the extent to which ICT infrastructure and platforms are developed within each country.

Immediately after SIDs 3, UN DESA's United Nations Project Office on Governance (UNPOG) conducted customized research on e-government for promoting sustainable development in SIDs and co-organized a series of capacity development events in collaboration with DPADM and other institutions. These included a capacity development workshop during the 2014 UN Public Service Forum, a session on "ICT and E-Government in SIDs: Responding to the SAMOA Pathway Call for Action" during the Sixty-ninth Session of the United Nations General Assembly in November 2014, a training workshop for SIDs in February 2015 in collaboration with National Information Society Agency (NIA) of the Republic of Korea and a workshop on sustainable development and disaster risk management using e-government in March 2015.

2.6 2030 Agenda for Sustainable Development Goals (SDGs)

Although ICTs have always been important to development thinking and practice (as encapsulated in, for instance, the term "ICT4D"), a new global program initiated at the UN General Assembly in September 2015 has expanded this connection.

The global development plan known as the 2030 agenda, or the "Sustainable Development Goals" (SDGs) (United Nations General Assembly 2015b), replaces eight "Millennium Development Goals" pursued in the developing countries during the period 2000–2015. The SDGs expand the focus as well as the reach of development assistance, for they apply to all countries, not just those in the developing world, and include environmental, economic and infrastructural goals. Each of these 17 goals has targets for every country to achieve by 2030.

The 2030 SDGs provide a "comprehensive, far-reaching and people-centered set of universal and transformative Sustainable Development Goals and Targets" (United Nations General Assembly 2015a) supported by a host of associated initiatives: the Declaration of Principles and Plan of Acton adopted by WSIS (discussed below); the Addis Ababa Action Agenda on Financing for Development[18]; the establishment of Broadband Commissions for Sustainable Development[19]; the activities of the Commission on Science and Technology for Development; and the Internet Governance Forum—among others.

Goal 9 (c) aims to significantly increase access to information and communication technology and to provide universal and affordable access to the Internet in least developed countries by 2020 (United Nations General Assembly 2015a; United Nations General Assembly 2015b, p. 25). This paragraph, while referring

[18]http://www.un.org/esa/ffd/wp-content/uploads/2015/08/AAAA_Outcome.pdf.

[19]http://www.broadbandcommission.org.

only to LDCs, exemplifies a significant change in the mindset of member states: it identifies the importance of ICT infrastructure improvement as the agent for change which provides an international plan of action for improving the ICT infrastructure. An improved infrastructure with broadband that facilitates a faster Internet is the platform required for developing e-government.

At the same General Assembly resolution 70/125[20] reaffirmed the WSIS vision of a people-centered, inclusive and development-oriented information society, assessed progress to date, identified gaps and challenges and made recommendations for the future.

Through extensive consultation, the WSIS Forums of 2015 and 2016 have mapped 11 ICT "lines of action" to assist in achieving the 17 SDGs. These include Internet access and affordability, infrastructure, innovation, information access, increased efficiency, early warning, disaster risk management, knowledge sharing and data.[21] Each of these has pertinence to the Pacific Islands context, and will significantly influence "ICT4D" activities in the region in the coming decades.

2.7 Conclusion

The potential of ICTs to contribute significantly to global development is evident in the extent to which they are embedded in the current global development architecture. E-government holds the potential to deliver public services efficiently, effectively and transparently, and has grown through extensive global dialogue sustained by the main intergovernmental organizations operating at global level. The UN and the World Bank have taken leading roles in the articulation of policy frameworks (in the case of the UN) and connectivity and capability (in the case of the World Bank) toward transformative e-government. Partly as individual governments and partly as a regional bloc (as will be discussed in Chap. 3), Pacific Island countries have attracted support for ICT infrastructure and for the development of capability in policy and application, and there has been growing PICs participation in global and regional meetings sponsored by ICANN, IGF and other technical and consultative groups.

The rapid development of ICTs in the past two decades has been accompanied by intergovernmental action at global level on a number of fronts. The most fundamental issue has been, and remains, determining who has "authority over"—who controls—the Internet. Countries can see the merits of ensuring that the Internet remains a "global public good" at the same time as they ensure that their national interests remain protected. Simultaneously with this perception, dialogue about ownership and control has been about usage, including use by government in an

[20]Outcome document of the high-level meeting of the General Assembly on the overall review of the implementation of the outcomes of the World Summit on the Information Society.

[21]http://www.itu.int/net4/wsis/sdg/.

expanding range of applications referred to as e-government. UN agencies such as the ITU have played a major convening role in this global policy dialogue, although other global agencies as well as global telecommunications actors have also been influential. The UN's Division for Public Administration and Development Management, for instance, has recommended that SIDs base integrated and comprehensive e-government strategies on seven "key success factors", commencing with political commitment, which should lead to adequate legal frameworks, a coordinating body, national e-government policy, sufficient budget, plus awareness and necessary government reform.

Small states, especially the small states in the Pacific region, have not been major actors in global policy settings. Some, but not all, have been represented at the global meetings and only a few among them have made statements in plenary sessions. Some struggle to meet voluntary reporting opportunities. Fiji, for instance, was the 690 only PIC to respond to a survey for the Final WSIS Targets Review: (ITU 2014). However, participation in global policy processes has steadily increased, as global agencies have tailored their programs for small island states and in some instances, for the Pacific Islands. The global agencies have shared knowledge of e-government gained from their experience in developing countries across the globe. Chapter 3 will consider in more detail the ways in which these global initiatives have influenced implementation within the countries of the Pacific Islands region.

References

Beschorner N (2015) ICT development impacts—pacific potential. Paper presented at the ICT Ministerial Meeting, Tonga, June 17

Beschorner N, Kuek SC Narimatsu J (2015) Information and communication technologies for jobs in the Pacific Island Countries. Washington: World Bank. https://openknowledge.worldbank.org/handle/10986/22963. Accessed 24 Oct 2016

International Telecommunications Union (2013) National ICT policy: knowledge-based report. http://www.ipu.org/conf-e/109-3.htm. Accessed 17 Jun 2016

International Telecommunication Union (2014) Final WSIS Targets Review: Achievements, Challenges, and the Way Forward, International Telecommunication Union, Geneva. http://www.itu.int/en/ITU-D/Statistics/Documents/publications/wsisreview2014/WSIS2014_review.pdf. Accessed 24 October 2016

Inter-Parliamentary Union (2003) The contribution of new information and communication technologies to good governance, the improvement of parliamentary democracy and the management of globalization. Resolution adopted unanimously by the 109th Assembly. Accessed 17 June 2016

Inter-Parliamentary Union (2016) World e-Parliament Report 2016. IPU, Geneva. http://www.ipu.org/pdf/publications/eparl16-en.pdf. Accessed 24 Oct 2016

Inter-telecommunications Union (2013) Report on the WSIS Stocktaking 2013, Geneva. http://www.itu.int/net/wsis/stocktaking/docs/reports/stocktaking_report_2013.pdf. Accessed 24 October 2016

Inter-telecommunications Union (2014) Report on the WSIS Stocktaking 2014, Geneva. http://www.itu.int/dms_pub/itu-s/opb/pol/S-POL-WSIS.REP-2014-PDF-E.pdf. Accessed 24 October 2016

Jansen J, de Vries S, van Schaik P (2010) The contextual benchmark method: benchmarking e-government services. Gov Inf Q 27(3):213–219
Lee YB (2014) E-Government for promoting sustainable development in Small Island developing states (SIDs): background paper for the special event of the 69th session of the United Nations general assembly second committee. http://workspace.unpan.org/sites/Internet/Documents/UNPAN93784.pdf. Accessed 24 Oct 2016
Obi T (ed) (2015) 2015 Waseda—IAC International E-Government Ranking Survey. Waseda University Institute of e-Government, Tokyo
Organisation for Economic Co-operation and Development (2005) Paris Declaration on Aid Effectiveness. Paris: Organisation for Economic Co-operation and Development, Paris
Pacific Islands Forum. (2002). Communications action plan—2002 review. Pacific Islands Forum Secretariat, Suva. http://www.forumsec.org/resources/uploads/attachments/documents/Forum%20Communications%20Action%20Plan%202002%20Review.pdf. Accessed 10 Oct 2016
Risse T (2004) Global governance and communicative action. Gov Opposition 39(2):288–313
Pacific Islands Forum Secretariat (2003) Pacific Islands input to the World Summit on the Information Society. PIFS, Suva
Strickland E (2015) Navigating the undercurrents: an analysis of Pacific Island regional ICT policy. (Ph.D). University of Queensland, Brisbane
United Nations Department of Economic and Social Affairs (2014) United Nations E-Government Survey 2014: E-Government for the Future We Want. United Nations, New York
United Nations General Assembly (2015a) Resolution A/RES/70/184. Information and communications technologies for development. http://www.un.org/en/ga/search/view_doc.asp?symbol=A/RES/70/184. Accessed 24 Oct 2016
United Nations (2015b) Transforming our world: the 2030 agenda for sustainable development, A/RES/70/1 C.F.R.
United Nations. Secretary General (2005) Mauritius strategy for the further implementation of the programme of action for the sustainable development of Small Island developing States. United Nations, New York. http://unctad.org/en/Docs/a60d401_en.pdf. Accessed 14 Oct 2016
Working Group on Internet Governance (2005) Report of the Working Group on Internet Governance. ITU, Geneva. http://www.wgig.org/docs/WGIGREPORT.pdf. Accessed 14 Oct 2016
World Bank (2016) World Development Report 2016: digital dividends. The World Bank, Washington. www.worldbank.org/en/publication/wdr2016. Accessed 17 Oct 2016
World Summit on the Information Society (2003) World Summit on the Information Society. Plan of Action. Geneva. http://www.itu.int/net/wsis/docs/geneva/official/poa.html. Accessed 24 October 2016
World Summit on the Information Society (2003a) Declaration of principles. Building the information society: a global challenge in the new millennium. Document WSIS-03/GENEVA/DOC/4-E. http://www.itu.int/net/wsis/docs/geneva/official/dop.html. Accessed 24 Oct 2003
World Summit on the Information Society (2003b) Plan of action. Document WSIS-03/GENEVA/DOC/5-E. http://www.itu.int/net/wsis/docs/geneva/official/poa.html. Accessed 10 Oct 2016
World Summit on the Information Society (2005) Tunis commitment. Tunis: http://www.itu.int/net/wsis/docs2/tunis/off/7.html. Accessed 24 Oct 2016

Chapter 3
E-Government Policy Processes at Regional and National Levels in the Pacific: Actors, Institutions and Networks

Graham Hassall

Abstract This chapter explores how policy processes at regional and national levels assist Pacific Island countries (PICs) and their various development partners translate the broader developments and possibilities advocated in global forums into workable solutions for each small states' unique sets of circumstances. After first identifying and explaining the roles of the actors, institutions and networks engaged in e-government at regional level, it traces regional e-government policy dialogue over the past three decades. The chapter then examines policy processes at national level. Although policy dialogue at regional and national levels is related, it is only at national level that legislative frameworks are established, policies are implemented and major expenditure is incurred. Regional-level policy dialogue identifies needs and assists the coordination of capacity building and development assistance, but the roll-out of e-government in PICs depends to a great extent on the leadership and coordination capacity in individual countries.

3.1 Regional ICT Policy Processes: Actors, Institutions and Networks

Chapters 1 and 2 introduced the global context of small island states. This chapter explores how policy processes at regional level assist Pacific Island countries (PICs) and how their various development partners translate the broad developments and possibilities advocated in global forums into workable solutions for each small state's unique sets of circumstances. The first section introduces many actors, institutions and networks that collectively engage in regional policy dialogue. It highlights the extent to which the small island states of the Pacific—as well as the international agencies and private sector actors interested in the region—gather in

---

G. Hassall (✉)
Victoria University of Wellington, Wellington, New Zealand
e-mail: graham.hassall@vuw.ac.nz

R. Cullen and G. Hassall (eds.), *Achieving Sustainable E-Government in Pacific Island States*, Public Administration and Information Technology 27,
DOI 10.1007/978-3-319-50972-3_3

collective dialogue spaces as a matter of efficiency. The second major section of the chapter shifts the focus to national level. Although policy dialogue at regional and national levels is related, it is only nationally that legislative frameworks are established, policies implemented and major expenditure incurred. Regional level policy dialogue identifies needs and assists the coordination of capacity building and development assistance, but the roll-out of e-government in PICs depends to a great extent on the leadership and coordination capacity in individual countries. The political economy of each country then becomes a factor in the speed and extent of e-government implementation.

PICs cooperate on a wide range of policy issues through regional institutions and policy networks formed at various levels. These institutions and organizations include:

- those established by the PICs themselves (and those established before independence) to cover a wide range of policy domains;
- those established by Asia-Pacific organizations to service PICs, such as the Asian Development Bank, and other organizations/programs formed by international bodies to reach the Pacific region, such as the multi-stakeholder Pacific Region Infrastructure Facility; and
- those stakeholders specifically focused on telecommunications, such as Pacific Islands Telecommunications Association and the Asia Pacific Telecommunity.

3.1.1 CROP Agencies

A considerable number of inter-governmental working groups, professional bodies, private sector associations and regional and international development agencies work collaboratively to achieve policy coordination in the Pacific region. Nine of the major actors are designated as members of the Council of Regional Organizations in the Pacific (CROP), and the meetings of their representatives, on a wide range of issues, produce decisions intended to guide activities within individual countries. The Council does not direct the work of its member agencies, but does have a strong agenda-setting role. Other bodies, while not CROP members, may be equally important or influential and are invited into CROP consultations and activities on an ad-hoc basis. The most prominent of these are set out in the following table (Table 3.1).

CROP agencies vary greatly in the reasons for their establishment, their size and their mandates (Graham 2008). In matters of ICT and e-government, they have either led or have participated in extensive regional dialogue on ICT-related matters since 1999. However, a kcy question concerns the efficiency and effectiveness of regional, versus national, planning. As will become evident, policy dialogue at regional level has proven suitable for setting standards and undertaking needs assessments, but lack of formal decision-making authority or convening power at regional level concerning ICTs and the roll-out of e-government has meant that

Table 3.1 CROP agencies and other regional development agencies

CROP agencies	Pacific Islands Forum; the Pacific Community; University of the South Pacific; Pacific Islands Development Program; South Pacific Tourism Organization; Pacific Power Association; Forum Fisheries Agency; Pacific Aviation Safety Office; Secretariat of the Pacific Regional Environment Program
Other regional organizations and networks	PASO, Fiji School of Medicine, Pacific Islands Law Officers Network, Pacific Islands Private Sector Organization, Pacific, PITA, Pacific Islands Chapter of the Internet Society, Internet Governance Forum
Asia Pacific organizations	Asian Development Bank, UNESCAP, Asia Pacific Telecommunity, Asia Pacific Top Level Domain, Asia Pacific Regional At Large Organization

project implementation, especially when involving legislative action and the installation of infrastructure, has required bi-lateral arrangements between PICs and development partners or vendors.

3.1.1.1 Pacific Islands Forum

The 18-member Pacific Islands Forum Secretariat (the independent PICs plus Australia and New Zealand), established in 1971 with its seat in Suva, Fiji, provides policy support and coordination to the Pacific Island heads of government, who meet annually to made decisions on matters of mutual concern.[1] The Secretariat has divisions for economic governance, political governance and security, strategic partnerships and coordination. As the coordinating agency for all policy initiatives at regional level, the Forum has taken an active interest in such ICT and e-government challenges as improving access to communications technology, reducing the cost of providing ICTs, establishing higher bandwidth to the global ICT "backbone", removing inappropriate regulatory environments in order to foster higher levels of investment and strengthening ICT skills. The Forum's Regional Security Committee has addressed the challenge of cyber-fraud,[2] and the Forum Leaders' meeting has deliberated for some years the possibility of a regional ICT Council. While the Forum had its own ICT adviser for a time, from 2004 its focus shifted to the "Pacific Plan" for regional collaboration (which included consideration of an 'ePacific Plan'. The current Framework for Pacific Regionalism,[3] which seeks not merely increased cooperation among the PICs, but a "pathway" toward

[1]French Polynesia and New Caledonia were admitted as full members of the Pacific Islands Forum in 2016.

[2]http://www.itu.int/en/ITU-D/Regional-Presence/AsiaPacific/Documents/Events/2014/September-COP/Nola_Faasau.pdf.

[3]Pacific Islands Forum (2015).

"deeper integration" (Pacific Plan Review 2013), continues to include consideration of regional ICT collaboration and leadership.

3.1.1.2 The Pacific Community

The 26-member Pacific Community (SPC) established in 1947 in Noumea, New Caledonia to help restore stability after the Second World (known until 1997 as the South Pacific Commission and until 2016 as Secretariat of the Pacific Community) includes, in addition to 22 Pacific Island countries and territories, four founding-member metropolitan powers (Australia, France, New Zealand and the United States of America).[4] SPC currently generates knowledge, policy options and technical assistance in the fields of economic development, fisheries, land resources, public health, social development, human rights education, geoscience, statistics, climate change and education.[5] Whereas the SPC, as a scientific and technical organization does not have a specific e-government program, its many services have direct bearing on the e-government capacity of the PICs it serves. Its Geoscience Programme, in particular, focused during 2010–15 on enhancing ICT capability in the fields of Ocean and Islands governance, water and sanitation and disaster reduction, drawing on its expertise in such areas as geographic information systems (GIS) and remote sensing and data management.

Until 2010 the geoscience program had the status of a CROP agency and was named the South Pacific Applied Geoscience Commission (SOPAC); it is now the SPC GeoScience Division. First established in 1972 as a UNDP program before becoming, in 1990, an independent inter-governmental organization geared towards promoting sustainable development in its member countries, SOPAC led GIS and Remote Sensing work in the region,[6] conducted ICT outreach on behalf of PIFS and implemented early phases of the Pacific region's ICT policy. From 1990 until its absorption into SPC and SPREP in 2010, SOPAC played a leading regional role in ICT development. The Forum Leaders' Meeting communiqué for 2004, for instance, noted:

> Leaders welcomed the completion of work by the South Pacific Geoscience Commission on the Environmental Vulnerability Index (EVI) and reaffirmed its importance as a tool for measuring and managing environmental vulnerability of countries and the need to incorporate such a tool in future sustainable development processes. Leaders supported the promotion of the EVI at the upcoming meeting to review the Barbados Programme of Action, as the first step towards its global operationalisation and the application of the EVI

---

[4]The Netherlands and United Kingdom withdrew in 1962 and 2004 respectively when they relinquished their Pacific interests; Australia, France, New Zealand and the United States of America are members of SPC but are regarded as "donor" countries rather than "developing" countries, and as such are not among the countries considered in this book.

[5]http://www.spc.int/.

[6]Notably between 1993 and 2010.

regionally and nationally as a complementary environmental management tool for sustainable development. (PIF 2014)

By 2004 SOPAC was installing web-based map servers (e.g., in Fiji, NLC boundaries, in Tonga the cadastral (mapping) of the Tongan island of Tongatapu; and in Papua New Guinea, mining exploration licenses in 14 PICs as part of an EU-funded “Reducing Vulnerabilities in Pacific ACP states”[7] project, which continues to the present.[8] Upon its expiration in 2010, SOPAC uploaded its entire body of knowledge as the digital collection SOPAC Compendium.[9]

3.1.1.3 The University of the South Pacific

A third significant CROP agency with long-standing engagement in ICT development is the University of the South Pacific. USP, which is co-owned by twelve PICs, has actively promoted access to the Internet. In May 2004, for instance, it supported the general thrust of the Fiji Government’s draft policy “e-Fiji: the Future Online”, but called for increased emphasis on the application of ICTs to human and social development, for reform of regulatory and legal frameworks and for reduction in the cost of telecommunications through cessation of monopoly practices. In that year, also, it successfully challenged the Fiji Telecommunications Company’s high-cost monopoly over Internet access and established a physical connection with the Australian Academic and Research Network (AARNET).

The University’s pursuit of ICT capability gained momentum with the opening in 2010 of the Japan Pacific ICT Centre at the USP’s main campus in Suva. The Centre is mandated “…to be a Regional Centre of Excellence for ICT and to spearhead research, new learning technologies, development and leadership in ICT to ensure that the South Pacific can participate in the global information society” by advancing ICT knowledge through learning, teaching, workshops and conferences and quality research by working closely with stakeholders by ensuring a relevant and sustainable solution to deal with ICT issues in the region and by drawing on international knowledge in ICT.

Since 2010 the Centre has run courses, workshops and other activities designed to disseminate international knowledge in ICT to stakeholders and Pacific societies.[10] It accommodates a World Bank-funded Pacific ICT Regulatory Resource Centre (PIRRC) and the Pacific Island Telecommunications Association.

[7]In 2000 the European Union signed the “Cotonou Agreement” concerning trade and development with 78 states from Africa, the Caribbean and Pacific, the terms of which some PICs are yet to agree with.

[8]http://gsd.spc.int/reducing-vulnerability-eu-b (Council of Regional Organisations in the Pacific (CROP) Information and Communications Technology Working Group, 2005). (Council of Regional Organisations in the Pacific. (CROP) Information and Communications Technology Working Group, 2005).

[9]http://ict.sopac.org/compendium/.

[10]http://www.jica.go.jp/project/fiji/002/materials/pdf/japan_pacific_01_01.pdf.

The PIRRC, which collects key industry statistics, issues periodic reports on the state of telecommunications competitions in PICs and provides advisory services to regulators in member countries, has struggled with funding uncertainties but continues to operate.[11] Another organization accommodated at the Japan Pacific ICT center, the Pacific Computer Emergency Response Team, (PACCERT) ceased operation in 2015 after Pacific governments failed to pay their member contributions. As a relatively new actor in the region, the Japan Pacific ICT Centre is yet to develop the leadership and attract the resources required to achieve its purposes.

3.1.2 UN bodies

Alongside the regional organizations introduced above, a range of United Nations-related organizations and programs contribute to the development ICT and e-government capacity in the Pacific Islands. In some of what follows, the regional programs of institutions introduced in Chap. 2 are considered in more detail.

3.1.2.1 International Telecommunications Union

The International Telecommunications Union's (ITU) role in facilitating policy dialogue at global level, particularly in relation to the World Summit for the Information Society and to the ICT development aspirations of small island developing states (SIDS), have been considered in Chap. 2. From 2009 to 2013 the ITU-EC project "Capacity Building and ICT Policies, Regulatory and Legislative Frameworks in the Pacific Island Countries" (ICB4PAC), assisted all 14 independent PICs with the development of their national ICT policy. All countries but Marshall Islands and Samoa were assisted with cybercrime policy and all, except Cook Islands, Kiribati, Samoa, Solomon Islands and Vanuatu, were assisted with licensing policy issues.

ITU projects specifically tailored for the Pacific region are noted in Box 3.1. Most of the ITU's regional projects reached all 14 independent PICs, and were valued at between US$ 500,000 and 1 million each (apart from training-only programs, which each cost approximately US$ 100,000). Collectively, this represents a level of capacity building which the small states could not have otherwise received. As noted in Box 3.1, the ITU has begun the second phase of ICB4PAC to

[11]"The efforts to establish the Centre dates back to 1996. In March 2006, the Pacific ICT Ministerial Forum held in New Zealand, recognised the importance of pooling scarce resources in the regulation of telecommunications services, and directed officials to approach the World Bank and other development partners. After a study was commissioned by the World Bank in 2008, the recommendation to establish the PIRRC at the Japan-Pacific ICT Centre was endorsed at the Pacific Regional ICT Ministers Meeting on 18 June 2010, in Tonga": https://www.usp.ac.fj/news/story.php?id=910.

assist PICs, in particular with national ICT policies, Universal Access Fund, review of licensing and cyber-crime legislation.

Box 3.1: ITU projects specifically tailored for the Pacific region

- Centre of Excellence Network in the Asia-Pacific Region (2007–14)
- Capacity Building and ICT Policy, Regulatory and Legislative Frameworks Support for Pacific Islands States (2008–13)
- Rural/Outer Island Communications in the Pacific (2008–12; Marshall Islands, Nauru, Papua New Guinea, Tonga)
- Implemented Assistance in Telecommunication/ICT in ASP Region (2009–11)
- Assistance in Telecommunication Policies and Regulations in the Asia-Pacific Region (2012–14)
- Ongoing Development of Satellite Communications Capacity and Emergency Communications Solutions for the Pacific Islands (2014–17)
- Implementing the Climate Change Adaptation Component of the Satellite Communications, Capacity, and Emergency Communications Solutions Project for the SIDS of the Pacific (2014–18)
- ITU NBTC Training Program 2016 (2014–18)
- ITU NBTC Training Program (2015)
- ICB4PAC II (2015–17)

3.1.2.2 United Nations Economic and Social Commission for Asia and the Pacific (UNESCAP)

In the ICT field, the Bangkok-based United Nations Economic and Social Commission for Asia and the Pacific (UNESCAP) has focused on economic connectivity, social connectivity and disaster risk reduction. Given the small size of PICs when compared to the Commission's Asian member countries, the ICT needs of the former have not been at the center of ESCAP's work programs, and it is not uncommon for reports on Asia-Pacific to lack detailed consideration of individual Pacific Island countries.[12] A 2004 report on regional cooperation to bridge the digital divide in accomplishment of the Millennium Development Goals (UNESCAP 2004) for instance, noted without further analysis the existence of the Forum Communications Action Plan, the Pacific Islands ICT policy and Strategic Plan, and the Pacific 14D Initiative, which focused on telehealth, distance learning

[12]For example, United Nations. Economic and Social Commission for Asia and the Pacific (2007), which doesn't have any Pacific Island coverage.

and universal access through community tele-centers. However, with the establishment of UNESCAP's Asian and Pacific Training Centre for Information and Communication Technology Development[13] in Incheon in the Republic of Korea in 2006, an eight-module curriculum tailored to the needs of PICs was developed, covering the Linkage between ICT Applications and Development; ICT for Development Policy, Process and Governance; e-Government Applications; ICT Trends for Government Leaders; Internet Governance; Network and Information Security and Privacy; ICT Project Management in Theory and Practice; and Developing Diverse Funding Structures or Options for funding ICT4D (Sovaleni 2009).

3.1.2.3 Asia Pacific Telecommunity

The 38-member Asia-Pacific Telecommunity (APT) (plus 4 Associate Members and 134 Affiliate Members) and comprising governments, telecom service providers, manufacturers of communication equipment and research and development organizations, with headquarters in Bangkok, was founded in 1979 as a joint initiative of the United Nations Economic and Social Commission for Asia and the Pacific (UNESCAP) and the ITU to foster telecommunication services and information infrastructure throughout the Asia-Pacific region (Asia-Pacific Telecommunity 2013). Its activities in the fields of policy and regulation, radiocommunication, standardization, human resource development and ICT development, are principally training workshops and standard-setting meetings involving the relevant ministries of member countries. Additionally, the APT has assisted members to prepare for global conferences such as ITU Plenipotentiary Conference, World Telecommunication Development Conference, World Radiocommunication Conference, World Summit on the Information Society (WSIS), World Telecommunication Standardization Assembly and ITU meetings.

3.1.2.4 UNDP Pacific Office

In the Pacific region, the United Nations Development Programme, the UN's principal development agency with offices in Fiji, Papua New Guinea and Samoa, has included ICT and e-government projects in its country programs. The earliest of these was the Fiji-based e-Pacifika Project to facilitate the development of national ICT strategy plans for 12 countries. Although this outcome was not achieved within the project's timeframe, it constituted "a good starting point" (UNDP Multi-Country Office Samoa 2008 p.4).

---

[13] UN-APCICT - http://www.unapcict.org/.

UNDP also helped establish the PeopleFirst Network in Solomon Islands, comprising some 14 community-owned e-mail stations in schools or clinics in rural areas, capable of sending messages at 2 Kilobytes per second (2Kbps) to a main station in Honiara, from which they were forwarded to their recipients via satellite. In 2004 UNDP Samoa, UNESCO and InternetNZ combined to establish the Pacific Internet Partnership, a program to fund travel fellowships to enrich technical knowledge in the Pacific. Although the UN agency later withdrew from the program it continued as an umbrella for organizing inaugural Pacific Internet Governance Forum in New Caledonia. UNDP initiated an online "Pacific Solution Exchange" in partnership with ADB, PIFS and USP, which has gradually built a knowledge base on ICTs and climate change.[14]

3.1.2.5 UNESCO

The United Nations Educational Scientific and Cultural Organization, which has an office in Samoa, has fostered the development of ICTs in PICs since the mid-1990s. Although the organization has focused on the development of national information policies and the cultivation of "information for all", this objective is underpinned by access to ICTs and their use by government (Zwimpfer Communications 2002). In later years UNESCO focused on the use of ICTs in the education sector, promoting ICT literacy (Pernia 2008) and the development of parliamentary websites (Hassall 2007).

3.1.3 Internet Society—Pacific Chapter

The Pacific Chapter of the Internet Society (PICISOC) (www.picisoc.org) emerged in 1999 on the impetus of informal meetings in the 1990s of IT managers and administrators from such CROP agencies as the Forum Fisheries Agency and SOPAC. The organization was formalized as part of the global Internet Society (ISOC), complete with office bearers and an annual meeting and conference (PACINET). PICISOC is to be distinguished from other regional actors in the ICT sector in that it is a non-governmental organization having open membership, which results in the organization's events becoming informal but professional meeting spaces for officials, business leaders, IT professionals, academics and others who share an interest in ICTs whether from a technical, policy, or commercial perspective. By 2013 PICISOC had some 600 members from more than 20 countries.

[14]http://www.solutionexchange-un.net/pacific/.

3.1.4 Additional Asia-Pacific Institutions

Additional Asia-Pacific institutions that are active in ICT policy dialogue include the Asia Pacific Network Information Centre (APNIC), the Pacific Islands Telecommunications Association (PITA) and the Asian Development Bank, each of which will be described briefly.

The Brisbane-based Asia Pacific Network Information Centre, which is the 'names and numbers' organization for the Asia Pacific, works with the domain-name system on the numbering and technical side of the Internet. APNIC introduced the Information Society Innovation Fund for Asia, a grants and awards program aimed at stimulating creative solutions for meeting ICT development needs in the Asia-Pacific region. The WSIS Stocktaking Report for 2013 noted that funds were allocated to 28 projects in 14 economies, and that:

> The successful projects showcased innovation, cooperation and technical knowledge, plus potential for creating social change in their communities. Projects reflect current technical and social discussions, including relevant issues such as exploring the opportunity for developing economies to get ahead in the IPv6 challenge and the deployment of wireless networks to serve isolated communities with cheaper and accessible communication services. (WSIS Stocktaking Report 2014, p. 117)

The Pacific Islands Telecommunications Association, a non-profit membership organization was formed to "improve, promote, enhance, facilitate and provide telecommunications services within Member and Associate Member countries in Melanesia, Micronesia & Polynesia and Australia and New Zealand".[15] PITA Membership is open to any telecommunications entity operating in the Pacific.[16] Suppliers of ICT equipment or services, as well as regional and international organizations can be associate members; and PITA has a Memorandum of Understanding (MoU) with numerous organizations in associated fields, including some already mentioned (ITU, APT, SPC, SOPAC, APNIC), but also the Commonwealth Telecommunications Organization (CTO), the Pacific Telecommunication Council (PTC), the Asia Pacific Satellite Communications Council (APSCC) and the Internet Corporation for Assigned Names and Numbers (ICANN). PITA also works closely with the PIFS and the Australian government. These MoUs and close links between PITA and a range of regional and international institutions highlight the varied links within the ICT ecosystem, as well as the reliance on regional organizations for the support of international institutions for their existence.

The Asian Development Bank,[17] established in Manila in 1966 and owned and financed by its 67 current members, is another regional institution that has made a substantial contribution to the growth of ICT connectivity in the Pacific. It has

[15]http://www.pita.org.fj/index.cfm?action=articles&cmd=view&id=F0848588-DFD6-07E2-E8BA0E7A53CF0875, accessed October 10, 2016.

[16]http://www.pita.org.fj/index.cfm?action=articles&cmd=view&id=F0848588-DFD6-07E2-E8BA0E7A53CF0875.

[17]www.adb.org.

undertaken e-readiness assessments in some member countries; actively promoted strategic alliances and partnerships with existing ICT initiatives at national, regional and international levels; and provided financial assistance for the establishment of public-private sector partnerships. It has regional and national programs and reports results with a high degree of transparency. Some of the ADB's regional assistance projects in the fields of ICTs and e-government are listed in Box 3.2.

Box 3.2: Asian Development Bank—Regional ICT projects 2001–2016

ADB5990	ICT Assessment in the Pacific (2001)
ADB6107	Preparing a Pro-poor Subregional E-Government Project for Providing Basic Service Delivery to Remote Communities in the Pacific (2003)
ADB6178	E-Rural Demonstration in the Pacific (2004)
ADB43046	Information and Communication Technology-Based Inclusive Growth and Poverty Reduction in the Pacific (2009–14)
ADB43071	Strengthening the Capacity of Pacific Developing Member Countries to Respond to Climate (2009–13)
ADB42291	Creation of the Pacific Information Superhighway with the University of the South Pacific Network (2009–13)
ADB42291	Higher Education in the Pacific Investment Program (2011-16)
ADB44383	Strengthening Regulatory Capacity for Information and Communication Technology Development in the Pacific (2011–15)
ADB47264	Strengthening Public Sector Management in the North Pacific (2013-16)
ADB46510	Pacific Private Sector Development Initiative (2013-2019)
ADB47114	Pacific Information and Communication Technology Investment Planning and Capacity Development Facility (2013–18)
ADB48240	Information and Communication Technology for Development Initiative Facility in Asia and the Pacific (2014-017)
ADB48345	Sector and Thematic Analyses in Policy Development (2014–18)
ADB48242	Knowledge for Solutions - Knowledge Portal (Subproject 3) (2014–18)
ADB49342	Implementing Information and Communication Technology Tools to Improve Data Collection and Management of National Surveys in Support of the Sustainable Development Goals (2015–19)
ADB48333	Applying Space-Based Technology and Information and Communication Technology to Strengthen Disaster Resilience (2015–2017)

Source: ADB website, www.adb.org

These ADB-led development projects had individual values between US$70,000 and (in the case of the Pacific Private Sector Development Initiative) US$38 million, making the ADB one of the most significant actors in the cultivation of e-government capability in the region. They often drew on development funds from other ADB members and partners, such as Canada's Climate Change Fund and the Asian Clean Energy Fund, as well as funds provided by Australia, India, Japan and New Zealand. Whereas some ADB-funded projects were for feasibility studies and readiness assessments, others helped with policy development, while yet others were for direct provision of computers, power supply systems, technical expertise, online knowledge-sharing portals, human resource management information systems and infrastructure. The Pacific Private Sector Development Initiative, covering the period 2013–2019, involves developing innovative financial products and branchless banking models, secure transactions reforms and business registries. Another current project, Implementing Information and Communication Technology Tools to Improve Data Collection and Management of National Surveys in Support of the Sustainable Development Goals (2015–19), seeks to improve data collection methods of national statistical systems in PICs, which are necessary in tracking progress with the Sustainable Development Goals. Internally, the ADB has worked to ensure that the knowledge products it generates through research and technical assistance are widely disseminated. The Bank's business plan for 2016–18 includes a range of ICT projects.[18]

3.1.5 Regional coordination

The actors, institutions and networks introduced in the preceding section collectively provide ICT knowledge diffusion, policy advice, innovation and adaptation in the Pacific Islands' context. However, reliance on cooperation, on "soft multilateralism" (Strickland 2015, p. 154), rather than on some more binding form of commitment and mutual institutional obligations, has rendered policy-processes at regional level unpredictable. The regional policy dialogue path has been characterized by ruptures and disruptions, as persistent national interests compete with the broader goals of regionalism. Furthermore, whereas regional agencies and regional consultations involving members of the executive branch of the various PICS might easily meet to agree on normative policy settings, implementation resides with individual governments and final decisions are in the hands of political leaders who don't necessarily respond to the policy advice given them. The following section reviews these "ruptures and disruptions" in two phases, the first from the meetings of Pacific Islands Communications ministers between 1999 and 2004 and the second from 2005 until completion of the MDGs at the end of 2015.

[18]http://www.adb.org/sites/default/files/institutional-document/178131/robp-pacific-2016-2018.pdf.

3.1.5.1 1999–2004—First Phase—Pre-Pacific Plan

Efforts to coordinate regional information policy commenced in 1999, when the Pacific Islands Forum adopted a "Vision for the Pacific Information Economy" together with a regional approach to telecommunications regulation and tariff levels, in collaboration with the APT, the Pacific Islands Telecommunications Association, and the ITU, other CROP agencies, and five donor countries (Australia, New Zealand, France, UK and the USA). The Ministerial Communique and Communication Action Plan, 1999 emphasized regional cooperation, development of ICT at national level to allow "all citizens to take part in the opportunities brought home by the global information economy", build an 'information economy' in the region and "enhance the region's competitiveness in the global economy". After stating, "Cooperation at a regional level will support national development efforts", the Communique added:

> Facilitating the construction and expansion of an interconnected and inter-operable information infrastructure across the region; Encouraging technological cooperation between Forum member countries in the development of the information infrastructure; Promoting free and efficient flow of information; Furthering the development of human resources; and encouraging the creation of policy and regulatory environments favorable to the development of the Pacific information economy (p. 2; cited in Strickland 2015, p. 156).

To coordinate this initiative, the CROP agencies established in 2001 a working group (WG) for ICTs through the merging of IT-PACNET (a technical cooperation group involving IT managers within the CROP agencies that had commenced in 1995) and the Information Sector Work Group (which commenced in 1998). The members of the new ICT WG were the Forum Fisheries Agency, The Pacific Chapter of the Internet Society, The Pacific Islands Development Program, Pacific Islands Telecommunications Association, Pacific Islands Forum Secretariat, the South Pacific Geoscience Organization, the South Pacific Board of Educational Assessment, the South Pacific Regional Environment Programme, the South Pacific Tourism Organization, UNDP and the University of the South Pacific. At its inaugural meeting the Working Group identified three objectives:

- To Review, clarify and provide advice on the development of priorities in the information sector;
- To Determine areas of complementarity and overlap and any potential gaps in the coverage of existing and proposed regional activities and initiatives in ICT; and
- To Recommend CROP implementation procedures for enhancing coordination and cooperation, and for ensuring that priority areas are adequately addressed.[19]

The establishment of the CROP ICT WG coincided with other regional assessment initiatives. In late 2001 the Governments of Australia, France and New

[19] At its 2004 AGM, held during the PICISOC Conference, the ICT WG agreed its mandate would be limited to monitoring National ICT implementations within the PIIPP framework.

Zealand sponsored a Pacific Information and Communication Technologies Needs Assessment and Strategy Planning Workshop, at SPC in Noumea, New Caledonia. The workshop, jointly organized by SOPAC, PIFS and SPC, with some 100 representatives of PICTs and regional organizations in attendance, drafted a policy and strategic plan that was refined by the new CROP ICT WG and issued in 2002 as the Pacific Islands Information and Communication Technologies Policy and Strategic Plan (PIIPP), entitled "Information and Communication Technology for every Pacific Islander". The PIIPP set out policies in pursuit of four "guiding principles": (1) human resources, (2) infrastructure development, (3) cooperation between stakeholders, and (4) appropriate policy and regulation (Pacific Islands Forum Secretariat 2002)—a list similar to the five "components of the Development Dynamic" reported 2001 (Accenture, Markle Foundation and UNDP 2001).

The guiding principles and the policies and strategic plan articulated in the document provided a comprehensive framework for action at regional and national levels, and envisioned the participation of national governments, regional organizations, educators, NGOs, the private sector, broadcast and print media and user groups. PIFS tasked the CROP ICT WG with the plan and its implementation, review and updating (Council of Regional Organizations in the Pacific 2002). A "Pacific Governance Project" on ICT issues was requested to facilitate information exchange between countries through PITA and CROP, and to undertake a Pacific data/stocktake. However, the CROP ICT WG was expected to meet only once per year, which was certainly too infrequent for it to gain momentum with its assigned responsibilities. The UNDP/UNOPS program, as noted above, developed the "e-Pacifika Project" to facilitate the development of National ICT Strategy Plans for Fiji and other Pacific Island countries.

The Working group organized an intergovernmental "Pacific Islands regional ICT consultation" in Suva in April 2003, in preparation for the World Summit for the Information Society, a global dialogue convened under United Nations auspices (described in Chap. 2). The Tokyo Declaration prepared at an Asian regional consultation before this Summit included an "Islands Paragraph", pleading the special circumstances of SIDS:

> These countries, vulnerable to environmental hazards, and characterized by small, homogenous markets, high costs of access and equipment, human resource constraints exacerbated by the problem of "brain-drain", limited access to networks and remote locations, will require particular attention and tailored solutions to meet their needs. (Para. 11, sec. 2, p. 4 of SWIS/PC-2/DOC/6-E)

In summary, the period 1999–2004 was one of agenda setting and stocktaking. International agencies, regional organizations, national governments and other stakeholders were engaged in policy dialogue, and expectations were high about the potential for ICTs to improve the performance of Pacific economies and delivery of government services. The agreed need was for faster development of high-speed networks, appropriate hardware and software procurement and maintenance, legal and regulatory frameworks conducive to fostering competitive markets "where appropriate" and the cultivation of expertise in ICTs. However, despite these signs of progress there was one major impediment: PIFS was unable to secure the technical and financial assistance necessary to formulate national plans and move to implementation. The next few years were to show that all of these depended on other key factors: political will, leadership and financing.

3.1.5.2 2005–2015: Second Phase—Production and Publishing of First Pacific Island Regional Digital Strategy

In 2004 the Pacific Islands Forum initiated a "Pacific Plan" to ensure that the region remains one of "…peace, harmony, security and economic prosperity" and called for the establishment of "stronger and deeper" links between the Pacific Island states. The Pacific Plan was to include a "digital strategy" and consultations took place among agencies and organizations to maximize cooperation for the development of ICT initiatives for the Pacific region. An Eminent Persons Group identified four goals for the full application of ICTs in the region:

- Improved access to ICTs, especially to and for rural and remote areas;
- Reduced costs of ICTs;
- Increased access to the global backbone; and
- Increased human resources capability to apply and use ICTs.

The *Pacific Regional Digital Strategy*[20] endorsed at the October 2005 Pacific Islands Forum Meeting established a set of objectives and strategies designed to address these needs (the list of objectives appears to have been expanded to include a fifth, "removing inappropriate regulatory environments in order to foster higher levels of investment"), but notably, whereas the Pacific Plan acknowledged the emerging significance of ICTs to the region it did not mention e-government.[21] As noted by Strickland,

> PIFS was designated as the lead agency for implementation, with SPC and other regional organisations playing a collaborative role. The CROP ICT Working Group called upon CROP agencies to support implementation of the Strategy (DS) and stressed the need for donors and agencies to harmonise their activities … The ICT Ministers Meeting designated a task force to oversee implementation of the Digital Strategy and adopted a six-step process for exercising this oversight. SPC expressed the view that the greatest impediment to implementation of the strategy would likely be lack of action at the country level, as this had been the case with past regional ICT-related initiatives. Its experience with delivery of services to member countries and territories included implementation of the Pacific Skye Edge hub (VSAT) connecting PICs to the Internet via satellite, a low-cost high-speed Internet access aimed at rural and remote communities. SPC also promoted the "One Laptop per Child" pilot project in Solomon Islands, in collaboration with the Government's Distance Learning Project. The laptops were low cost and thought suitable for rural areas. Software was designed to enhance literacy and numeracy skills. Other regionally-funded

[20]http://www.pmoffice.gov.ck/index.php?option=com_content&view=article&catid=25%3Athe-project&id=86%3Apacific-regional-digital-strategy&Itemid=18.

[21]Network Strategies 2010, p.i.

infrastructure projects initiated included scoping for the Pacific Island submarine cable network, and the South Pacific Information Network (SPIN). (Strickland 2015, p. 172)

The 2006 "Wellington Declaration" of Forum ICT Ministers (Pacific Islands Forum Secretariat 2006) stated: "ICTs while not an end in themselves, have a key role as a basis for economic development, while also promoting and enhancing social cohesion, cultural enrichment and environmental conservation" (para. 15), and noted seven priorities:

- Human resource development (including training and the establishment of systems to assist HRD);
- Price reductions of telecommunications services;
- USOs/telecommunications services (telephone and Internet) to outer islands/ outer lying areas;
- ICTs in education—ensuring school students have access to computers and the necessary teaching methods;
- Infrastructure development;
- Improved networked economies through e-government and e-commerce; and
- Development of policy and regulatory frameworks.
- The ministers agreed to set up a taskforce to pursue seven initiatives:
- develop qualitative and quantitative indicators to assist with establishing each country's progress towards agreed goals;
- encourage countries to provide data and assess e-readiness (infrastructure, services, platforms and user capacity and costs);
- encourage countries to develop or refine ICT policies with particular emphasis on increasing stakeholder engagement and providing copies to the Pacific Islands Forum Secretariat for dissemination on its website for reference by other countries;
- encourage countries also to provide at least e-mail capacity to Government agencies and a basic web site to gain experience in e-government;
- Seek World Bank, Asian Development Bank and ICANN assistance with regulatory and policy capacity development;
- recommend a regional approach to the purchase or bulk lease of satellite capacity to increase affordability of access by remote communities; and
- Develop complementary policies and legislation at national level and cooperate regionally to combat the problem of "spam".

SOPAC supported pursuit of these initiatives by coordinating PIC and donor activities in respect of PIC e-readiness assessments (with UNDP and the CROP ICT WG), ICT Policy development (with the Republic of China), e-government (with UN-APCICT and the Korean government) and tele-center support (with the Pacific Telecenter Online Community) (Sovaleni). However, although SOPAC conducted an e-readiness survey of 14 Forum Island Countries in 2008, using Harvard University's holistic *Readiness for the Networked World* model (Harvard University. Center for International Development 2000) there were no subsequent annual ministerial meetings, nor was a regional ICT Council established, nor was the 2005 strategy adequately followed up and implemented. It was at about this time that an influential report on UNDP-funded ICT projects in several Pacific Island countries concluded: “the intended outcomes for these projects are basically irrelevant to the actual situation on the ground” (Boase 2009, p. 26). The failure to identify priorities and align projects with local needs is characteristic of many struggling ICT/e-government projects in the region (Budden 2006, p. 14).

Pacific Region Infrastructure Facility

In 2008 Australia, New Zealand, the World Bank and the Asian Development Bank established the Pacific Region Infrastructure Facility (PRIF) “to assist Pacific island countries develop and maintain critical economic and other infrastructure.” The Facility was to provide up to A$200 million funding over four years, initially focused on Kiribati, Samoa, Solomon Islands, Tonga, Tuvalu and Vanuatu, “to help develop competitive local private sectors to deliver infrastructure maintenance and construction services” (Pacific Islands Forum 2008), including transport, energy, communications and water and sanitation infrastructure.

According to Strickland,

> The PIFS member countries which had liberalised, or were willing to, were then encouraged and facilitated to attract private sector investment through this Pacific Islands Regional Infrastructure Fund by the World Bank, ADB, Australia and New Zealand. The potential for funding through PRIF was a lever and incentive to national institutions to enact the policy and regulation changes which the international institutions had been seeking from the inceptions of this policy process. From PRIF’s founding there was increased activity seen for projects and initiatives linked the private sector interests being facilitated in the region, including particularly new submarine Internet cables. (Strickland 2015, p. 173)

In 2008, alongside PRIF’s funding, other funding and contributions from international institutions also picked up dramatically, with a range of research published by international institutions and donor countries, such as the ITU “Capacity Building and ICT Policy, Regulation and Legislative Frameworks Support for Pacific Island Countries (ICB4PAC)” program; and ICT-related capacity building workshops funded by a range of international institutions, including the World Bank and ADB funded (Network Strategies 2010) which, Strickland notes, tended to focus on liberalization of legislation and changes to regulations (Access to affordable ICT; Efficient and effective utilization of ICT for sustainable development and Adoption of ICT as a national priority in PICTs (Strickland 2015, p. 173). The Pacific Leaders, at the 40th Pacific Islands Forum held in Cairns in August 2009, called for a review and update of the Pacific Plan Digital Strategy (PPDS). This 2010 report, the *Framework for Action on ICT for Development in the Pacific* (FAIDP) is considered in more detail below.

By 2010 there was general agreement that the regional Digital Strategy had not brought the desired outcomes. A PIFS-funded report on the Strategy found that whereas liberalization had expanded reach, availability and affordability of mobile services, telecoms penetration in PICs remained low when compared to developed countries (Network Strategies 2010). The pace at which ICT policies and legislation were being developed was hindered by capacity and resource constraints (Network Strategies 2010, p. i) and "converged applications (such as e-government, e-commerce, e-health and e-education) are either non-existent or in their infancy" (Network Strategies 2010, p. ii). Regional coordination of these efforts, furthermore, was a major challenge:

> Partners, donors and other agencies and organizations have made substantial and important contributions to capacity building and the promotion of ICT in the Pacific since the Digital Strategy. This has resulted in considerable progress in achieving the objectives of the Strategy although there is still much work-in-progress on key aspects such as national ICT policies and planning, accessibility of ICTs to rural populations, and regional ICT coordination and harmonisation. With respect to the latter there is certainly a need for improved regional coordination as current channels appear to be ineffective, resulting in gaps in information and knowledge about current and planned activities among partners and potential inefficiencies. Opportunities for stakeholder consultation and engagement appear to be very ad hoc, leading to considerable potential for overlap. This situation is exacerbated by the paucity of information and statistics available on ICT progress at the national level and the lack of published information on objectives, developments and outcomes of regional initiatives.[22]

The "Tonga Declaration" on "ICT for development, governance and suitable livelihoods" that resulted from a June 2010 meeting of Pacific ICT Ministers in Nuku'alofa endorsed the *Framework for Action on ICT for Development for the Pacific and tasked* SPC to work with USP, PICTs and development partners to develop a three year rolling Strategic Action Plan and to coordinate its implementation (Secretariat of the Pacific Community 2010b). As part of the WSIS process, in 2011 SPC hosted an inaugural "Pacific Internet Governance Forum" (Secretariat of the Pacific Community 2011). A Digital Strategy "2.0" was published, along with its program framework and major changes in regional responsibility for the policy.

The WSIS Stocktaking Report for 2012 noted:

> In the Pacific region, the Framework for Action on ICT for Development in the Pacific (FAIDP) outlines a new approach to developing and improving ICT services in order to support development, strengthen governance and improve the livelihoods of communities in the region. It acknowledges that national ICT policies and plans implemented through integrated and well-coordinated approaches provide the principal means for ICT to contribute meaningfully to sustainable development. It promotes a 'whole sector' approach, based on the concept of 'many partners, one team'. This approach recognizes that numerous stakeholders contribute to ICT development in the region and accepts them as equal partners (International Telecommunication Union 2012, p. 71)

Policy meetings in 2010 and 2011 agreed to fund more research and investigate infrastructure financing.[23] The 2010 *Framework for Action on ICT for Development*

---

[22]Network Strategies 2010, p. iii.

[23]"ADB International Development - Tonga Fiji Submarine Cable Project in Tonga," (n.d.); "ADB Pacific Regional ICT Connectivity Project, Phase 2 (Solomon Islands and Vanuatu)," World Bank 2011.

in the Pacific (Secretariat of the Pacific Community 2010a), issued in response to a request from the Pacific leaders, shifted the emphasis from regional action to supporting the initiatives of individual PICs over the next three years; PICs focused on obtaining funding for initiatives and for capacity building assistance from multilateral organizations: regulatory frameworks, emergency communications, access for rural and remote areas, ICT for disabled persons, cybersecurity and international connectivity capacity.

Box 3.3: ICT Timeline—Pacific Islands

1995	CROP IT managers group established
1998	Information Sector Work Group established
2001	ICT working group established—merging of IT-PACNET—a technical cooperation group involving IT managers within the CROP agencies
2002	February—Pacific Islands ICT Policy and Plan (PIIPP) endorsed at ministerial level at Forum Communications Policy Meeting
2002	e-Pacifika program implemented by United Nations Development Programme (UNDP) and United Nations Office for Project Services (UNOPS) with Japanese funding
2003	Pacific report prepared in advance of WSIS Geneva
2004	Pacific Plan; USP breaks FINTEL monopoly and connects to AARNET
2004	Sub-regional Symposium on ICTs for Development in Pacific Islands Developing Countries
2005	Pacific Islands ICT Policy and Strategic Plan endorsed by Pacific Island Forum Leaders
2006	Pacific Islands Forum ICT ministers Wellington meeting - focuses on implementation of the digital strategy and establishes a task force to advance the digital strategy
2008	Pacific Regional Infrastructure Facility established
2008	Gondwana-1 submarine cable joins Noumea to Sydney
2008	Picot-1 submarine cable extends from Noumea to Mouly, Poindimie and Xepenehe in New Caledonia
2009	40th Pacific Islands Forum calls for Pacific Plan Digital Strategy (PPDS) to be reviewed and updated
2009	American Samoa-Hawaii (ASH) submarine cable joins American Samoa to Hawaii
2009	PIPE Pacific Cable-1 (PPC-1) submarine cable joins Madang (PNG) to Sydney and Guam
2010	Tonga Declaration: 'Framework for Action on ICT for Development in the Pacific' (FAIDP) SPC/ITU
2010	Japan ICT Center opens at the University of the South Pacific
2010	Honofua submarine cable joins French Polynesia to Hawaii

2010	HANTRU1 submarine cable joins Marshall Islands to Federated States of Micronesia
2011	1st Pacific Internet Governance Forum; SPC Framework For Action On ICTs; and Universal Access Workshops in Noumea
2011	Digital Strategy 2.0 launched
2014	SIDS III in Apia, Samoa, results in "SAMOA pathway"
2014	Interchange Cable Network 1 (IICN1) submarine cable joins Port Vila to Suva
2015	Third Pacific ICT Ministerial meeting and Pacific ICT Officials meeting - Tonga
2015	SDGs launched—with significant expectations for ICT development and use

2015 ICT Ministers Meeting

The ITU and EU-sponsored Third Pacific ICT Ministerial meeting and Pacific ICT Officials meeting in Tonga in 2015 established links with the United Nations Project Office on Governance (UNPOG), and with UN DESA's Division for Public Administration and Development Management (DPADM) to broaden the regional strategy around ICT and e-government for SIDS; to update the earlier Pacific Regional Digital Strategy of 2005–2010; and the Framework for Action set out in FAIDP. In addition to its need to review the FAIDP, another of the meeting's aims was to leverage commitment to the sustainable development of SIDS made by Heads of State gathered at the Third UN Conference on Small Islands Developing States in Apia in 2014, and to consider how to increase use of ICT infrastructure for development, as called for in SIDS III's S.A.M.O.A. Pathway. The UN's Project Office on Governance noted in its presentation to the meeting that the development of ICT in the Pacific region had for some time been "isolated and uncoordinated with agencies working in silos and no clear cohesive architectural design that is cost efficient", and that the "current status of ICT infrastructure development in Pacific SIDS is much behind of the average of the world" (Lim 2015, p. 26). UNPOG's recommendation was that "SIDS focuses their efforts on the establishment of ICT/e-government infrastructures first".

The ICT Ministerial meeting agreed to a program of training throughout the region to enhance e-government capacity, share best practice and encourage cooperation, a policy of using ICTs to learn how to apply them more effectively. But this approach assumes that a regional approach is the solution to the problems outlined above, and that a regional 'voice' will be stronger than that of individual countries and national ICT/e-government policies, also advocated by the ITU. There is little evidence to show that either approach would be more successful. This book, in covering both regional policies and initiatives and highlighting case studies and examples from individual countries in the region, seeks to throw some light on this issue.

ICTs in the Pacific Plan

The ICT ministers' meeting of 2015, together with the impetus provided by the SIDS III Conference in Samoa, revived the improvement of ICT capacity in the Pacific as a regional policy priority. Information and communication technology was one of five "priority issues" selected from 68 submissions for closer consideration by the region's leaders. (The others were (i) increased economic returns from fisheries and maritime surveillance; (ii) climate change and disaster risk management; (iii) West Papua; and (iv) cervical cancer.) A subsequent statement of priorities concerning the "Framework for Pacific Regionalism" announced by the chair of the Pacific Islands Forum in January 2016 noted:

> The priority initiative pertaining to information and communication technology (ICT) asks for an assessment of the merits in establishing a regional ICT Advisory Council. This assessment is being led by the University of the South Pacific with support from the Pacific Islands Forum Secretariat. (Pacific Islands Forum. Forum Chair 2016)

Crucial stakeholders in this area, including the Asian Development Bank and the World Bank, are contributing to the analysis of ICT priority issues, and the CROP ICT Working Group will provide high level guidance to the assessment.[24] The inclusion of a broad group of stakeholders allows for a considered and well-informed response to Forum Leaders' acknowledgment of the numerous challenges that exist in realizing the benefits of ICT in the region. The recommendations of the assessment were presented to leaders in 2016.

3.2 National ICT Policy Processes

The preceding section detailed progress by regional organizations toward coordination of ICT initiatives in the Pacific region. Equally important is the development ICT policy and e-government policy by the political and public sector leaders in each country. Most PICs now have articulated ICT policies, with a few (Nauru, Niue, Solomon Islands and Tuvalu) still in the draft phase. The ITU claims some credit for stimulating this process. After its 2010 survey found that only 5 of the 14 countries had a national ICT policy, the ITU convened a workshop "for decision-makers" in June 2010, and by April 2013, with the ITU's in-country assistance, 13 countries had policies in place (ITU 2013, p. 2).

Policy frameworks have the important role of establishing the direction a government wants to go and how it intends to get there. Establishing an ICT environment conducive to the operation of e-government requires decisions concerning a range of matters, including Internet and telecoms infrastructure and services, protection of personal data and privacy, technical development concerning Internet protocol and domain names, security and authentication and ICT regulatory and

[24]http://cropict.org.fj/.

market issues (International Chamber of Commerce Commission on E-Business 2011).

Most PICs have taken decisive steps toward selection of connectivity infrastructure—although the high costs of laying undersea cables between small states with their small economies had slowed this process.[25] Additional decisions must be taken on hardware, including computers, peripherals and their location. Levels of integration must be envisaged, and these can be both vertical and horizontal, and within government and beyond it. Integration also involves decisions concerning the establishment of a central government portal, shared platforms and the extent of information sharing: will government use a whole of government approach? Is it working toward more open government in matters such as budget preparation or planning processes? Progress with broadband delivery and other ICT services in a country, including e-government, asset ownership, access, governance, funding and usage requires parliaments to pass laws on telecommunications, radio spectrum, competition, consumer protection and content. Privacy acts are required to ensure that public servants do not access data except as needed, and all such laws and policies need to be monitored over time and revised in the light of experience.

Finally, the policy framework should identify intended areas of transformation. These usually focus on efficiency, productivity and effectiveness. Transformation shifts from one-way to two way communications and eventually to full service, where transactions with government can be fully completed online. Table 3.2 depicts the status of ICT policy in PICs and indicates the agency most responsible for them.

3.2.1 Political Will and Leadership

Discussion with political and public sector leaders in some PICs has made clear to us the importance of strong leadership toward e-government. For political leaders, the development of ICT capacity and capability may be important, but nonetheless competes for attention and resources with such equally important sectors as health, education, agriculture, climate change, energy and transport, among others. Furthermore, the potential offered by e-government may not be well understood by some in governments, and thus given lower priority (West 2007).

Leadership in the public sector is equally important. Leadership of ICT policy has for the most part been assigned to a government's ministry of communications, but leadership of e-government has proven more challenging with only a few countries establishing, to date, effective whole-of-government approaches. This has meant that the evolution of a facilitating environment (i.e., national policy, licensing

[25]The term "infrastructure" refers to physical backbone, spectrum resources, interconnection, access, and to such technologies as GSM, WCDMA, LTE and LTE-A, WiMax, Satellite and Cable.

Table 3.2 Status of ICT Policy in PICs and the agency responsible

Country	Policy status	Policy documents	Policy actors
American Samoa	Yes		
Cook Islands	Yes	Cook Islands National Information and Communication Technology Policy, 2015-2020	National ICT Committee
Federated States of Micronesia	Yes	FSM National ICT Policy	Department of Transport and Télécommunications
Fiji	Yes	National ICT Policy 2004, Policy Directions and Strategies for the Development and Growth of Information and Communication Technology, Fiji Information Technology Development Policy, e-Fiji the future online, Creating Information Economy for Fiji	IT Advisory Council
French Polynesia	Yes		
Guam	Yes		
Kiribati	Yes	National ICT Policy 2011	Ministry of Transportation and Communication
Marshall Islands	Yes	National ICT Policy 2012	
Nauru	In progress		
New Caledonia	Yes		
Niue	In progress		Niue Island Information, Technology & Communication and Development Council (NiDC)
Northern Mariana Islands	Yes		
Palau	Yes	Palau National ICT Policy 2013	National ICT Policy Task Force
Papua New Guinea	Yes	National Policy on Information and Communication 2008; National ICT Act 2009; NICTA Act 2009; Vision 2050	Interagency ICT Task Force
Samoa	Yes	ICT Policy and Strategy Plan 2004; National e-strategy; National ICT Policy; The Pacific Plan's Digital Strategy)	National ICT Steering Committee

(continued)

Table 3.2 (continued)

Country	Policy status	Policy documents	Policy actors
Solomon Islands	In progress	National Communication Technology Policy, The Pacific Plan's Digital Strategy	ICT Working group
Tokelau	No	National Strategic Plan	
Tonga	Yes	National ICT Policy 2008; National ICT strategic plan	Ministry of Meteorology, Energy, Information, Disaster Management, Climate Change and Communications
Tuvalu	In progress	Tuvalu National ICT policy, "Information and Communication Technologies for every Tuvaluan Citizen"	Ministry of Communications, Transport and Tourism
Vanuatu	Yes	The National Information and Communication Technology Policy 2013	Ministry of Public Infrastructure and Utilities
Wallis and Futuna	Yes		

and regulation) for the provision of online services has been somewhat dependent on the vision, leadership and resources within individual government departments and agencies, rather than on directions, standards and values for e-government set at whole-of-government level (Table 3.3).

In addition to the challenge of enacting legislation that is adequate and timely, governments have to allocate political and administrative responsibility for their oversight. While in the majority of cases this has been given to the Ministry of Communication, Cook Islands and Vanuatu retain it in the Office of the Prime Minister.

Samoa established a multi-stakeholder ICT committee chaired by the Prime Minister, and in Fiji an IT Advisory Council operates. In Samoa, committee members were divided into working groups to consider what steps were required for infrastructure, a communications sector plan, law and regulation and capacity building. Departmental heads were invited to advise the committee on the types of services they anticipated developing, and a list of currently-used applications was assembled. Fiji was an early leader in e-government in the Pacific (Hassall 2005). Having established an Electronic Data Processing Centre in 1966, services such as billing, payroll and civil registration were brought together under Information Technology and Computing Services in 2000, and a first e-government program was implemented in 2006. Although a military take-over of a democratically elected government occurred that year, Fiji's initiatives in e-government continued. In Solomon Islands, in the absence of any movement to establish a whole-of-government approach, the Ministry of Finance established an ICT

Table 3.3 Government agency responsible for coordination of e-government

Country	Ministry
American Samoa	US Government
Cook Island	ICT Policy and Planning Office of the Prime Minister
Federated States of Micronesia	Department of Transportation, Communication and Infrastructure
Fiji	Ministry of Tourism, Trade and Communications
French Polynesia	Ministry of Economic Recovery, the Blue Economy, Digital Policy and Investment
Guam	Department of Administration, Office of Technology
Kiribati	Ministry of Communications, Transport and Tourism
Marshall Islands	Ministry of Transport and Communication
Nauru	Ministry of Transport and Telecommunications
New Caledonia	Direction of Technology and Information Services
Niue	Ministry of Infrastructure
Northern Mariana Islands	
Palau	Ministry of Public Infrastructure, Industries and Commerce
Papua New Guinea	Ministry of Communication and Information Technology
Samoa	Ministry of Communications and Information Technology
Solomon Islands	Ministry of Communication and Aviation
Tokelau	Ministry of Telecommunications
Tonga	Ministry of Information and Communications
Tuvalu	Ministry of Works Communications and Transport
Vanuatu	Ministry of Public Infrastructure and Utilities
Wallis and Futuna	French Government

Support Unit to provide hardware and software support for the Ministry and for such other government departments as decided to innovate.

A fundamental policy question confronting government in a number of PICs has been whether or not to open telecommunications markets to competition and, if so, on what terms. Communications has long been associated with national security and there has also been a strong belief that small markets could not sustain multiple sets of infrastructure. By convention, Post and Telecommunications had been regarded as a service that only government should, or could, provide. However, such views have been legally and commercially challenged through the persistent approaches of private companies—such as Digicel and Vodafone—which had revolutionized telecoms markets in other parts of the world and which view Pacific markets among their next frontiers. Digicel first entered the South Pacific market in Samoa in November 2006 and Papua New Guinea in July 2007. By 2016 the company had expanded to Vanuatu, Tonga, Fiji and Nauru.[26] Vodafone entered the

[26]http://www.digicelpacific.com/en/about/our-locations.

Samoan market in 2007 and is also present in Fiji, Papua New Guinea, Solomon Islands, Tonga, Guam and the Commonwealth of the Northern Mariana Islands.

One common source of resistance to the introduction of new, cross-agency platforms that e-government leaders face is a preference for existing ('legacy') systems. After a department has spent scarce financial and human resources establishing, maintaining and refining information systems that are critical to their core business, the prospect of shifting to new systems not of their choosing, has not often been viewed favorably. Samoa's government does not wish to have its data depend on a private company's network (i.e., Digicel) and is intent on acquiring a second undersea cable.

Such institutional and political resistance to reform appears to have taken place in Papua New Guinea. At independence in 1975 the colonial-era and government-owned Department of Posts and Telegraphs became the Postal and Telecommunications Services Division of the Department of Public Utilities, before being divided into Telikom PNG, Post PNG and PNG Telecommunications Authority in a corporatization exercise in 1996. Telikom, a 100% state-owned business, established B-Mobile in 2002 to trade in the mobile phone sector. When PNG's National Executive Council decided in 2005 to formulate an ICT policy based on introduction of competition, the political leadership appeared to be divided, with some advocating reform and others seeming to resist and to protect the business position of the state-owned enterprises (Duncan 2011).

Traditional views of governments about communications administration and access were also challenged by the development banks, whose calculations suggested that greater competition in telecommunications markets would expand market reach, boost economic activity and growth and allow for more online government service delivery.[27] Development assistance flowed to governments and private sector actors willing to embrace policy settings by which governments became creators and regulators of markets rather than direct service providers.[28] Governments could choose to privatize their telecommunications entities, establish state-owned enterprises or enter public-private partnerships. In several cases entry to new markets has been through acquisition of a local company or by partnership with a local company. The Samoa government established Computer Services Limited by privatizing SamoaTel. Government commissioned a report by the consulting firm Leading Associates to make recommendations on how to deregulate the market. At the time the government desired nation-wide coverage which it did not believe a new provider could achieve.

Whether government pursues a "universal access" ICT policy and if so, by what means, is a distinctive characteristic of a country's ICT and e-government policy. The consequence of accepting this principle—as espoused by WSIS declarations—

[27]Evidence for these effects is documented in studies by the development agencies as well as local studies (Stanley 2008; Pacific Institute of Public Policy 2012).

[28]See, for instance World Bank Projects WB075739 (Samoa 2002-2011), WB143465 (Vanuatu 2013), WB128013 and WB13229 (Marshall Islands 2013) referred to in this chapter.

is that government must ensure ICTs reach well beyond the more easily-serviced urban areas, to the remotest provinces and islands—and at affordable prices. Should government leave access to market forces, as in Solomon Islands, or should it make coverage a condition of commercial licensing, as in Samoa, Vanuatu and elsewhere? Choice of policy in this matter will greatly affect the role of the national regulator tasked with ensuring that service providers meet the terms of their licenses.

3.2.2 *Sequencing of Infrastructure, Legislation and Applications*

Sequencing of e-government is important. Infrastructure, including hardware and software, is best planned once the needs of government agencies are known, understood and assessed. In practice, e-government capacity has emerged in most Pacific countries non-sequentially. Networks and computer banks in some instances were built before needs assessments were complete, before existing software environments assessed and before aspirational services were considered. This tends to happen in contexts where human and material resources are limited and where governments have responded to the opportunities placed in front of them, whether these be by international development agencies, development banks, bilateral partners or even by the private sector; and a combination of technical assistance, grants, loans and public-private partnerships play a role. In Papua New Guinea, for instance, officials in the Ministry of Communications faced a constant challenge in maintaining appropriate sequencing, as installation of hardware often ran ahead of needs assessments.

Infrastructure, particularly fiber-optic connectivity via submarine cables or connectivity via satellites, requires considerable financing and negotiation with international partners. Between 2000 and 2016, more than 62,000 km of submarine cable were rolled out in nine projects across the Pacific Ocean, bringing faster and cheaper connectivity to American Samoa, Federated States of Micronesia, Fiji, French Polynesia, New Caledonia, Marshall Islands, Papua New Guinea, Samoa and Vanuatu.[29]

In 2016 new plans were made for undersea cables to CNMI, Fiji, Guam and Samoa, (and yet other projects intended to islands within countries, such as within Fiji from Viti Levu to Vanua Levu as part of the World Bank's Tui Samoa Cable project). In the case of the North Pacific, Telecommunications company Docomo Pacific will link the CNMI islands of Saipan, Tinian and Rota to Guam by undersea fiber-optic cable (Atisa), a US$25 million project to replace and upgrade cable damaged 2015 by Typhoons Dolphin and Soudelor (Telecom Announces $25 Million Undersea Cable For CNMI 2016). Other projects currently being developed

[29] http://submarinecablemap.com.

include the Hawaiki Cable, intended to connect New Caledonia, Vanuatu, Fiji, Samoa and Hawaii to Australia, New Zealand and the west coast of the United States; the Solomons Oceanic Cable Network to link Honiara, Auki and Noro with Sydney; Tui-Samoa to link Samoa and Fiji; and the Interchange Cable Network 2 (ICN2) to link Honiara to Lugainville in the north of Vanuatu.

In the North Pacific, Guam has been a critical node for submarine cables since the beginnings of modern communication links and continues to expand its links between Asia and North America. In 2014 a consortium of global communications companies announced a US$50 million undersea cable system (SEA-U.S.) with landing sites in Manado (Indonesia), Davao (Philippines), Piti (Guam), Oahu (Hawaii) and Los Angeles (Guam Telecom 2014).

Satellite connectivity has also expanded. For remote islands such as Nauru and Niue, satellite is the more feasible option: until 2016 the local company Internet Niue (renamed 'Rocket Systems') provided the country's villages with free Internet using revenue from the .nu domain, but was forced to introduce a fee when demand for expanded services outpaced this source of revenue. (Radio New Zealand International 2016). In other countries satellites are relied on to reach outer islands.

Several governments have begun linking their departments via local area broadband networks. In 2013 the Samoan Government accepted Chinese assistance to establish a 'National Broadband Highway' between government offices in Apia, allowing government departments to communicate with each other without providing access to the Internet (which requires an additional subscription through a private ISP (BlueSky, CSL or IPacifika), should a department have such need). Tonga attempted to install a LAN between its government buildings in 2000 and Solomon Islands made plans to develop a Wide Area Network in Honiara. In Papua New Guinea the major government buildings were linked by a fiber-optic network by 2014 and the government has been working toward an Integrated Government Information System (IGIS), which officials regarded as the first necessary step toward developing an e-government policy. The system was trialed with the civil registry and with the National Executive Council before being expanded.

As noted above, the goals for e-government include increasing transparency, increasing the efficiency and effectiveness of government and providing these outcomes through high quality regulation (Lee 2014). The development banks and agencies have thus given considerable assistance to PICs for the development of legal and institutional frameworks to support expanded ICT and e-government environments. In 2013, for instance, the World Bank undertook projects in the Marshall Islands to strengthen the legal, regulatory and institutional environment so as to "…transition from a monopolistic to a liberalized telecommunications market" through reform and restructuring of the majority state-owned National Telecommunications Authority (NTA), among other activities. A World Bank project that commenced in Vanuatu in 2013 sought to strengthen oversight institutions. The Samoa Telecommunications and Postal Sector Reform Project from 2002 to 2011 assisted with privatization of Samoa Tel and additional strengthening of regulatory policy pertaining to interconnection, convergence and "downstream effects of privatization".

Table 3.4 National projects funded by ADB

Country	Number of projects	Focus
Fiji	3	Improving infrastructure services; implementation of the Fiji ICT Strategy
Kiribati	1	Land use management
Marshall Islands	1	Health management information system
Micronesia	1	Infrastructure planning and implementation
Papua New Guinea	5	Maritime and waterways safety, road transport, microfinance, DIGICEL mobile telecommunication expansion, public expenditure review and rationalization
Samoa	1	SchoolNet and community access
Solomon Islands	3	Education, broadband policy and capability, renewable energy for telecom networks, economic recovery support
Tuvalu	1	Education
Vanuatu	3	Financial services, secured transactions reforms

In the past decade, the Asian Development Bank has assisted nine PICs to finance at least twenty ICT projects.[30] Papua New Guinea has benefited from five of these, and Vanuatu and Fiji three each (Table 3.4).

These projects have delivered management information system plans and strategies; land management systems including GIS systems; ICT assessments in specific sectors, particularly education; technical assistance; workshops; telecommunications infrastructure policy; and computer hardware and software.

Subsequent chapters will show e-government applications to be emerging across the Pacific to meet the needs of payroll, taxation and customs, government procurement, business registration, health, education, agriculture, fisheries, immigration, citizenship, passports, land titling and registration and electoral rolls. Small islands such as Niue have built Internet improvement into their tourism sector strategic plan for 2014–17 (Radio New Zealand International 2014). Papua New Guinea's Forestry Authority has established, with support from the Japanese International Cooperation Agency, satellite surveillance to detect illegal logging. (Illegal Logging In PNG Being Monitored By Satellite System: Forest Minister calls problem 'a thing of the past' 2015) Provident funds have introduced mobile services for their customers. Increasing numbers of government departments and agencies have developed 'outward facing' websites.

[30]These are in addition to the regional projects outlined in Chap. 2.

Box 3.4: ADB Projects Assisting the ICT Capacity and Capability in PICs

2000	Marshall Islands: Reviewing the Health Management Information Systems (HMIS)
2003	Fiji: Implementation of the Fiji ICT Strategy
2004	Kiribati: Supporting Land Use Management on Kiribati (Christmas) Island
	Tuvalu: Education Sector Reform and Development Project
	Vanuatu: Secured Transactions Reforms
2006	Fiji: Improving Infrastructure Services (formerly Power Sector Development Road Map)
2007	Papua New Guinea: Support for Public Expenditure Review and Rationalization
	Vanuatu: Expanding Access to Financial Services
	Samoa: SchoolNet and Community Access Project
2009	Papua New Guinea: DIGICEL Mobile Telecommunication Expansion Project
2010	Papua New Guinea: Microfinance Expansion Project
2011	Solomon Islands: Renewable Energy for Telecom Networks
	Federated States of Micronesia: Strengthening Infrastructure Planning and Implementation
2012	Papua New Guinea: Regulating and Sustaining Road Transport
2013	Solomon Islands: Proposed Loan and Grant Solomon Islands: Broadband for Development Project (Report and Recommendation of the President to the Board of Directors)

Vanuatu: Expansion of Rural Financial Services

3.3 Conclusion

In 2006 the Pacific Islands Forum's advisor on e-government wrote:

> If we reflect on the three factors of bureaucratic and political climate, economic and social conditions, and technology, we must also conclude that much development of all three must occur before the full potential of e-government can be realized in the Pacific. I think it is fair to say also that most of the challenges of e-government, commencing with the basic acceptance that it has value, through the processes of blending with current processes, the resource limitations—both human and capital—to the ultimate sustainability including citizen acceptance and participation are ahead of us—or perhaps not even "on the radar". (Budden 2006, p. 54)

This chapter has reviewed the complex regional governance arrangements in the Pacific Islands that Budden mentioned and noted their progress since that time. Following a period of preliminary attempts beginning in 1999 to devise ways to coordinate at regional level, a second phase was spurred by the 2004 launch of the 'Pacific Plan' and from 2005 a 'Pacific Regional Digital Strategy'. However, the slow rate of progress led to a review of this strategy in 2010. The option of a regional ICT council has been resurrected but is yet to be formalized.

The 'real world' implementation of e-government is thus invariably messier than abstract models suggest or anticipate and technical considerations invariably vie with a country's political economy in decision-making processes. Departments that have labored to establish their data systems are often initially reluctant to forego these and join in whole-of-government information systems. Departments of personnel, finance, health, immigration, labor and revenue, each collect data on a common set of citizens and create considerable overlaps of information, yet each is often reluctant to forgo these separate databases through concerns about technical difficulties and about harmonization of data and data sharing—which, if not controlled by adequate legislation, runs the risk of information leaking both inside government departments and beyond them. And if not well-designed technically, they run the risk of incorrect data entry, although this challenge can be met by establishing a core database that contains an individual's name and common identification number, which can then be drawn on by departments for their more specific information needs. Papua New Guinea is introducing a registration number but most Pacific countries have not commenced consideration of the idea.

Although there has been considerable dialogue at regional level clarifying the potential of e-government for the PICs and setting development frameworks and timetables, the task of implementation has always been the responsibility of national governments, where political and administrative decisions are made. The main challenges have been funding and timely acquisition of infrastructure to provide access to the Internet. Yet even when public sector officials gain the attention of MPs, ICT infrastructure and policy remain just one of many urgent items on a government's agenda, alongside health, education, transport, energy and other crucial sectors. While government in the Pacific Island Countries has not yet been 'transformed', it has in some sectors experienced incremental improvement in little more than a decade and this, given the region's challenges with capacity, is good progress.

References

Accenture, Markle Foundation, & UNDP (2001) Creating a development dynamic: final report of the digital opportunity initiative

Asia-Pacific Telecommunity (2013) Ad-hoc group for work program: working methods for policy and regulation forum for Pacific (PRFP). Paper presented at the 37th session of the management committee, Vientiane, Laos

Boase R (2009) Final report. outcome evaluation of the information & communications technology for development (ict4d) project of the UNDP multi-country office (MCO) based in Samoa prepared for the UNCP MCO based in Samoa and the Governments and people of Cook Islands, Niue & Samoa. Retrieved from Vancouver:
Budden J (2006) E-government in the pacific: an opportunity for regional synergies? Reg Dev Dialogue 27(2):53–63
Council of Regional Organisations in the Pacific (CROP). Information and Communications Technology Working Group (2005) Information and Communications Technology Working Group meeting Warwick Hotel, Sigatoka, Fiji 26–27 Apr 2005 Retrieved from Suva
Council of Regional Organizations in the Pacific (2002) Pacific umbrella initiative "Pacific Islands information and communication technology for development 2002–2012" (Pacific 14D Initiative)
Duncan R (2011) Telecommunications in Papua New Guinea. In: Findlay C (ed) The impacts and benefits of structural refoms in the transport, energy and telecommunications sectors in APEC Economies. APEC Policy Support Unit, Singapore, pp 433–445
Graham K (2008) Models of regional governance: Is there a choice for the Pacific? In: Graham K (ed) Models of regional governance for Pacific Island States: the future architecture of Pacific regionalism. University of Canterbury Press, Christchurch, pp 19–38
Guam Telecom joins consortium building SE Asia-U.S. cable link: $250 million project to lay 10,000 mile undersea fiber-optic cable (2014) Pacific Daily News
Harvard University. Center for International Development. Information Technologies Group (2000) Readiness for the networked world: a guide for developing countries. https://cyber.harvard.edu/readinessguide/guide.pdf
Hassall G (2005) ITC policy and practice in the Fiji Islands: national and regional dimensions. In: Marshall S, Taylor W, Yu X (eds) Encyclopaedia of developing regional communities with ICT. Idea Reference Group, Hershey, London, Melbourne & Singapore, pp 471–474
Hassall G (2007) Pacific Islands parliamentary websites: a preliminary survey. In: Prasad BC, Roy DC (eds) Development problems and prospects in Pacific Islands. Nova Science Publishers, New York, pp 121–136
Illegal logging in PNG being monitored by satellite system: forest minister calls problem 'a thing of the past'. (2015, July 22). The National
International Chamber of Commerce commission on e-business, I. A. T (2011) An inventory of ICT policy positions and practical guidance. International Chamber of Commerce, Paris
International Telecommunications Union (ITU) (2013) National ICT policy: knowledge-based report. Retrieved from Geneva
International Telecommunication Union (2012) Report on the WSIS Stocktaking 2012. Retrieved from Geneva
Inter-telecommunications Union (2014) Report on the WSIS Stocktaking 2014, Geneva. http://www.itu.int/dms_pub/itu-s/opb/pol/S-POL-WSIS.REP-2014-PDF-E.pdf. Accessed 24 October 2016
Lee YB (2014) E-Government for promoting sustainable development in small island developing states (SIDS): background paper for the special event of the 69th session of the United Nations General Assembly second committee. Retrieved from New York
Lim J-H (2015) E-Government for sustainable development in SIDS. www.itu.int/en/ITU-D/Regional-Presence/AsiaPacific/Documents/Events/2015/August-RDF2015/Session-3/S3_JaeHong_Lim.pdf. Accessed 24 Oct 2016
Network Strategies (2010) Review of Pacific regional digital strategy: final report for the Pacific Islands Forum Secretariat—Part A - Technological capacity. Retrieved from Suva
Pacific Institute of Public Policy (2012) Net benefits. Internet uptake in the Pacific. Retrieved from Port Vila, Upgrading the coconut wireless
Pacific Islands Forum Secretariat (2002) Policy and strategic plan. Information and communication technologies for every Pacific Islander. PIFS, Suva

Pacific Islands Forum Secretariat (2006) Forum information and communications technologies ministerial meeting, 30 March 2006, Wellington, New Zealand, Wellington Declaration. Retrieved from Suva

Pacific Islands Forum (2008) Thirty-ninth Pacific Islands Forum, Alofi, Niue, 19–20 August 2008—Forum Communiqué. Retrieved from Suva

Pacific Islands Forum (2015) The framework for Pacific regionalism. Retrieved from Suva

Pacific Islands Forum. Forum chair (2016) Famework for Pacific regionalism: update on the implementation of regional priorities. Retrieved from Suva

Pacific Plan Review (2013) Report to Pacific leaders. Retrieved from Suva

Pernia EE (2008) Strategy framework for promoting ICT literacy in the Asia-Pacifc region. Retrieved from Bangkok

Radio New Zealand International (2014) Niue tourism chair resigned over poor strategic plan—Plan did not embrace "growth' in use of NZ monies

Radio New Zealand International (2016) After brief shutdown, Niue WiFi provider up and running: company has a month to resolve issues with government

Secretariat of the Pacific Community (2010a) Framework for action on ICT for development in the Pacific (FAIDP)

Secretariat of the Pacific Community (2010b) Pacific regional information and communication technology (ICT) ministers' meeting, Nuku'alofa, Tonga, 18 June 2010. Retrieved from Suva

Secretariat of the Pacific Community (2011) Pacific internet governance forum "Developing the future together": Report for sponsors and supporters of the inaugural Pacific internet governance forum (Noumea, New Caledonia, 9–10 Apr 2011). Retrieved from Noumea

Sovaleni S (2009) ICT@SOPAC. Retrieved from Suva

Stanley L (2008) The development of information and communication technology law and policy in Papua New Guinea. Pac Econ Bull 23(1):16–28

Strickland E (2015) Navigating the undercurrents: an analysis of Pacific Island regional ICT policy (PhD). University of Queensland, Brisbane

Telecom announces $25 million undersea cable for CNMI (2016) Saipan Tribune, p. March 11

UNDP Multi-Country Office Samoa (2008) Terms of reference for evaluation of ICT for development in the Cook Islands. Niue and Samoa, Retrieved from Apia

UNESCAP (2004) Meeting the challenges in an era of globalization by strengthening regional development cooperation. Retrieved from Bangkok

United Nations. Economic and Social Commission for Asia and the Pacific (2007) Assessment of the status of the implementation and use of ICT Access Points in Asia and the Pacific United Nations development account project on knowledge networks through ICT access points for disadvantage communities

West DM (2007) Digital government: technology and public sector performance. Princeton University Press, Princeton

World Bank (n.d.) ADB international development—Tonga Fiji submarine cable project in Tonga. Washington

Zwimpfer Communications (2002) Internet infrastructure and e-governance in Pacific Islands Countries. A survey on the development and use of the internet. UNESCO, Wellington

Chapter 4
Regulation, Access and Connectivity in Pacific Island states

Brian Louey-Gung

Abstract This chapter explores one of the essential pre-conditions of successful e-government implementation, a reliable and affordable telecommunications market, and considers whether an independent regulatory body overseeing a competitive telecommunications market is essential to achieving that outcome. The chapter discusses the primary regulatory principles that are necessary to support e-government and the functions of a telecommunications regulatory regime: price regulation, spectrum management, licensing, universal services and access, cyber-security, competition and consumer protection. Based on these principles, national ICT policy areas that should be given priority are identified. The chapter then focuses on the economic situation facing Pacific Island countries (PICs) and its impact on telecommunications markets. It includes a simple study of the cost of supporting an independent ICT regulator indexed against population and gross domestic product. A number of other key issues emerge from this discussion: the affordability of regulation; connectivity (also a critical market issue for PICs); the management of international telecommunications gateways; and some issues related to regulations and markets affecting PICs. The chapter concludes with some comment on possible solutions that would assist PICs to develop robust telecommunications markets to support e-government and economic growth.

4.1 Introduction

What are the most important elements of a rollout of e-government services? According to Gulati et al. (2014) the most important predictors of the success are a nation's financial and human resources, followed by having a suite of national ICT and e-government policies. A further critical finding is that an independent regulatory body overseeing a competitive telecommunications market is strongly

B. Louey-Gung (✉)
Independent IT and Telecommunications Strategy Consultant, Suva, Fiji
e-mail: brian.louey-gung@alumni.unimelb.edu.au

R. Cullen and G. Hassall (eds.), *Achieving Sustainable E-Government in Pacific Island States*, Public Administration and Information Technology 27,
DOI 10.1007/978-3-319-50972-3_4

associated with greater e-government success (Gulati et al. 2014). These results are not surprising. A commitment to industry liberalization and an independent regulator[1] identifies a government that believes in the economic benefits that can arise from a focused strategic approach to the use of ICT (World Bank 2009); and dedication to these strategies is demonstrated in the creation of national ICT and e-government policies. However, even with such a strategic focus, it is essential to have access to sufficient financial capital and human resources.

What does this mean for Pacific Island countries (PICs), several of which are classified in the World Bank's lowest economic development category and house the world's smallest national populations? Providing the basic needs of food, fresh water and shelter is a major challenge, and for many Pacific Island residents merely obtaining a tertiary education requires a move to another country—generally to one of the twelve PICs affiliated with the University of the South Pacific, to one of the newer national universities in the region or to Australia or New Zealand. Put simply, many PICs do not have substantial financial or human resources.

Ultimately, this question must be answered for each country individually; however, there are some common principles that apply, not just to PICs, but to all e-government deployments. The second section of this chapter discusses the regulatory principles that are essential to support e-government. For e-government to be successful, business and consumers must be comfortable using the technology and confident that they are not being overcharged or vulnerable to scams. The principles go directly to protecting consumers and to preventing, or at least managing, instances of market failure. Resultant regulatory functions include:

- Price regulation—to ensure that prices are fair and reasonable.
- Spectrum management—to ensure that all wireless operators have access to a fair proportion of the spectrum resource.
- Licensing—to ensure compatible and reasonable technical standards and to ensure that operators actions support government policies on such things as customer service standards, universal access and the collection of important statistical information.
- Universal access—to ensure affordable access to basic ICT services for the entire community, including those outside towns and cities.
- Cybersecurity—to ensure business and consumers are safe when operating online.
- Competition and consumer issues—to address more general issues of market failure and unfair behavior by operators (ITU 2014).

A bottom-up approach to these functions is adopted in this chapter to indicate the national ICT policy areas that should be given priority, irrespective of whether the market is liberalized or there is an independent regulator. The discussion then

[1]A regulator is an institution that oversees and regulates a market, in this case a telecommunications market, to ensure that the market operates in a fair and efficient way, and complies with any relevant legislation.

focuses on the economic situation facing the PICs. It looks at a simple study of the cost of supporting an independent ICT regulator indexed against population and gross domestic product. While the study is not sufficiently detailed to suggest an 'ideal' level of expenditure, it does illustrate that PICs, with their extremely small populations, have to spend many times the *per capita* cost of other countries if they are to have effective independent national regulatory bodies. PICs must individually decide if an independent ICT regulatory body is a sensible use of their scarce national resources when much of their population lacks secure food, housing and education. The question is whether the twin goals of a liberalized ICT market and an independent regulatory body are a realistic ambition for only the larger of the PICs.

An example relevant to all PICs is provided to illustrate a number of the regulatory concepts discussed in this chapter. A consequence of the geographical isolation of PICs and their relatively small populations is that the entity that supplies international telecommunications connectivity will have a monopoly, or near monopoly over this essential service. International connectivity is highly relevant to e-government for at least three reasons: (1) for government to government interactions; (2) for access to cloud services, which are increasingly being used for e-government in developed countries; and (3) because inter-island distances in the PICs are often so large that international satellite facilities are the only feasible option for island-to-island domestic communications. Irrespective of the competition status of the rest of the industry, there is a strong argument that international telecommunications need to be regulated in the best interests of the country.

4.2 Key Regulatory Principles for E-Government

Much has been written as to how e-government can be used to enhance the development and implementation of regulation, such as facilitating dialogue with stakeholders, educating the public on regulatory requirements and easing the cost of compliance. Little has been written, however, about the critical role of effective and appropriate regulation to facilitate e-government. E-government success can be hampered in many different areas that can be made subject to regulation if and where necessary. Much of this regulation is based on competition theory, but is still relevant in the monopoly markets of many PICs.

4.2.1 ICT/Telecommunications Regulators in PICs

As at the end of 2015, independent ICT regulatory authorities operated in six of the PICs (Fiji, Kiribati, Papua New Guinea, Samoa, Solomon Islands and Vanuatu—see Fig. 4.1); Cook Islands, Tonga and Palau were in the process of forming independent regulatory authorities (in Tonga the Ministry of Communications currently holds that role). Various different models have been adopted, as in the rest of the world, suggesting that there is no single correct way to create a regulator. Palau is notable in being the only PIC that liberalized its telecommunications industry prior to establishing a suitable regulatory regime and an independent regulator.

Box 4.1: ICT/Telecommunications Regulators in PICs

Fiji
The primary ICT regulator, the Telecommunications Authority of Fiji, is responsible for all ICT issues with the exception of competition and consumer issues, which come under the Fiji Commerce Commission.
http://www.taf.org.fj/ (formed 2008); http://www.commcomm.gov.fj/ (formed 2010)

Kiribati
The Communications Commission of Kiribati is the sole regulatory authority for ICT issues.
http://www.cck.ki/ (formed 2005 as the Telecommunications Authority of Kiribati)

Papua New Guinea
The National Information and Communications Technology Authority (NICTA) is the sole regulatory authority for ICT issues.
http://www.nicta.gov.pg/ (formed 2010)

Samoa
The Office of the Regulator is responsible for ICT, postal services and the Energy industries.
http://www.regulator.gov.ws/ (formed 2006)

Solomon Islands
The Telecommunications Commission of Solomon Islands is the sole regulatory authority for ICT issues.
http://www.tcsi.org.sb/ (formed 2009)

Vanuatu
The Telecommunications and Radiocommunications Regulator is the sole regulatory authority for ICT issues.
www.trr.vu (formed 2008)

4.2.2 ICT Regulation in Developed Economies

A 2012 survey conducted by the ITU for the production of the ICT Regulatory Toolkit (ITU 2014) asked 158 ICT regulatory authorities to identify their five key regulatory responsibilities.

Table 4.1 Regulatory responsibilities identified by ICT regulatory authorities

Responsibility	% respondents	E-government relevance
Price regulation	88.6	Unreasonably high prices deter the use of e-government services
Spectrum allocation and assignment	88.6	Essential for mobile coverage and long-distance communications
Licensing	83.5	Technical license conditions ensure operational compatibility between different e-government services Operator license conditions impose compliance with e-government policies
Universal service/ access	79.1	Provides affordable access for all citizens to e-government services
Cyber security	39.2	Ensures e-government services are safe to use

Source ITU-*info*Dev ICT regulation toolkit (ITU 2014)

A sixth responsibility—competition behavior—was not among the responses, but overlaps strongly with price regulation, licensing (customer service standards) and cyber security. Competition regulation ensures customers and citizens are treated fairly when using (and paying for) the telecommunications services that are used to access e-government services (Table 4.1).

This portfolio of responsibilities of a typical modern ICT regulatory body has relevance for the success of e-government—although some links are not always obvious.

4.2.3 Price Regulation

Unreasonably high ICT prices will make e-government too expensive for the government itself, for its internal uses and for other users such as businesses and citizens that try to access e-government services. This situation will increase resistance to the development of e-government services, and when the services are deployed users will be less willing to make use of them. So reasonable ICT pricing is essential to the success of e-government.

But what is a 'reasonable' price? According to mainstream economic theory, market forces determine optimal (also known as 'efficient') pricing, which is considered to be 'reasonable'. The principle applies at all levels of the market, including retail, wholesale and interconnection. Regulation should be required only where the market has failed to efficiently set prices or where such a failure is reasonably anticipated. Unfortunately, the proliferation of regulators around the world with a pricing mandate testifies to the market's failure (realized or anticipated) to set efficient prices. This is directly relevant to most PICs, where small market sizes make market failure more likely.[2] For example, even though several PICs markets have been successfully liberalized with multiple market entrants, supply of international connectivity remains an effective monopoly, controlled by the submarine cable operator or the main satellite operator.

The large majority of price regulation work focuses on wholesale pricing, the consensus being that retail price regulation is not necessary where wholesale prices are reasonable (ITU ICT Regulation Toolkit 2014). The most common exceptions are the imposition of price caps, or the principle of geographically uniform retail prices, generally applied under Universal Access policies. The technique of price capping will normally allow prices to rise in line with an appropriate economic indicator such as CPI. The three main techniques for wholesale price regulation are cost-based pricing, retail-minus pricing and benchmarking. Each has strengths and weaknesses, and the most appropriate one will depend on individual circumstances. A detailed discussion of these techniques is beyond the scope of this chapter, but a summary is provided below.

4.2.3.1 Cost-Based Pricing

Cost-based pricing has been in use for at least 30 years and is probably the most common method for regulating situations where competition is being introduced into a monopoly market. During this time, many academic studies contributed to developments in the methods used. However, improvements will often impose an increased requirement for data collection, making it more expensive to implement. Cost-based pricing, as the name implies, determines a price based on the cost of providing the product or service, allowing a reasonable margin for profit. Unfortunately, there are many complicating factors underlying this simple principle. Two of the more contentious are:

- Cost allocation: how are costs allocated (such as the cost of running the telephone exchange) to all the services that use the same piece of equipment? The cost of running the exchange is relatively straightforward to determine; the allocation of those costs to all the services supplied by that exchange can be done in numerous ways which have very different outcomes.
- Weighted Average Cost of Capital (WACC): All major businesses borrow money to fund infrastructure rollout, and WACC is the assumed average interest rate they pay for those loans. Again, this might seem to be a straightforward concept, but it can be calculated in many ways, with small differences in WACC being responsible for major price differences.

[2]Regulators should not automatically impose price regulation, especially for new services. It is not unusual for regulators to allow the market to set prices for new services and only regulate if necessary.

A further complication is that cost models must be adapted or created for each new technology. For example, cost models of 3G mobile technologies are substantially different to those used for modelling 2G technology, and 5G technology may require radically different cost models compared to the latest 4G models. Common methodologies, which can be applied to most technologies, include:

- Long Run Incremental Costs (LRIC) and many variants such as LRIC+ and TSLRIC (Total Service Long Run Incremental Costs);
- Fully Allocated Costs (FAC);
- Stand Alone Costs (SAC); and
- The Building Block Approach (based on actual incurred costs and forecasts of expenditure and demand specific to a particular business) (Incenta 2015).

Each methodology has known strengths and weaknesses, so deciding on the most appropriate one to apply in a given situation can be controversial. It is not unusual for a government to specify a particular methodology in legislation, narrowing the regulatory options, although this may restrict the regulator's ability to quickly adapt to technology and other change, leading to poor regulatory outcomes.

4.2.3.2 Retail-Minus Pricing

Retail-minus pricing assumes that the retail market is fully competitive—an assumption that is uncertain even in the small PICs with liberalized markets. The cost of providing the retail service (including a reasonable retail profit) is estimated and deducted from the retail price, giving the regulated wholesale price.

An increasingly common problem with this approach is that it assumes that wholesale services need to correspond strongly with retail offerings. For broadband services such correspondence is increasingly unlikely, as international capacity, for example, may be wholesaled as a certain capacity (Mbits per seconds) available 24 h a day, irrespective of how much is actually used, whereas retail broadband is generally offered with a monthly data cap.

4.2.3.3 Benchmarking

Benchmarking is attractive because it can provide a regulated price without needing to strongly justify cost data, forecasts, cost allocation schemes and other potentially controversial aspects of alternative pricing methodologies.

It gathers data from markets similar to the one in question and makes adjustments where necessary to account for significant differences in cost drivers and other factors. This determines a benchmark range for the market in question. The regulated price is then chosen from within this range—depending on government policy, objectives and other socio-economic factors.

Benchmarking's most important aspects are the selection of markets to include in the benchmark and the selection of the final value from within the benchmark range. Both can be highly controversial. The selected markets ideally should be broadly similar in the cost drivers:

- The regulatory environment: e.g. the pricing and costing methodologies applied by the regulator, or quality of service regulation.
- State of market development: e.g. Degree of competition; number of competitors in the market; extent of vertical and horizontal integration; range of services offered.
- Socio-economic environment: e.g. population distribution; population density; income levels; terrain.

The selection of the final value from within the benchmark range will generally reflect government policy. For example, a policy of encouraging new entrants may see a regulated interconnection price chosen from the lower end of the benchmark range.

4.2.4 The Need for Pragmatism

Above all, a regulator operating in the PICs environment needs to maintain a high level of pragmatism regardless of the chosen pricing method. All methods described have different strengths and weaknesses, and all present problems with the collection of data. The selection of a pricing method will depend as much on what data and resources are available as which method might theoretically provide the best answer given the circumstances.

In performing a regulated price study, the regulator must also be aware of government policy—for example, a policy to attract more competition in either the regulated market, upstream or downstream markets. To encourage new entrants, the regulator may choose to apply price discrimination for either a limited time or until a new entrant achieves a certain market position.

Other challenges for which there are no easy solutions include:

- How best to take account of donor funding: such funding does not expect a commercial return and significantly reduces the risks faced by the operators. However, reducing the risk premium may make prices so low that alternatives are priced out of the market. For example, cost-based wholesale prices for a largely donor-funded facility may be so low that other facilities cannot compete.
- In countries that have multiple international gateways it may be necessary to prove that they all have Significant Market Power (SMP) (see Sect. 4.3).
- All countries want to encourage investment, so it is important that investors are certain that regulated pricing will not unduly reduce the value of their investment. This means that pricing principles, once established, should not be changed capriciously.

4.2.4.1 The Way Forward for PICS

The comparative details and arguments for and against each methodology can be readily found,[3] but are not central to this discussion. There are two salient points to be made with respect to PICS: the first is that implementation of any existing pricing methodology is very costly. Examples abound of individual cost determinations exercises extending over several months, if not a year or more, with expenditure on consultations and rebuttals incurred by several key stakeholders.[4]

While developed-economy regulators engage external expertise for pricing exercises, the risk of engaging external contractors is greater for PICs as they have little if any pricing expertise on staff to properly manage the exercise in the best interests of the country. In addition, the cost may be so high as to require donor funds, adding more complexity to the process.

PICs (and other developing economies) need a price regulation methodology whose cost is reasonably containable and is readily able to be updated for technology's advances. This could be a new methodology, or a modification on the principles of benchmarking or LRIC. Unfortunately, no such new or modified methodology exists, or is being developed—while much development is occurring, it seems to be adding complexity in the quest for perfect pricing, rather than simplifying and reducing cost.

There appears to be no generally accepted price regulation methodology whose cost of implementation is reasonably containable, leaving PICs to cope with either LRIC or benchmarking. In these circumstances, it is recommended that legislative or other hard limits be placed on the pricing process in order to contain costs. Rather than the 'best' pricing outcome, PICs may have to compromise on achieving a 'good' pricing outcome.

The second salient point is affordability. Irrespective of the chosen pricing methodology, PICs may find that the regulated price is not affordable for many of their citizens, in which case other strategies must be devised—such as providing and promoting community-based facilities for access to e-government services rather than expecting individuals to use their personal equipment.

Ultimately, PICs need a pricing approach which incurs much less cost than current developed-economy approaches but which will still provide reasonable outcomes.

---

[3]For example, see websites of the ICT regulators in the UK (OFCOM: http://www.ofcom.org.uk/) or Australia ACMA: http://www.acma.gov.au/.

[4]One extreme example is a recent regulatory pricing decision by the New Zealand Commerce Commission which took nearly three years and involved nine distinct consultations and 23 rounds of industry submissions and cross submissions (Commerce Commission 2016).

4.2.5 Spectrum Management

Spectrum management is relevant to e-government in that it is an essential component of international and domestic telecommunications connectivity. While an increasing number of PICs are accessing high capacity international bandwidth via submarine cable, and some larger population centers are being provided with fiber-optic connectivity to homes and businesses, it is financially impractical to extend such access to the vast majority of PICs citizens. Where submarine cables are not present, international connectivity continues to be provided by satellite. Domestically, many remote communities are reachable only by satellite services, and even in fiber-connected larger population centers most people will access local telecommunications services using mobile telephone technology. Thus, for the foreseeable future, access to spectrum underpins the basic telecommunications infrastructure in most PICs and is essential to the development of e-government.

Fortunately, spectrum resources in PICs are not under pressure and the task of ensuring access to sufficient spectrum is relatively straightforward. PICs have historically followed global international technical standards for radio equipment and spectrum assignments, facilitating access to the global economies of scale present in the current mobile and satellite telecommunications industries. There are no indications that this practice is going to change and no sensible reason to do so.

4.2.6 Licensing

E-government is supported by licensing policies in two distinct categories: technical standards and licenses to operate. Licensing practices have been used extensively in the past to maintain strict government control over technology choices and market entrants in both developed and developing economies, but in more recent times the trend is towards technically flexible licensing and open market entry. Given that PICs are market and technology followers, it is prudent for PICs to adopt such practices so that market players can take full advantage of global market economies of scale and introduce new technologies and services as soon as practicable.

4.2.6.1 Technical Standards

Technical standards can be imposed through a licensing regime whereby the minimum standards of equipment and services are specified by government fiat. The ICT industry strongly supports technical standards, which are considered critical to interoperability; it also encourages avoidance of costly battles over technologies through the work of the ITU (2016a) and industry bodies such as the

GSMA,[5] which promotes global equipment and service compatibility (GSMA 2016). These standards must be formally adopted by countries in order to have legal force. Governments may apply additional criteria for local markets and also have the option of ratifying only a subset of available standards. For example, Australia and many other countries in the Asia-Pacific region have adopted the APT-700 MHz band plan as the template for spectrum allocation, whereas the USA has a different 700 MHz plan, both of which have been ratified by the ITU.

Such standards support e-government by ensuring technical compatibility and interworking of equipment. In practice, PICs have little choice but to accept technical standards determined by global bodies, as they have minimal development and manufacturing capabilities themselves. However, there are mechanisms through which PICs can influence the development of global technical standards, albeit such efforts require substantial resources and expertise.

4.2.6.2 Operating Licenses

Operating licenses authorize a person or commercial entity to conduct a business in a particular field or to sell particular products or services. Within the ICT industry, operating licenses are often required to run a telecommunications network, a TV or radio station, or, increasingly, an Internet service provider. The value of operating licenses to e-government is that license conditions can be used to:

- provide guidelines for the quality of customer services that should be provided and give the regulator the legal power to intervene if and when necessary—with mandatory standards, monitoring and enforcement powers;
- support universal access policies through mechanisms such as mobile coverage targets and contributions to a universal service fund; and
- mandate regular collection and reporting of important statistical information—to gauge the appropriate time to roll out e-government initiatives and to monitor the progress of those rollouts.

License conditions can also be used to limit the extent to which a single entity (person or company) can dominate the media and telecommunications industries in a country. This is considered to be necessary in democratic countries where a single entity dominating media outlets could bias news reporting.

Most PICs' markets are very small and will struggle to attract multiple market players and open market entry may seem irrelevant; however, it may be worthwhile to enable such practices in legislation so that they can be activated if and when appropriate at a future date.

[5]The GSM (Groupe Spéciale Mobile) and the GSM Association which comprises mobile operators, is focused on ensuring that mobile operators have timely and affordable access to appropriate spectrum to meet the rapidly growing demand for mobile broadband services (GSMA 2016).

4.2.7 Universal Access and Service

Three pre-requisites for successful e-government targeted at citizens are:

- the majority of citizens are able to access e-government services;
- the access is affordable; and
- citizens are comfortable interacting with the e-government services.

Universal access and service policies address these requirements.

Universal service policies pre-date mobile telephony by many decades and originally referred to the provision of affordable fixed telephone services to individual households, usually mandating a defined basic voice service that was price controlled or subsidized. Universal access policies are focused on similar issues, but for shared community facilities, such as public telephones. Over recent years the focus of Universal Service and Access has broadened to include mobile telephony, which is particularly relevant to PICs, where the provision of fixed facilities to all communities, let alone individual households, would be prohibitively expensive. Mobile facilities, on the other hand, are much less expensive to provide in terms of coverage and access to acceptable handsets. There has also been a marked shift in user preference from fixed facilities to mobile across the world. Those PICs with active policies have applied the universal access approach through incentives to mobile operators to expand their service coverage, facilitating the spread of personal mobile services and enabling Internet connected community facilities such as schools, libraries and medical centers.

Affordability is directly addressed by pricing regulation, but it is open to the government to implement additional affordability measures through universal access policies. For example, it is common for universal access or service policies to ensure that telecommunications prices for remote users remain affordable, despite the significantly higher costs of providing services to such communities. While various methods are used to maintain affordability, the outcome is that remote services are effectively subsidized, either internally by the operator or explicitly by the government through a universal access grant. Vanuatu, for example, has a universal access policy that is often recommended to other PICs, which requires telecommunications services to be available to 98% of the population at an agreed uniform price.[6] This is achieved by imposing a levy on providers and making a subsidy available to those providers (to be allocated by the regulator through a transparent and accountable process) who service hard-to-reach geographical areas (Vanuatu 2013). This is referred to as a supply-side approach. By contrast, Kiribati's universal access policy takes a demand-side approach; projects funded by its Universal Access Fund from levies authorized by its universal access rules (Kiribati Communications Commission (CCK) 2015) and the Kiribati

[6] This must include voice, text messaging, and data download speed minimum of at least 21 Mbps and upload speed of at least 12 Mbps.

Communications Act 2012 may be used to develop markets, government facilities and industries in areas which would not otherwise support a competitive market.

The Solomon Islands Telecommunications Act 2009 makes provision for a similar pro-active Universal Access Policy; this was to be supported by funding from the World Bank as part of its Solomon Islands Telecommunications and ICT Development Project. The evaluation report of the project notes that the funds allocated for a Technical Advisor were not used and activity pertaining to the Universal Access and the provision of ICT services to rural areas was dropped since a universal access policy was not deemed to be necessary "in view of the increased coverage of ICT services in rural areas during the project execution phase" (World Bank 2016). The report noted that 90% network coverage had been achieved during the project and 66% of the eligible population were mobile subscribers. However, it should be noted that coverage, affordability and uptake are not identical and that pursuance of the mandated UA policy may be necessary in the longer term.

The third element of citizen comfort in using e-government services is a challenge to PICs where many citizens in remote communities do not yet have electricity, let alone modern communications. Yet it is in these remote communities that both government and citizens have much value to gain from the implementation of e-government. In addition to increasingly efficient solar power systems and rapidly extending mobile coverage, citizen training and familiarization courses are being offered in some PICs to make citizens aware of the opportunities and benefits of e-government. Governments also need to address issues of data security and privacy to ensure ongoing trust in e-government systems.

4.2.8 Cybersecurity and Other Data Protections

An absolute requirement for successful e-government is that all users have confidence in the systems. Such confidence is achieved only when all of the factors involved in e-government are working well and are protected.

At the technical level, all government and associated business computers, networks and terminals (including smart phones) should:

- be protected from online attacks by regularly updated firewalls and anti-virus programs. The government can only ensure that its own systems are protected; it is impractical to mandate protection for all users. However, the government can promote the use of protective products and ensure suitable products are readily available (many are free);
- have backup and recovery systems in place so that if an attack should occur, service can be maintained and/or readily restored; and
- ensure confidential data (personal and business) is strongly encrypted and physical security of storage devices is high.

At the human operator level, formal policies must:

- define access permissions to different levels and types of private and confidential data;
- provide suitable training to all people with such access, including in the application of strong personal passwords; and
- monitor systems for breaches of confidentiality and, where breaches have occurred, investigate and apply penalties.

And finally, it may be necessary to create and distribute information for citizens on how to use e-government systems. Personal training could be provided at community centers such as schools and libraries.

Government and business access to Computer Incidence Response Team (CIRT) facilities (in the Pacific region more commonly referred to as Computer Emergency Response Team or CERT) is strongly recommended (ITU 2016b). The idea of a Pacific CERT to supplement the work of the AsiaPacificCERT (APCERT)[7] and the Asia Pacific Network Information Center was first advocated in 2007 by various groups including the Pacific Islands Chapter of the Internet Society (PICISOC), and endorsed by the Pacific ICT Ministers meeting in Tonga in 2009. PacCERT was formed in 2012, with the assistance of the ITU and the Japanese International Cooperation Agency (JICA) and established at the Japan-Pacific ICT Centre at the University of the South Pacific; all 22 PICs and territories are constituent members. Currently PacCERT is 'in hiatus' but services could be negotiated from CERTs in Australia, New Zealand, the USA, or any other Pacific region country with CERT facilities.[8] In Australia, for example, two CERTs are in operation, the government funded CERT Australia (2016), while the other is a commercial operation with many corporate and Pacific clients.

4.2.9 *Competition and Consumer Issues*

An essential element of the success or failure of e-government systems is the link provided by telecommunications carriers; consumers must be comfortable using the telecommunications services to access and interact with e-government systems. The relationship between the carrier and the consumer is complex, encompassing a large range of regular interactions, ranging from the carrier's advertising and promotions to one-to-one transactions of billing and customer services.

The Australian Telecommunications Consumer Protection Code (Communications Alliance 2016) provides excellent examples of the types of consumer issues that may need to be addressed by a regulator. They include:

---

[7]APCERT is an organisation comprised of CERTs established in developed countries around the Asia Pacific, including Australia, China, Taiwan, Hong Kong, Japan, New Zealand, and several others.

[8]Tonga recently announced the opening of a national CERT in the capital Nuku'alofa. see http://www.mic.gov.to/news-today/press-releases/6159-pm-launched-tonga-national-cert.

- advertising claims must be able to be proved on request;
- advertising must not be deliberately misleading;
- customer contracts must be written in plain language, readily understandable by the majority of the population;
- bills must be accurate and clearly explained; and
- complaints must be handled politely and resolved in a timely manner.

It is worth noting that all of the aforementioned issues apply irrespective of whether the market has been liberalized, whether there is a monopoly telecommunications carrier (as is the case for many PICs), or whether there are competing telecommunications carriers (as, for example, in Fiji, Palau, Papua New Guinea, Samoa, Tonga and Vanuatu.)

While competition in telecommunications is not essential for successful e-government, it can be argued that competition will result in more widespread availability of services and more affordable services, thus providing an important boost for the take-up of e-government. Telecommunications competition raises a large number of additional regulatory issues, involving the management of significant market power, information advantages and negotiating power. These are best illustrated through an examination below of the regulatory management of international connectivity in PICs.

Competition issues become even more problematic where one market entity is well established in the face of new entrants. The established entity will naturally want to protect its market position and will take whatever actions are legally available. Government and regulators can act in many different ways to moderate the extent to which a company in a powerful market position can leverage its market power against smaller players. Two common examples are regulated infrastructure sharing and mandated number portability.

4.2.9.1 Infrastructure Sharing

Infrastructure sharing describes the practice of competing carriers sharing infrastructure such as mobile towers or even radio spectrum. Infrastructure sharing potentially reduces the cost of rolling out infrastructure by a substantial amount, as well as reducing the number of mobile phone towers to be built, minimizing the amount of visual pollution.

Despite these many benefits to carriers and the community, infrastructure sharing may not be attractive for a well-established carrier facing a new competitor seeking to roll out its network quickly and cheaply: it would be seen as simply helping the new competitor while gaining little in return if government policy is to encourage the new entrant. In this case the regulator, backed by suitable legislation, may regulate and mandate infrastructure sharing in defined situations.

4.2.9.2 Number Portability

Number portability enables a mobile phone customer to change operators/providers while keeping their mobile phone number. This can be achieved technically by many different methods; in best practice implementations, the porting of numbers is so prompt that the customer can be active on the new operator's network within 10–15 min of making the request and satisfying the necessary identity checks. In other implementations, it may take several days for porting to occur, during which time the customer may have no mobile service.

Number portability is considered to be very important where a new operator/provider is entering an established market—having to change a telephone number is a significant barrier to changing a mobile operator. In such cases, it may be necessary for the government to mandate number portability. However, in a more balanced market, number portability may simply impose an unnecessary additional cost on the industry. Fiji, for example, has two mobile operators but does not have number portability. It is difficult to find reliable statistics, but anecdotally many mobile customers in Fiji already have two services—one from each operator, so imposing number portability in Fiji at this time may do little to lower the barriers for migration between operators.

In the small telecommunications markets of most PICs, if it is government policy to encourage new entrants, it would be prudent to give the regulator sufficient power to moderate the efforts of the established telecommunications providers seeking to protect their market share and impede new entrants.

4.3 Key Issues

4.3.1 The Affordability of Regulation

Given that an ICT regulator has a number of well-defined functions to perform, including the monitoring and imposition of complex international standards, there are certain basic requirements of the office that may bring incur considerable costs. The question then arises as to whether these are affordable in the context of small markets. The author undertook a survey of World Bank countries with established independent regulators that have published staff numbers and costs in their annual reports. Seventeen countries were chosen, based on availability of data and population size (in 2012)—all countries have a population of less than 10 million, except for Australia (23 million) and Malaysia (29 million). Australia was included as a representative developed country in the region; Malaysia was included as a comparatively high population developing country near the region. They provide a simple check that countries with significantly larger populations have broadly similar results.

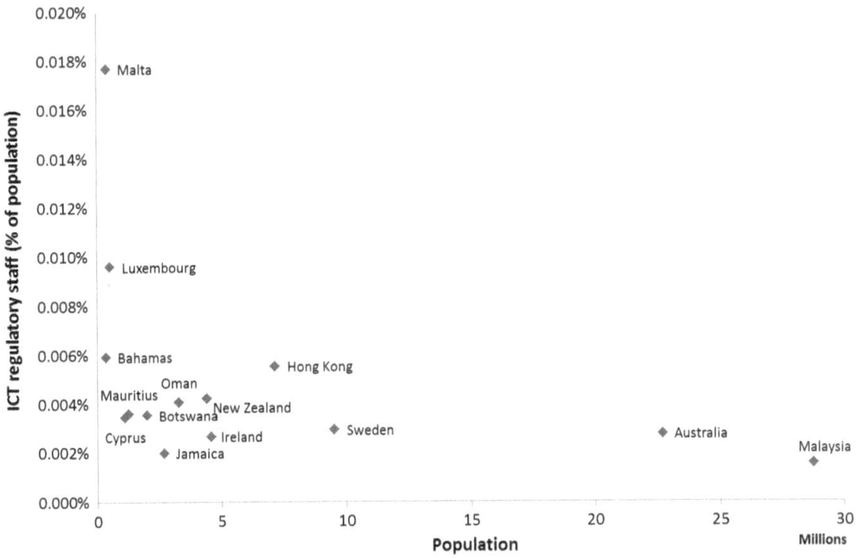

Fig. 4.1 ICT regulatory staff as a percentage of population (2012). *Sources* World Bank for population and GDP; Regulator staff numbers and operational expenditure: 2012 Annual reports

Figure 4.1 shows that the number of ICT regulatory staff is at or below 0.005% of the population for most countries, with only the smallest populations—Bahamas (population 372,000), Malta (418,000) and Luxembourg (531,000)—clearly above this figure. Apart from Hong Kong (7.1 million), all other countries fall under 0.005%. For scoping purposes, 0.005% has been chosen as an upper indicator of the number of ICT regulatory staff required for a functional regulatory office.

Figure 4.2 shows ICT regulatory expenditure is at or below 0.03% of GDP for most countries, including two of the four smallest in the sample by population: Bahamas (population 372,000) and Malta (418,000) are above, but Luxembourg (531,000) and Iceland (320,000) are below.

4.3.1.1 Implications for PICs

Table 4.2 shows the application of these two scoping benchmarks for investment in ICT regulators, of 0.005% of population and 0.03% of GDP, for the PICs providing this data. The five most populous countries—Papua New Guinea, Fiji, Solomon Islands, Vanuatu and Samoa—as well as Kiribati have established independent ICT regulators. The Papua New Guinea regulator operates within the informal benchmarks outlined above; Solomon Islands and Vanuatu regulator staff are well within the staff benchmark but expenditure is some 4–5 times higher than the GDP benchmark derived from comparable countries; Samoa exceeds the staff

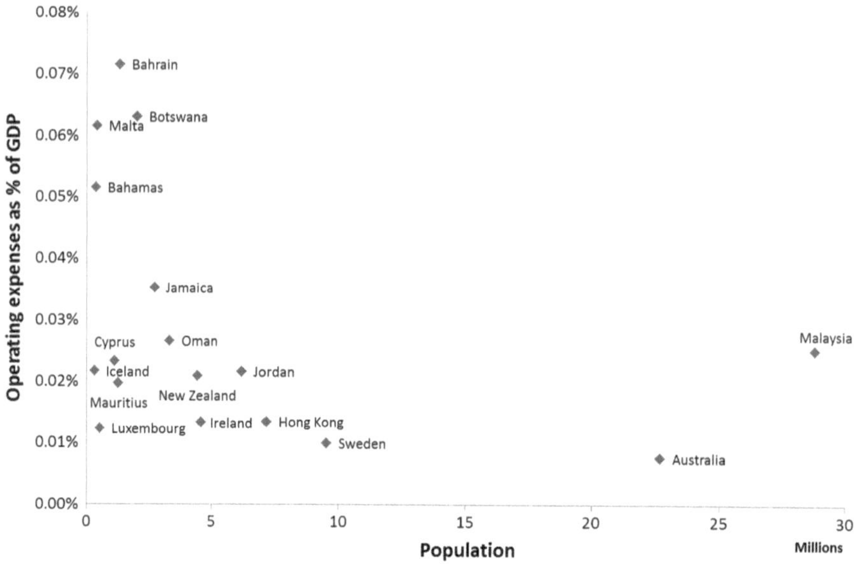

Fig. 4.2 ICT regulatory expenditure as a percentage of GDP (2012)

Table 4.2 Application of scoping benchmarks for regulatory expenditure in PICs

Country	Population ('000)	Staff (0.005% pop)	Expenditure (US$[a]) (0.03% GDP)
Papua New Guinea	7167	359	6,130,000
Fiji	875	43	1,260,000
Solomon Islands	550	27.5	307,500
Vanuatu	247	12	336,000
Samoa	189	9	254,000
Tonga	105	5	154,000
Federated States of Micronesia	104	5	115,000
Kiribati	101	5	80,000
Marshall Islands	53	3	55,332
Palau	21	1	116,000
Tuvalu	10	0.5	GDP not available

[a]GDP (USD PPP) World Bank (downloaded 4 Feb 2014)

benchmark[9] (expenditure as percent of GDP data was not available); data for Fiji and Kiribati was not available.[10]

This comparison suggests that the more populous PICs can probably operate their ICT regulators within the staffing levels, if not the total expenditure, of more developed countries. In the case of Solomon Islands as much was spent on consultancy as on staffing (TCSI 2012), although this was only the Regulator's second year of operation, and the figure may cover establishment costs. However, the figures for staff numbers and expenditure benchmarks for the smaller countries quite clearly show the challenge facing those countries—to create an effective national regulator with 5 or fewer staff and/or a budget of not much more than US $100,000 to US$150,000[11]—when, especially in the early years, that budget should allow for at least one expert with substantial international experience and responsibility for capacity building.

Pragmatically, five staff is not sufficient to cover all of the specialized areas of regulation; and US$100,000 is hardly sufficient to attract an expert with substantial international experience, let alone fund multiple other staff members. Again, pragmatically, no matter how this problem is approached, smaller Pacific Island countries wishing to maintain an independent national regulatory authority[12] will have to pay comparatively more in regulatory costs than developed economies. The question then becomes one of determining the most cost-effective manner to access the benefits of ICT.

4.3.2 Connectivity

4.3.2.1 Connecting the Pacific to the Rest of the World

Pacific Island countries are characterized by communities isolated by vast stretches of ocean. While some countries, such as Nauru, comprise one main island that houses almost its entire population, the population of others, such as Kiribati, is spread across island groups that are separated by thousands of kilometers of ocean.

The only viable telecommunications technologies capable of bridging these distances are satellite and submarine cable. (see the Telegeography Submarine

---

[9]The Samoan regulator also regulates the energy industry.

[10]Data for the Pacific Islands regulators was provided in-confidence and is not reported explicitly here.

[11]For example, the scoping benchmarks for Tonga and the Federated States of Micronesia, as the next largest countries by population. Tonga has since announced that it will be setting up an independent ICT regulatory agency.

[12]Reductions in regulatory costs may be achieved by sharing regulatory expenses with other countries. This may be via a shared regulator or by sharing specific regulatory functions, such as spectrum management. A more detailed discussion of these options is beyond the scope of this paper.

Cable Map at: http://www.submarinecablemap.com.) These technologies deliver all international connectivity for Pacific Island nations and a significant amount of domestic long-distance connectivity—where standard terrestrial long-distance technologies such as microwave are defeated by inter-island distances or inaccessible terrain.

Connection to a submarine cable is made via a cable landing station, and connection to a satellite service is made via a satellite ground station. Comparisons between the different technologies can be made on a number of different criteria (see Table 4.3).

Table 4.3 Comparison between different technologies used for international telecommunications in the Pacific Islands region

Costs incurred by PICs to access and to use international telecommunications services	Submarine cable landing stations incur a very high capital cost which provides access to very high bandwidth. Cost per unit of bandwidth used is relatively low. However, for many PICs, the bandwidth made available is far in excess of any reasonable estimate of market demand into the foreseeable future Satellite ground stations are much more scalable to market demand, from relatively inexpensive VSAT units appropriate for small communities to major ground station facilities that can match the capacity of submarine cable landing stations. However, the cost per unit of bandwidth used is relatively high Ground stations for non-geostationary satellite services serving larger communities must use tracking antennas, incurring extra cost. For smaller communities, it may be possible to use broad beamed non-tracking antennas, but only lower bandwidth services will be available
Locations	Bringing a submarine cable to shore is a complex task that is heavily influenced by undersea topography. Accordingly, the number of locations that are suitable for landing stations can be quite restricted and on some islands suitable locations may be lacking. Substantial regulatory protections are then required to ensure that close-to-shore activities do not damage the cable in shallow waters Finding a suitable location for a satellite ground station is much easier as it need only be located within the satellite footprint, have line-of-sight to the satellite, and sufficient space for the satellite antenna
Areas served	Submarine cables can serve only places where the cable is brought to shore, including areas that can be connected to that location by terrestrial telecommunications infrastructure Satellite services can provide service to any location within their footprint, which in some cases is the entire Pacific region. (See later section on C-band spectrum)
Weather effects	Satellite services are adversely affected by rain and heavy cloud. Tropical rain showers reduce satellite throughput, and it is not unusual for service to be completely interrupted. Antenna diversity can alleviate this effect, by placing an alternative antenna sufficiently distant so that rain is unlikely to affect both at the same time. Many PIC islands are physically too small to implement diversity Submarine cables are not affected by weather

(continued)

Table 4.3 (continued)

Latency[a]	Latency adversely affects interactive online services, with the low latency of submarine cables preferred in comparison to the higher latency of satellite services. Even the acknowledgement requirements of the Internet (through the TCPIP protocol) need to be carefully managed to accommodate the latency inherent to geo-stationary satellite services. Mainstream services affected by high latency include: • ordinary telephone conversations; • audio and video conferences; • online role-playing games; • high frequency trading Latency over a submarine cable depends on the length of the cable, which rarely is laid in a straight line due to undersea topography. Latency over satellite services depends on the height of the satellite orbit. Typical latency figures are: • Australia to Fiji using the Southern Cross cable: 20 ms • MEO satellite system O3b: about 50 ms • geostationary satellite: about 250 ms

[a]Latency, in the context of telecommunications, is the time interval between a signal being sent and the signal being received

4.3.2.2 C-Band Satellite Spectrum

C-band satellite[13] transmission is essential for PICs as it is the only technology that is able to provide continuous affordable modern communications to small, very remote island communities. This is due to its ability to offer an extremely wide coverage footprint—several C-band satellite services offer a single 'global' beam that covers all parts of all PICs. Small communities are able to access these services at an affordable price because they share the beam with not only many other small communities, but with larger population centers as well.

Other satellites operating at higher frequencies, such as O3b Networks (2015) and the proposed Kacific service (Kacific 2015), can offer greater data speeds than C-band, but their coverage footprint is limited to less than 1000 km in radius. This is large enough to individually cover only a small number of PICs islands, and few entire PICs. Thus, a small community will be offered service only if it is lucky enough to fall within a coverage footprint aimed at a larger, commercially viable, population center. Many small PIC communities are not close enough to larger population centers to be able to take advantage of such fortuitous coverage.

However, C-band satellite is under threat. The ITU's World Radiocommunication Conference of 2015 (ITU 2015) agreed to the allocation of frequencies in the lower part of the C-band spectrum to (3.4–3.6 GHz) to meet the ever-increasing demand for mobile services even though several powerful lobby groups claim that satellite

[13]C-band is the name given by the ITU to a portion of the electromagnetic spectrum in the microwave range of frequencies, (between 4 and 8 GHz) that is allocated to satellite communications.

services, especially in tropical and sub-tropical regions will suffer serious interference as a result (CASBAA 2015; Intelsat 2015). While PICs have plenty of spectrum and can avoid interference caused by shared uses of the spectrum by simply not authorizing those uses, the worldwide market for C-band satellite can be expected to decline: alternative technologies are available for all users outside the PICs, so over time those alternatives will gain more of the market. A declining global C-band market will see prices rise, and eventually manufacturers can be expected to stop making C-band satellite equipment as the market becomes too small to remain economically viable.

It is possible that alternatives will be available before C-band satellite goes into a serious decline. Two potential services are currently under development: Google Loon, which is highly speculative and not targeted at PICs; and OneWeb, which is still very early in its development. Until a viable alternative is developed for the Pacific it is essential that the PICs become active in defending satellite use of C-band spectrum.

4.3.3 International Gateway Wholesale Operations

The satellite ground stations and submarine cable landing stations that are the visible technical structures of international connectivity for all Pacific Island nations are known as international gateways. Most Pacific Island nations have only one international gateway, which accesses one of several commercial international satellite services.[14] Some Pacific Island nations have two international gateways—a submarine cable landing station to provide the main connectivity, while satellite delivers backup and domestic long-distance services.[15]

Regulators use a number of guiding principles to oversee the wholesale operation of an international gateway operator that is considered to have significant market power (SMP) in its dealings with its customers (the retail service providers).

First, all customers of the gateway operator should be treated in a non-discriminatory manner. This is especially important if one of the retail providers is closely related to the gateway operator. For example, Amalgamated Telecom Holdings Ltd (ATH) in Fiji is sole owner of the submarine cable gateway operator Fintel, sole owner of Fiji's monopoly fixed network provider and major retailer Telecom Fiji Ltd and 51% owner of Vodafone Fiji, one of two mobile telephony operators and retailers in Fiji. ATH therefore has 'effective total control' of Fintel and TFL and a controlling interest in Vodafone Fiji (Reddy 2012). Reddy asserts that regulatory oversight is needed 'now more than ever' and is critical to ensure the protection of end users from anti-competitive pricing regimes (Reddy 2013). Put

[14]Where there is only one international gateway it accesses a geostationary C-band satellite service such as from Intelsat, SES New Skies or APStar.

[15]Where submarine cable is not a viable option, it is increasingly common for main connectivity to be provided by the O3b satellite service, while the original C-band satellite service is retained for backup and domestic long-distance services.

simply, Fintel's dealings with its associated companies TFL and Vodafone Fiji must be non-discriminatory in comparison to its dealings with its other Fiji customers.

Second, despite the need to be non-discriminatory, a government policy imperative to support the entry of new competitors may justify asymmetric regulation, but only during the start-up phase. For example, a new entrant may be granted a limited period of favorable pricing or other conditions. The period may be for a specified time, or until a specified market threshold is achieved, such as market share targets.

Third, the balance of power between the retailers and the gateway operator should be broadly equal. Regulation may be necessary to moderate the market power of the gateway operator, but it should not shift the power balance too much towards the retailers. If the gateway operator knows the amount of traffic being carried by each retailer (and therefore the total amount of incoming and outgoing traffic) but individual retailers only know their own traffic figures, selectively releasing such information may affect consumer attitudes in the retail market and therefore affect negotiating strength.

Finally, the services and prices offered to the retailers should allow the retailers sufficient flexibility to differentiate their retail offerings and should not unduly favor one approach to the retail market over any other. The wholesale price might be structured as a price per customer per month plus a price for each unit of usage. A high price per customer coupled with a low unit price for usage will favor a retailer targeting high usage customers. Conversely, a low price per customer coupled with a high unit price for usage will favor a retailer with a large proportion of low usage customers.

Gateway operator interactions with retail service providers can be divided into the initial set-up, day-to-day operations and financial processes. The regulator should maintain an interest in the following areas of initial set-up (offered as an example of the detail required in all operator/retail interactions):

- Physical space in the gateway building required to house customers' interconnection equipment. How many customers will need to be accommodated?
- Ancillary services in the gateway building such as uninterruptable power supply, air conditioning and security.
- Specifications required to be satisfied by the customers' equipment, covering diverse aspects ranging from physical size and power requirements to the interconnection protocols for voice and data connectivity.
- Procedures for delivering, installing, testing and commissioning customers' equipment into the gateway building.
- Acceptance testing of the initial service offering.

4.3.4 Regulation of Companies with Significant Market Power

Companies and other market entities are considered to have significant market power (SMP)[16] if they can act fairly independently of their competitors. In the extreme case of a company holding a monopoly, there are no competitors and so all monopolies have SMP. SMP can be manifested in a number of ways. For example, a company with SMP might be recognised by one or more of the following behaviors or market situations:

- It can set its prices substantially independently of the prices of its competitors, generally charging a significant premium for the same or similar product or service offered by other suppliers.
- It controls an essential facility that its competitors must access, such as the only submarine cable landing station or the only satellite ground station.
- It has a substantially higher market share than any of its competitors (although this is rarely considered to indicate SMP by itself and other factors must also be present).
- It significantly influences the access that competitors have to essential facilities, such as sites for mobile phone base stations.

Companies with SMP are not constrained by normal market pressures. In the absence of regulation, they can act as monopolists, charging unreasonably high prices and favoring some customers over others. It is well-recognized that such behavior is most likely to militate against the long-term national interest (OECD 1996). One characterization of SMP regulation is that it is a proxy for the discipline of a competitive market.

Regulators should closely monitor the behavior of companies with SMP to ensure that such companies do not take excessive profits at the expense of the overall economy and the community. The focus should be on both the operational and financial aspects of the relationship between SMP operators and their customers. Scrutiny of operational aspects addresses the fairness of customers' access to facilities and services; scrutiny of financial aspects addresses the fairness of the prices charged by the gateway operator, including wholesale prices.[17]

---

[16]The European Commission determines that an entity has Significant Market Power if it "enjoys a position of economic strength affording it the power to behave to an appreciable extent independently of its competitors, customers and ultimately consumers" (European Commission 2002).

[17]Regulatory scrutiny of wholesale prices is addressed under the more general discussion of price regulation in the main discussion.

4.4 Conclusion

This chapter began with the assertion that the most important predictors of a successful e-government rollout are a nation's financial and human resources, with a further significant element being the creation of a competitive telecommunications market overseen by an independent regulatory body. Those countries that have deregulated their telecommunications markets and established an independent regulator are noted in Sect. 4.3; others have found it more challenging to introduce a competitive market, largely because of size and isolation and because they have retained any necessary regulatory mechanisms within government (e.g. Nauru). Monopoly operators, sometimes part-owned by government, remain the model in many PICs. However, most PICs now have a National ICT Policy which articulates the need for a competitive market, a regulatory regime and a Universal Access Policy, (as well as essential supporting legislation) all of which are seen as essential steps towards the effective uptake of e-government. Some form of regulation is essential in a deregulated market. But therein lies a critical issue for PICs—the affordability of regulation.

There is no obvious answer to the affordability problem. Three possible approaches are presented below to seed a necessary discussion in and among the PICs.

4.4.1 Regional Regulation

Could a regional regulator work for the PICs? It would facilitate the creation of substantial economic benefits while alleviating the shortage of essential expertise in the region. In addition, there are working models in the European Commission and the Eastern Caribbean Telecommunications Authority that could be modified for the PICs. However, irrespective of the details of these (or other) potential models, it must be recognized that substantial effort expended over several decades was required to address national political and cultural issues.[18] The important question is whether the PICs (or a subset of the PICs) have the intent and are willing and able to expend the necessary effort to establish and support a regional regulator. The prospect of a regional regulatory authority for the Pacific has been raised in the past and quickly rejected. So it can be inferred that the political will is not present at the moment. However, a reduced variant of regional regulation could be one that provides only technical expertise, on issues such as spectrum management.

[18]The first substantial step towards the European Union was the formation of the European Coal and Steel Community with the signing of the Treaty of Paris in 1951. ECTEL was formed in 2000 as an outcome of the Organisation of Eastern Caribbean States, a regional organisation that commenced in 1981. ECTEL reference https://www.itu.int/ITU-D/treg/Events/Seminars/GSR/DSR/documents/Document17.pdf.

National governments would retain full control of policy agendas, but would have access to a pool of technical experts to provide policy advice. The regional body may also be able to undertake monitoring and enforcement activities such as detecting and resolving radio interference problems.

4.4.2 Outsourced Regulation

Some PICs are closely associated with industrialized nations, namely Australia, New Zealand, the United States, France or the United Kingdom. The United States Federal Communications Commission has jurisdiction in the United States Pacific territories (Guam and American Samoa) although Guam has its own regulatory body (the GUAM Public Utilities Commission.) It is possible that some smaller PICs associated with Australia and New Zealand may be able to outsource their ICT regulation, leveraging the value of this alliance. However, such countries would need to carefully negotiate and monitor the relationship to ensure that their interests were being properly served. To date there is no evidence that Pacific nations have contemplated such an arrangement, however, it would give PICs access to necessary high quality regulatory resources.

4.4.3 Individual Approaches to National Regulation

A competitive telecommunications market in the Pacific islands is generally pictured as a former monopoly fixed network operator competing with two or more mobile operators—one of which may have formerly been part of the monopoly operation. For example, Fiji has two regulatory agencies (The Telecommunications Authority of Fiji and the Fiji Commerce Commission) and a consumer advocacy organization, the Consumer Council of Fiji, all of which have a role to play. It is clear that effectively regulating such a market structure demands the full range of regulatory disciplines, and some of the more populous PICs (e.g. Vanuatu) have successfully liberalized their markets with full regulatory oversight. However, many of the smaller countries (e.g. Tuvalu, with a population of around 10,000 and a per capita income of less than US$4000 per year) are unlikely to be able to attract new operators to create such a market, especially in the early years of liberalization. In this situation, it is sensible to provide for only those regulatory disciplines that are going to be required while allowing sufficient flexibility to add the remaining disciplines later. For example, a market that has competing retail service providers selling the services of a monopoly wholesale network operator does not require regulations addressing interconnection or infrastructure sharing, but it may be essential to structurally separate the monopoly network operator's wholesale and retail arms. Indeed, even if a new operator is interested in entering the market, the additional regulatory cost may not be warranted, at least in the immediate term.

Applied sensibly, this type of staged national regulation is capable of easing the cost burden in the early years of market liberalization, until the market and broader economy have grown sufficiently to be able to afford full liberalization and associated regulation. It may be the only practical approach for many PICs. It will be some years before several of the PICs should even contemplate full liberalization and, for those with populations well under 10,000 full liberalization may not be a sensible aspiration.

The vital goal is for all Pacific Island citizens to have reliable and affordable access to modern telecommunications. This is the essential infrastructure that underpins the economic and social benefits that ICT development in general, and e-government in particular, can bring to small island developing countries that comprise the majority of PICs. To achieve this goal, PICs will learn from more economically developed nations, but ultimately may have to forge their own paths.

References

AusCert (2016) http://www.cert.gov.au

CASBAA (2015) C-band invasion. CASBAA, Hong Kong. http://www.casbaa.com/resources/regulatory/satellite-issues/c-band-invasion/. Accessed 15 July 2016

Commerce Commission of New Zealand (2016) Unbundled copper local loop and unbundled bitstream access services final pricing principle. The Commission, Wellington. http://www.comcom.govt.nz/regulated-industries/telecommunications/regulated-services/standard-terms-determinations/unbundled-copper-local-loop-and-unbundled-bitstream-access-services-final-pricing-principle/. Accessed 1 Aug 2016

Communications Alliance (2016) C628:2015 telecommunications consumer protections code. http://www.acma.gov.au/~/media/Networks/Regulation/pdf/C628_2015%20Telecommunications%20Consumer%20Protections%20Code%20pdf.pdf. Accessed 1 Aug 2016

Communications Commission Kiribati (CCK) (2015) Universal access rules. CCK, Betio, Tarawa. http://www.cck.ki/index.php/downloads/finish/25-universal-access/654-universal-access-rules

CERT Australia (2016) Home page. http://www.cert.gov.au Accessed 1 Aug 2106

European Commission (2002) Commission guidelines on market analysis and the assessment of significant market power under the community regulatory framework for electronic communications networks and services. Official J Eur Comm (2002/C 165/03). European Commission, Brussels

GSMA (2016) About us/What we do. GSMA, London. http://www.gsma.com. Accessed 26 Jul 2016

Gulati GJ, Williams CB, Yates DJ (2014) Predictors of online services and e-participation: a cross-national comparison. Gov Inf Q 31(4):526–533. doi:10.1016/j.giq.2014.07.005

Incenta (2015) Post 2020. TSLRC vs the building block approach. Report for Chorus. Chorus, Wellington. https://www.chorus.co.nz/file/71401/External-Report—Incenta-BBA-v-TSLRIC-final-for-publication.pdf. Accessed 15 Sept 16

Intelsat (2015) Protecting C-band spectrum. Intelsat, Luxembourg. http://www.intelsat.com/homepage/about-us/in-the-industry/c-band-spectrum/. Accessed 25 Jul 2016

International Telecommunications Union (ITU) (n.d.) The Eastern Caribbean telecommunications regulatory authority: ECTEL. ITU, Geneva. https://www.itu.int/ITU-D/treg/Events/Seminars/GSR/DSR/documents/Document17.pdf. Accessed 27 Jul 2016

ITU (2014) ICT regulation toolkit. Section 6.6. Functional aspects of regulation. ITU/InfoDev, Geneva. http://www.ictregulationtoolkit.org/6.6. Accessed 15 Jul 2016

ITU (2015) World radiocommunication conference allocates spectrum for future innovation. ITU, Geneva. http://www.itu.int/net/pressoffice/press_releases/2015/56.aspx#.WBbLNRR2XnY. Accessed 28 Oct 2016

ITU (2016a) ITU-T in brief. ITU, Geneva. http://www.itu.int/en/ITU-T/about/Pages/default.aspx. Accessed 25 Jul 2016

ITU (2016b) National CIRT capacity building. ITU, Geneva. http://www.itu.int/en/ITU-D/Cybersecurity/Pages/Organizational-Structures.aspx

Kacific (2015) High speed internet to the Pacific Islands, Indonesia, Timor-Leste, The Philippines, New Zealand and Papua New Guinea. Kacific Broadband Satellite, Singapore. http://kacific.com

O3b networks (2015) Home page. http://www.o3bnetworks.com. Accessed 1 Aug 2016

OECD Policy Roundtable (1996) The essential facilities concept. OECD, Paris. http://www.oecd.org/competition/abuse/1920021.pdf. Accessed 15 Sept 16

Reddy M (2012) Telecommunications market in Fiji: is there competition. Presentation to the PacINET meeting, Suva, November 2012. PICISOC. http://www.picisoc.org/wp-content/uploads/2012/05/Tel-Pap-USP-ICT-221112_Dr-Mahendra-Reddy.pdf. Accessed 16 Aug 2016

Reddy M (2013) Telecommunications market in a small economy: is there a need for price control? Asian Econ Financ Rev 3(7):843–857

Telecommunications Commission of Solomon Islands (2012) Annual report 2011. TCSI, Honiara. http://www.tcsi.org.sb/images/stories/pdfs/tcsiannualreport2011.pdf. Accessed 30 Sept 2016

Vanuatu (2013) Telecommunications & radiocommunications regulator. Universal access policy. TRR, Port Vila. http://www.trr.vu/index.php/en/telecom-industry/universal-access/universal-access-policy. Accessed 26 July 2016

World Bank (2009) Information and communications for development 2009. Extending reach and increasing impact. World Bank Group, Washington. https://issuu.com/world.bank.publications/docs/9780821376058. Accessed 1 Aug 2016

World Bank (2016) Solomon Islands—telecommunications and ICT development project. Implementation completion report. World Bank Group, Washington. http://documents.worldbank.org/curated/en/2016/06/26472498/p113148-implementation-completion-report-review 28.7.16. Accessed 28 Jul 16

Chapter 5
Mobile Technology in Pacific Island Countries: the Potential for M-Government

Amanda H. A. Watson, Joseph Kim Suwamaru, Ioana Chan Mow and Sarah Logan

Abstract The chapter focuses on the use of mobile phones in Pacific Island countries for government service delivery. The chapter argues that mobile telephone technology is worthy of attention in its own right, as compared to other technologies. This is because mobile phones have been transformational in the lives of Pacific people in ways computer-based technologies have not been, as most people have never had access to them; many do not have access to electricity or basic services such as banking, postal services and landline telephones. Despite limitations, such as small screen size, incomplete network coverage and limited data capacity, mobile phones are widely used in the Pacific. This chapter explores the extent to which mobile phones have played a role in e-government in the Pacific to date, as well as the potential for them to contribute in the future.

The Pacific region was one of the last in the world to gain widespread mobile network coverage, perhaps due to its remoteness, dispersed populations and vulnerability. The region has witnessed dramatic growth in the uptake of mobile technology since around 2005, fueled by competition, market liberalization, network expansion, introduction of pre-paid mobile services and sale of more affordable mobile devices. In what has been a substantial change in rural com-

The original version of this chapter was revised: Authors' affiliations have been updated. The correction to this chapter is available at https://doi.org/10.1007/978-3-319-50972-3_15

A. H. A. Watson (✉)
Australian National University, Canberra, Australia
e-mail: ahawatson@hotmail.com; amanda.watson@anu.edu.au

J. K. Suwamaru
Independent researcher, formerly at Divine Word University, Madang, Papua New Guinea

I. Chan Mow
National University of Samoa, Apia, Samoa

S. Logan
University of New South Wales, Sydney, Australia

R. Cullen and G. Hassall (eds.), *Achieving Sustainable E-Government in Pacific Island States*, Public Administration and Information Technology 27,
DOI 10.1007/978-3-319-50972-3_5

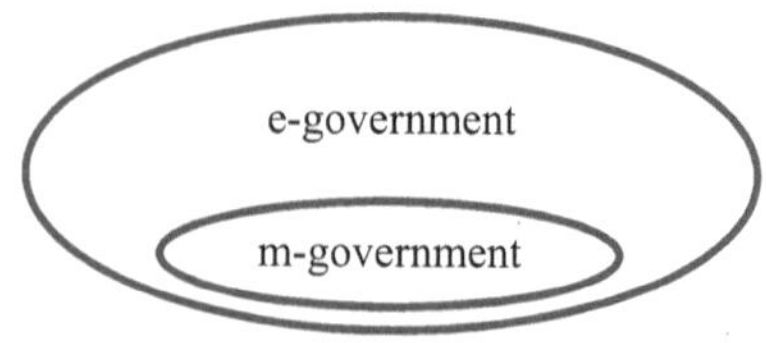

Fig. 5.1 M-government is a subset of e-government. *Source* Adapted by Amanda H. A. Watson from Akter et al. (2013)

munities over the last decade, mobile phone networks now reach many communities, even those on far-flung islands and in other remote locations. The chapter provides context, analysis and examples from the Pacific, many of which are previously undocumented, as well as other examples from the literature.

5.1 Introduction

The term e-government can refer to the use of a range of digital technologies, including the Internet, office computers, networked computer databases and other electronic devices. Thus, the use of mobile phones in government interactions, civic participation and service delivery, or m-government, can be thought of as a subset of e-government (see Fig. 5.1).

This chapter will briefly outline the context in terms of mobile phone network coverage in the Pacific Island Countries (PICs). It will then refer to the theoretical framework and explore the extent to which use of mobile phones can contribute to e-government as m-government. It will then introduce two main aspects of m-government, m-service and m-participation, providing relevant examples. The chapter will conclude with remarks about the extent to which these phenomena are evident in the PICs.

In short, the chapter argues that mobile phones have been transformational in the lives of people in PICs in ways that computer-based technologies have not been because most people have never had access to them. Despite their limitations, such as small screen size, incomplete network coverage and limited data capacity, this chapter conveys an important message about the contribution that mobile phones have made to date in e-government in the Pacific, as well as the potential for them to contribute in the future. The chapter provides analysis and examples from the Pacific, many of which are previously undocumented, as well as other examples from the literature. This chapter's authors have conducted research addressing the relevant issues in the Pacific region and thus the chapter draws on the available literature and their experiences, with particular focus on Papua New Guinea (PNG) and Samoa as illustrative examples. Demographic information regarding PICs, including land size and population figures, can be found in the first chapter of this volume.

5.2 Mobile Phones in PICs

The Pacific region was one of the last regions of the world to gain widespread mobile network coverage, perhaps due to its remoteness, dispersed populations and vulnerability. The region has witnessed dramatic growth in the uptake of mobile technology since around 2005, reaching 4.1 million users by the end of 2014, with annual growth rates above rates of other developing regions (GSMA 2015). According to Cave (2012), 60% of people in the Pacific own mobile phones and the number is increasing. Such growth has been fueled by competition, market liberalization, network expansion, introduction of pre-paid mobile services (ITU 2011), more affordable mobile devices (GSMA 2015) and a "growing youth population accelerating adoption" (Cave 2012).

A range of regimes exist in the Pacific, from fully liberalized environments, such as Samoa and Vanuatu, to monopolistic environments, such as the cases of Niue and Marshall Islands (Miniwatts Marketing Group 2016). Most of the other PICs are at various stages of introducing liberalization and competition, with plans for full liberalization (Network Strategies 2010). With the exception of Federal States of Micronesia and Kiribati, PICs with monopolistic regimes tend to be small island countries with small populations, such as Cook Islands, Marshall Islands, Nauru, Niue and Tuvalu, where it is feared that the size of the market might not be able to sustain competition (Miniwatts Marketing Group 2016).

While mobile telephony services tend to be provided by private companies, governments usually have a role in regulating the telecommunication sector and in some cases have partial or full ownership of service providers. For example, in Tonga, there are currently two mobile network operators, one of which is owned by the Tongan government, while the other is Digicel, a private company operating in the Caribbean and in six PICs: PNG, Samoa, Tonga, Vanuatu, Fiji and Nauru. The Tongan government also has an 80% share in Tonga Cable Limited, which was established in 2009 "to build and manage a submarine fibre optic cable to connect Tonga to an international network service" (Tonga Cable Ltd. 2016).

Box 5.1 Growth in mobile phone uptake in Samoa

The Polynesian nation of Samoa provides an example of the trends that have occurred in PICs in recent years. The first mobile network in Samoa was established in 1997, but until 2005 this was a monopolistic arrangement, with stagnation in growth of mobile services, characterized by limited coverage and high costs. In 2006, two new mobile providers entered the market, leading to substantial investment in mobile infrastructure and thus increase in mobile network coverage (Chan Mow 2014). The establishment of a regulatory framework (Chan Mow 2014), the setting up of an independent regulator, the de-regulation and opening up of the telecommunication market, and the privatization of SamoaTel were all made possible through the World Bank-funded Telecommunication Sector Reform Project, implemented from

2002 to 2013 (World Bank 2011). Since 2005, there has been a seven-fold increase in the number of mobile subscribers, from 22,000 in 2006 to 80,000 a year later (Australian Government 2008) and the price of phone calls has more than halved (Va'a et al. 2012).

Internet access in many PICs has been quite low until very recently, due to heavy reliance on access via fixed lines, exacerbated by inhibitive costs and limited bandwidth. Mobile phones have now made Internet access a reality for many previously unconnected populations, with 15% of Internet access in the region now through mobile phones (GSMA 2015). A noticeable trend is the increasing migration from second generation (2G) networks suitable for voice calls and text messaging (or SMS, short message service) to third generation (3G) networks which allow for Internet access, email and social media use. In some locations, primarily capital cities, such as Port Vila in Vanuatu and Port Moresby in PNG, fourth generation (4G) network coverage has been established, allowing for greater mobile Internet speeds (Watson 2015a). There is an increasing number of smart phones in the region (GSMA 2015). Nonetheless, network coverage remains predominantly 2G offered over global system mobile (GSM) and coded division multiple access (CDMA) networks, with many people in PICs having basic handsets, meaning that Internet access is not possible for these people (see Fig. 5.2).

In major cities, there is demand for greater capacity and faster data speeds coming from businesses, with a requirement for Internet access in offices and some even using mobile phone services as office phones. The general public in cities tends to consist of wage earners with disposable income and greater exposure to advertising through mainstream media; therefore, these people also want access to more digital services, compared to subsistence farmers in rural areas. To support

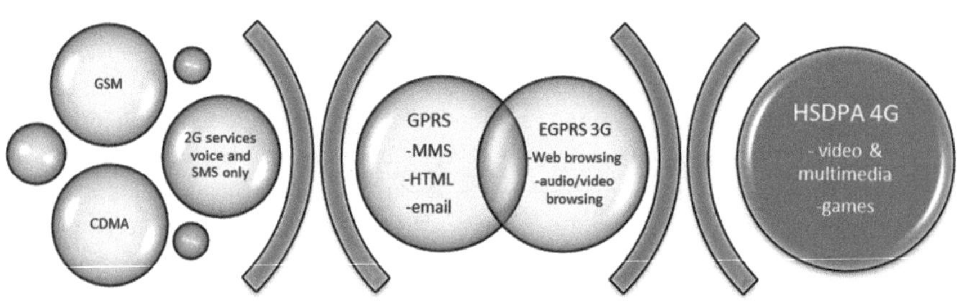

Fig. 5.2 Migration path from 2G to 4G driven by capacity, data speeds and return on investment. *Source* Joseph Kim Suwamaru

multimedia-based services, telecommunication companies have migrated towards 3G and 4G networks[1] (Hau'ofa 2014) in urban areas. Thus, mobile service providers segment the population, compete for corporate clients and offer 2G, 3G and 4G network offerings. Compared to customers who are making short voice calls and exchanging text messages, operators can charge more for data usage and thus differentiate offerings for improved return on investment (Hau'ofa 2014). For instance, in PNG, Digicel and bmobile Vodafone offer 3G service predominantly in urban areas. Similarly, in Vanuatu, a company named WanTok offers 4G service in the capital Port Vila, while two other companies, Telecom Vanuatu and Digicel, offer 2G and 3G coverage elsewhere in the country (Watson 2015a).

5.3 Theoretical Framework

This chapter, along with several others in this book, utilizes Heeks' ICT4D 2.0 Manifesto (Heeks 2009) as a useful theoretical framework. In the Manifesto, Heeks outlines how thinking and practice around the use of information and communication technologies (ICTs) for development (ICT4D) have progressed in recent decades. He also suggests ways forward. While Heeks focuses on ICT4D, he acknowledges that the wider development community has been similarly progressing towards more participatory approaches (Heeks 2009).

While only briefly touched on in the Manifesto (Heeks 2009), it is acknowledged that communication scholarship itself has also moved through phases regarding its understanding and approach. In the 1950s and 1960s, communication scholars were exploring mass communications theory, using media effects research, as "the mass media were thought to have a powerful and direct influence on individuals" (Melkote 2010; see also Flew 2007). Mass communications theory was linked to modernization theory in development circles and it was felt that the mass media could speed up the process of modernization by exposing people to new ideas and attitudes (Melkote 2010).

Subsequently, it was realized that media power is connected with other forms of power, such as political or economic power, and thus the political economy approach became important in the field of communication studies. This approach has been "the most influential framework" (Flew 2007) for discussions about the media over several decades (Flew 2007). Through the political economy approach and findings of industry-focused research, such as the concentration of media ownership explained by McChesney (2010), a notion of cultural imperialism evolved, linked with dependency theory (Flew 2007; Heeks 2009).

Heeks discusses ICT4D 1.0, spanning roughly a decade and a half from the mid-1990s, as a period in which there was an "invention-down approach – bringing

[1] 3G is offered over General Packet Radio Service (GPRS) and its enhanced version EGPRS; 4G is offered over High Speed Downland Packet Access (HSDPA).

new technologies into development contexts" (Heeks 2009). During that period, donors subsidized unsustainable tele-centers and missed the mobile phone phenomenon (Heeks 2009). By contrast, the approach advocated in ICT4D 2.0, from mid-/late-2000s onwards, is collaborative and participatory, with user engagement in poor communities undertaken throughout design and implementation phases (Heeks 2009). In addition to advocating development that is more inclusive and participatory, Heeks' approach also argues for a focus on existing, simple technology: "rather than wait for handset and bandwidth upgrades to allow mobile Internet access, what can be achieved for development through calls and SMS?" (Heeks 2009).

Globally, there is a growing body of literature on the use of mobile phones in development (for example, see Chib 2013; Donner 2008, 2015; Trucano 2013). There has also been a growing interest in the strategic use of mobile phones in the Pacific, a topic on which academic scholarship has begun to increase. While early studies at the time of mobile network expansion were primarily ethnographic, focused on attitudes towards and uses of the technology (see Watson and Duffield 2016), there is an emergent body of work exploring the use of mobile phones, particularly voice calls and text messaging, in development efforts and rural service delivery (regarding PNG, see Kaleebu et al. 2013; Logan and Gibbs 2015; Watson 2012a, 2014; Watson and Morgan 2014, 2015; Watson et al. 2015a; Watson and Poima 2015; Yamo and Watson 2014).

5.4 Can Mobile Technologies Enhance E-Government in PICs?

The ICT4D 2.0 era offers striking opportunities in PICs not possible during ICT4D 1.0. Indeed, the benefits of ICT4D 1.0 to e-government have been stymied by PICs' far-flung populations separated by masses of water, prohibiting them from taking advantage of the traditional tele-center model. However, the advent of mobile phones may change this. Mobile technology creates "virtual bridges that nullify physical distance" (Cardi 2014) and "mitigate the impact of geographical isolation" (Cardi 2014), potentially addressing the basic vulnerabilities of distance which characterize PICs. As the UN found, e-government holds the potential to address multiple challenges faced by PICs: e-participation and e-service components such as e-health, e-education and e-commerce can enable the governments of PICs to work and deliver as one and to advance sustainable development in all its dimensions (United Nations 2014).

Mobile phones offer particular solutions to e-government which are specific to the challenges PICs face. Most obvious, for example, are coordination of government activities and increased facility of communication, again as elsewhere (Mutula and Olasina 2014). Within these broader effects, it is important to note the constraints PICs face in taking up mobile technologies. The 2014 UN e-government

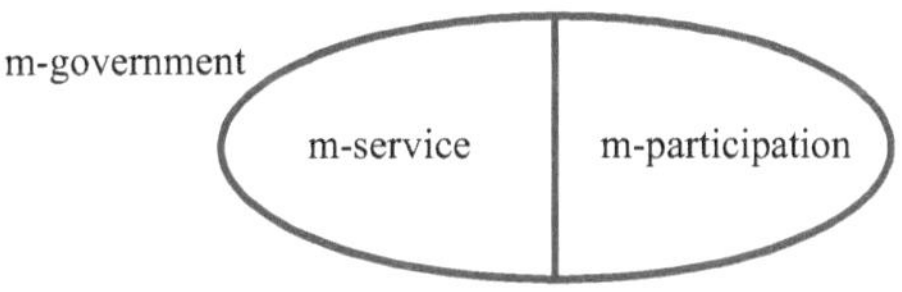

Fig. 5.3 M-service and m-participation as aspects of m-government. *Source* Amanda H. A. Watson

survey, for example, found that Small Island Developing States (SIDs) consistently rank lowly in the E-Government Global Development Index, with none of the 38 SIDs globally in the top 100 of those countries surveyed (United Nations 2014). It is also important to note that chief among the constraints is cost, both of infrastructure and of the technology itself. Given their small, generally poor and widely dispersed populations, PICs are less likely to be able to finance the construction of ICT infrastructure themselves and must rely on corporations or on donor support. Finally, given that mobile technology is inherently consumer driven, citizens themselves must have equal access to technology to enable their governments to make the most of its benefits.

Differences in benefits for m-government between simple SMS-based technologies and 3G technologies are evident—but the former is still dominant in the developing world. PICs will need to manage the unevenness of benefits versus cost, with regard to ICT4D 2.0 via mobile phones. As is arguably often the case, governments in PICs will have to face even more challenges than others in doing so, given their small, dispersed populations. Nonetheless, there may be even greater gains.

The chapter will discuss two main aspects of m-government: m-service and m-participation (see Fig. 5.3). The term m-service refers to use of mobile technology to deliver services. A typology distinguishes between three types of implementation using mobile phones in government service delivery: to provide an additional channel for existing government services; to extend traditional government services to people in previously under-served communities; and to provide new services (Raja et al. 2012). The examples presented in this chapter cover all three types of implementation. The term m-participation refers to the use of mobile phones for participatory initiatives, such as community consultation and community feedback mechanisms.

5.5 Roles of Mobile Phones in M-Service Delivery

Examples of m-service will be provided below from four sectors: health, education, disaster risk reduction (often referred to as DRR) and commerce. The health and education sectors are typically thought of in relation to government service delivery and both are receiving attention globally regarding strategic use of mobile phones (see respectively Chib 2013; Trucano 2013). In research conducted across seven provinces of PNG, survey respondents were asked about potential benefits of mobile phones in relation to healthcare and education, and were able to identify a number of perceived positives (Suwamaru 2014). The research findings showed that to date mobile phones have played supportive roles, rather than direct roles, in healthcare and education in PNG (Suwamaru 2014). Disaster risk reduction is of vital importance in the Pacific, as there is a high risk of exposure to natural disasters such as volcanic eruptions and cyclones, so this important facet of government service delivery has been included here. The section on commerce includes macro-economic growth, mobile financial services and other financial interactions, such as agricultural supply chains.

5.5.1 The Health Sector

5.5.1.1 Toll-Free Hotlines and Call Centers

In many PICs, citizens on dispersed islands or in far-flung communities can find it difficult to access healthcare services in a timely manner. While mobile phones cannot overcome distance or a dilapidated transportation system, they are becoming an indispensable medium to mobilize effort and support at a distance (Suwamaru 2014). Toll-free health hotlines or call centers may be able to enhance health service provision and potentially make health services more accessible to populations in selected healthcare systems. In PNG, a maternal health hotline in Milne Bay Province has been successful in saving the lives of women experiencing childbirth complications (Watson et al. 2015a). The toll-free number enables mobile phone calls between health workers, so that they can exchange real-time advice during time-critical medical emergencies (Watson et al. 2015a).

In Western Highlands Province of PNG, a toll-free call center has been established so that members of the public, as well as rural-based health workers, can call free-of-charge to speak to a health worker at any time (Watson and Poima 2015). Whereas in Milne Bay Province staff who are already on duty in the labor ward answer calls, establishing a call center is a more expensive option, requiring recruitment of new staff who work on shifts in it and have no face-to-face clinical shifts or other duties. As the service is also available to the public, extensive awareness activities are required to inform the public about the call center, its phone number and services. In Samoa, the two mobile phone companies, Digicel and Bluesky, sponsor a hotline for suicide prevention in partnership with non-government organizations Faataua le ola and Samoa Victim Support, so that people can call to ask for help.

5.5.1.2 Health Service Coordination

Mobile phone access across the Pacific has the potential to improve the convenience and speed of requests for medicines. Through mobile phone calls or text messaging, locations needing specific types and quantities of drugs can be identified and arrangements can then be made to source the appropriate drugs. This may reduce the need for expensive and time-consuming trips to ascertain the availability and status of drugs. In PNG, the Western Highlands Provincial Health Authority has a Closed User Group (CUG) (Yamo 2013; Yamo and Watson 2014), which allows officers to make unlimited calls and exchange unlimited SMS messages between one another. A trial was conducted in PNG in which voice-deactivated handsets containing tailor-made software were distributed to health workers so that they could report disease outbreaks via SMS (Rosewell et al. 2013; see also Box 11.2 in the health chapter).

5.5.1.3 Health Tips

In PNG, the non-government organization Population Services International established a service through which subscribers could receive weekly health tips as text messages. The service was promoted primarily through radio advertisements and at its peak it had around 40,000 subscribers. It was discontinued for a number of reasons, including difficulty with showing impact. In early 2016, the National Department of Health in PNG launched a new text messaging service, with support from the United Nations agency UNICEF, which provides text messages related to maternal health and pregnancy. In Samoa, SMS text messages are sent out to provide health advisories and health tips, particularly in relation to disease outbreaks such as chikungunya and dengue fever.

5.5.2 *The Education Sector*

In regard to aspects of mobile phone usage in school systems, research in PNG has found predominantly supportive roles rather than direct pedagogical applications (Suwamaru 2014). This may be true across the PICs because widespread availability of mobile phones is relatively new in these countries. As such, most examples to date regarding use of mobile phones in education are anecdotal or related to general usage, rather than explicitly related to m-government or m-service. One example of an m-service project in the education sector is the SMS Story project, outlined below.

Concerning aspects of mobile phone use in school systems, parents, teachers and students express varying opinions that reflect differing perspectives (Suwamaru 2013). Some parents appreciate mobile phones because they use them to remain in contact with their children who may be studying and residing far from home. Other parents have concerns about misuse of mobile phones, leading to disruptions in students' studies. Such competing views are not unusual because with any innovation there are likely to be positive and negative impacts (Rogers 2003).

In general, teachers felt that if used well mobile phones can be beneficial, whether for collaboration between teachers, between the school administration staff or between parents and students (Suwamaru 2014). Teachers have expressed appreciation for mobile phones but claim that responsible usage is important. Some teachers spoke of the untapped potential mobile phones have to support pedagogical efforts in PNG where logistics can be challenging. In the meantime, much of teachers' mobile phone use has been in supportive roles, such as delegation of tasks. For example, a teacher described how she uses the mobile phone to contact the subject head to organize a substitute teacher if she has to miss school to take her young child to a health clinic. Notwithstanding the foregoing, some teachers expressed strong opposition to the use of mobile phones during class times because students may not concentrate on their schoolwork (Suwamaru 2013; Watson 2011).

On the part of students, mobile phones are seen as school tools necessary for knowledge building (Suwamaru 2014). Some students claimed that mobile phones enabled them to access the Internet and generally served as useful study assistance tools, with helpful functions such as the calculator, alarm, clock, calendar and others. In PNG, students claimed that there are no alternative means of accessing the Internet. Nevertheless, several schools in PNG have banned mobile phones (Watson 2011); and in Samoa their use during class time is discouraged by the Ministry of Education to reduce distractions. Despite the pervasiveness of mobile phones, they are not used for mobile learning at any level of education in Samoa.

5.5.2.1 SMS Story

The aim of the SMS Story research project was to determine the effectiveness of sending daily mobile phone text message stories and lesson plans to teachers for improving children's reading ability in the early years of schooling in PNG (Kaleebu et al. 2013). The research was conducted by means of a control trial in which teachers in treatment schools received text messages, while teachers in control schools did not. The stories and lesson plans were designed to introduce children to reading English. Children's reading ability was assessed before and after the intervention. At the outset, treatment and control schools were comparable with respect to their characteristics and children's reading assessment results (Kaleebu et al. 2013). End-point reading assessment revealed a statistically significant difference between the results of the control and treatment schools, with children in treatment schools performing better than their counterparts in control schools across four of the five reading skills tested (Kaleebu et al. 2013). Thus, it was found that the intervention improved the reading ability of children (Kaleebu et al. 2013).

5.5.3 Disaster Risk Reduction and Disaster Response

Very important in the context of m-service to PICs is the facility mobile technologies have to deliver disaster risk reduction programs (for a more detailed discussion, see Chap. 10 in this volume and Noske-Turner et al. 2014). ICT4D 2.0

tools can assist in three phases of such responses (United Nations 2014). The first phase is in mapping, forecasting and preparing communities before a disaster occurs. The second phase is in early warning, whereby mobile phones can provide critical information quickly, for example in the case of the Vanuatu SMS alert system (Bolitho 2015). The third phase involves risk and disaster management during and after an emergency (United Nations 2014). Remote communities can use mobile technology to seek information that they would otherwise be unable to access (Watson 2012b). Milne Bay Province, a maritime province of PNG, has a maritime emergency call center to assist in alerting authorities to missing boats or other maritime emergencies (Digicel Group 2014).

Text messages are used as part of the early warning system in Samoa, via both mobile phone networks. Upon receipt of the signal of an approaching disaster (for example, a tsunami or cyclone) at the Meteorological Office, text messages are sent to key people in every village. Once they are informed, village bells are rung and sirens sounded to warn people of the approaching disaster. The Disaster Management Office and the Meteorological Office both have Facebook pages which they use for sending out alerts and updates in times of disaster such as cyclones, for example information on road closures and roads re-opening.

5.5.4 *Commerce and Economic Development*

Because the role of mobile phones in commerce and economic development has emerged across PICs in the last decade, this section of the chapter deals with the various related impacts, uses and services. While some of these examples may not appear to be explicitly linked to m-government or m-service, all of them have potential to impact, or indeed are already changing, the lives of people in far-flung, rural communities and are therefore worthy of mention.

5.5.4.1 Economic Development

Freer mobile markets have catalyzed improvements in communication infrastructure vital for both macro and micro economic development. Indeed, the Samoa declaration of 2014 frames the relationship between PICs and ICTs within a broader emphasis on private sector growth and economic sustainability, arguing for a focus on "promoting and enhancing the use of information and communications technologies for, inter alia, education, the creation of employment, in particular youth employment, and economic sustainability purposes" (United Nations Small Island Developing States Conference 2014).

The telecommunications sector itself has generated economic activity and contributed to growth. For example, in PNG during 2007 the mobile sector contributed 0.7% of total gross domestic product (GDP) growth (Australian Government 2008, p. 3). In the case of the GDP of Samoa, the "communications sector contribution rose sharply following the entry of Digicel in 2007 but has declined" (Minges and Stork 2015) since a peak in 2010.

Many PICs are characterized by dual economies, consisting of the formal sector and the informal economy, which can include small-scale farming, subsistence-based living and self-employment (Miniwatts Marketing Group 2016). In Malaita Province of Solomon Islands, sea shells are harvested by divers and processed into necklaces, which are sold across Melanesia, fetching healthy profits. Mobile phones are used along the value chain, from collection and processing of raw materials, to accessing markets, generating important income for rural entrepreneurs (Suwamaru 2013).

5.5.4.2 Mobile Financial Services

Governments typically do not deliver financial services to populations. Such services are generally provided by private institutions, which in turn are regulated by governments. Across large expanses of territory and ocean in PICs, the pressures of delivering financial services are magnified by the costs and security pressures of transportation of cash resources. Thus, people residing in many locations often do not have access to any financial services, such as bank branches. Financial services using even basic mobile technology can increase the viability of transactions across such distances, helping to improve markets (Donovan 2012). In addition, recent studies show that such services can facilitate the delivery of cash transfers, an important innovation in social service delivery (Aker et al. 2013).

Mobile phones are harnessed to support a range of mobile banking (m-banking) and mobile money (m-money) services to customers across PICs. Mobile networks enable commercial banks to offer m-banking services to customers, while non-bank m-money services are also available in some PICs (Vlies and Watson 2014). Services include checking of balances and transferring of funds. Offerings by banks have evolved from SMS-based services, such as BSP SMS banking, which commenced in PNG in 2008 (Watson 2013), to menu-driven systems (Vlies and Watson 2014) and possibly to an increasing number of smartphone applications in the future. Uptake of m-banking and m-money has not been as great in PICs as might have been expected. Reasons may include the absence of agent interoperability, resulting in limited efficiency and convenience for users (Fijian Election Office 2014), as well as other factors such as a continued reliance on and preference for cash (Suwamaru 2015). Nonetheless, mobile banking services involve over 25,000 transactions per day over the Digicel network in PNG (Hau'ofa 2014).

Digicel launched its mobile money product in Samoa in 2011. Take-up has been slow and it is estimated that only around 1–2% of subscribers are using the service. The main barrier appears to be a lack of confidence in mobile wallets compared to cash. Mobile banking services from the ANZ Banking Group Limited, the National Bank of Samoa (NBS) and Westpac can also be accessed from mobile phones, further reducing the need to use a mobile operator's mobile money services. According to the Central Bank of Samoa, during the 2011–2012 financial year,

0.8 million tala was created in mobile money accounts served by 57 cash-in/out agents (Minges and Stork 2015).

In some PICs, remittances from relatives working overseas make a contribution to local economies. While transfer of funds has always been possible, mobile financial services may be making such transactions easier in some cases. For Pacific Islanders living and working overseas, it is possible to transfer funds to relatives in some countries using the m-banking services offered by commercial banks or the Digicel remittance service which has been introduced in recent years in various PICs.

5.5.4.3 Income Generation

Mobile phones can also be considered as direct income generators, as exemplified by selling of call-credits and mobile phone accessories by various actors in markets and on street corners across the Pacific (Suwamaru 2014; Watson 2011). For example, small traders buy wholesale value call-credits from upstream suppliers, to sell to end-users who pay retail prices, or above-retail prices in many rural and remote localities. Signs advertising the sale of call-credits adorn many roadside stores and even residential houses.

5.5.4.4 Agriculture

Rural farmers supply urban markets with a wide variety of fresh garden produce and acquire information concerning supply and demand conditions in person, via mobile phones or through other means such as mass media and agricultural extension workers. Mobile phones can be used to coordinate logistics and enhance the local supply chain (Suwamaru 2015). For cash crops, such as coffee, cocoa and copra, mobile phones play essential roles in identifying upstream buyers with better prices (Suwamaru 2014). Farmers use "mobile phones to coordinate delivery and sale of local fresh produce in markets" (Suwamaru 2014) and business costs have been reduced through access to mobile telephony (Suwamaru 2014), for example farmers can access market prices (Suwamaru 2014). It is worth noting that not all agricultural producers in PICs can enhance their profitability through use of mobile phones, as certain enabling conditions, such as access to transport, may be needed in order for mobile telephony to be useful (Watson 2011).

In Samoa, the government launched an e-voucher system in partnership with Digicel Samoa, providing cash to farmers and fishermen after the devastation of Cyclone Evan (Minges and Stork 2015). To date, up to 4000 farmers and fishermen have received cash assistance through this program, with farmers receiving e-vouchers through their mobile phones and using the e-vouchers to purchase necessities, for example fertilizer at hardware stores. The Samoan Women in Business Development Incorporated (WIBDI) organization is developing a smart-phone application to help organic farmers by providing advice on harvest and pesticides (Minges and Stork 2015).

5.5.4.5 Advertising Through SMSs

Using text messaging for advertising has taken various forms, ranging from those SMSs originating from mobile phone operators to those representing private businesses and non-government organizations. Apart from the promotional material of mobile phone companies themselves, which was already sent to subscribers regularly, the first commercial advertisement via SMS in PNG was sent on the Digicel network during April 2014; the ease with which large numbers of people may be reached has resulted in an increase of SMS-based advertising in PNG since that time. For example, the following SMS-based advertisement for a retail store was sent to Digicel handsets in Port Moresby on August 9th 2014:

> HAUS DEPOT WAIGANI CENTRAL FUN DAY ON NOW! BBQ BALLOONS FACE PAINTING & MASSIVE DISCOUNTS IN STORE! SPEND K150 SAT-SUN GET FREE 2L COKE+SLICE OF PIZZA HURRY!!

In PNG, the National Information and Communications Technology Authority, NICTA, has developed an industry standard to provide regulatory safeguards regarding SMS-based advertisements. The code includes clauses that specify that unsolicited text messages should not be sent between 7 pm and 6 am and that citizens have the right to opt-out of SMS-based advertisements if they wish.

In Samoa, the advent of commercial advertising through mobile phone text messaging is a recent phenomenon, which first appeared at the end of 2014 for commercial advertising, although the mobile phone companies had already been using this avenue for advertising their own products, services and promotions for several years.

5.5.4.6 Pre-paid Electricity Top-Ups Via Mobile Phones

In PNG, many communities are not connected to the main electricity grid. For those with connection, a pre-paid system is in place, meaning that consumers pay for their electricity in advance, rather than at the end of a billing period. The electricity provider has a service for topping-up electricity meters directly from mobile phone credit and this is well-liked by rural villagers in locations with grid connection, who save on the time and cost involved in travelling to an urban center to purchase electricity (Watson 2011).

The use of mobile phones for topping-up electricity meters in Samoa is known as Cashpower and has been available at least since 2009 via the National Bank of Samoa. The service can be used only by customers who have a valid account with the bank. After registering with the Cashpower service, customers can top-up their electricity meters by sending a text message to a pre-defined number, as well as the amount of Cashpower top-up (see also Chan Mow 2014). This money is deducted from the customer's bank account.

5.5.4.7 Superannuation Balance Checks and Alerts

In research conducted during the early days of widening mobile network coverage in PNG, rural research participants had not used superannuation fund balance check services, although at least one was available by that time (Watson 2013). In Samoa, citizens can use their mobile phones to check their National Provident Fund (NPF) contributions and loan balances (Minges and Stork 2015). Further research could explore whether such services are frequently used and whether or not fund members find such services user-friendly and of benefit.

5.5.4.8 Tourism

The advent of mobile telephony has potential to contribute to business opportunities for small-scale tourism operators. For example, the combination of a global, online, local-travel-booking franchise network portal and mobile phone services has improved business for small-scale tourist operators in Samoa.

5.6 Roles of Mobile Phones in Participation (M-Participation)

Mobile phones have the potential to be harnessed in various ways for participatory initiatives in PICs—referred to as m-participation in this chapter and as m-governance in some literature. Such initiatives can assist in decision-making and information dissemination among leaders and citizens in a transparent and accountable manner (Logan 2012). Through m-participation, mobile phones could facilitate citizen-state engagement and contribute to increased transparency in PICs, as is possible elsewhere, with effects arguably intensified by the technology's capacity to overcome great distances (United Nations Development Programme 2012).

For those who have access to 3G or 4G coverage, a suitably advanced mobile phone handset and enough money to be able to purchase data, mobile phones could be used to access social media in order to find out about current events, such as government activities. Some experiences show potential for m-participation uses, which could lead to more transparency and accountability in decision-making and information dissemination.

5.6.1 Elections

The 2014 elections in Fiji witnessed increased use of text messaging over mobile phones and publishing on social media websites (Fijian Election Office 2014). In the 2015 elections in the Solomon Islands mobile phones were used in campaigns and in coordinating election operations (Haley and Zubrinich 2015). During political uprisings in PNG, mobile phones have been used to organize people to physically gather and demand action (Logan 2012). The permeation of mobile phones across PNG has been unprecedented, offering a means for some people to access and use social media (Logan 2012).

In the 2016 general elections in Samoa, both tablets and mobile phones were used for pre- and post-election surveys administered by an election monitoring team of observers from Australian National University and the National University of Samoa. The electronic surveys were pre-loaded on devices and filled in by election-monitoring observers as survey participants provided responses verbally.

5.6.2 Government-to-Government Communication

Esa'ala District in Milne Bay Province of PNG employs a Closed User Group (CUG) (Cobain 2014), which allows for voice calls and text messages between district officers, in order to improve local level government administration and service delivery. Supplementary to the CUG service, telecommunication companies also offer private dedicated lines (copper wire, fiber-optic or wireless) that enable organizations to exchange information within and between offices that are geographically dispersed, using email, networked computer databases and so on.

5.6.3 Civic Engagement

In many PICs, mobile phones offer the fastest mode by which contact can be made with government offices. The Open Parliament Project offers a service to PNG citizens so people can access information about parliamentarians via a website, through an automated text messaging service or through a call-back service. After about one year of operation, the automated service had provided information addressing over 9000 requests, while over 37,000 people from 173 countries had visited the website (Taule 2015).

5.6.4 Remote Data Collection

The use of mobile phones for data collection is a nascent field in the Pacific region. There are some opportunities for using mobile phones to collect data remotely, for example for conducting phone interviews or for conducting surveys via text messaging (Watson 2015b). This has not become mainstream to any large degree in any of the PICs, although the Solomon Islands Government has developed a proposal to test a standardized mobile data collection tool with five government agencies, for possible roll-out to all ministries (Solomon Islands Government 2015). The Phones against Corruption project (Watson et al. 2015b) represents the use of mobile phones to collect data and also, in this case, to address a governance issue. In Samoa, text messaging was used for a business survey conducted by the national statistics office.

5.7 Conclusion

Compared to computers and the Internet, mobile phones are much more widely available in PICs and as such have played a transformational role in people's lives. As mobile phone services have spread in recent years, people have adapted mobile technology and used it for unexpected purposes suited to local situations. The Pacific mobile revolution has provided potential for mobile phone services to be utilized in development efforts within the region. It may also be possible for governments of PICs to harness the reach of mobile phone networks to better deliver services to or collect information from widely dispersed populations (m-government). However, to date, strategic efforts to use mobile phones have been largely informal and ad hoc. An exception is the approach of the government of Vanuatu, which has been systematically exploring the potential benefits of m-government (Vanuatu Office of the Government Information Officer 2014). Another exception is the planned approach to data collection using mobile devices in the Solomon Islands (Solomon Islands Government 2015).

The chapter presents isolated examples of m-government in the PICs, which show potential to be replicable or scalable. It may be advisable for m-government trials to be conducted and rigorously evaluated, before the practice is widely adopted. Research into acceptance by potential users could provide valuable insights, for example exploring the extent to which citizens trust information received through mobile phones, as compared to other sources, such as local leaders and radio programs. Because many view their own mobile phones as private, personal devices primarily for social contact with relatives and friends, as well as for entertainment uses, incorporation into government service delivery may not always be welcomed. There may be some benefit in providing awareness for leaders and citizens on potential uses and benefits of m-government. Leaders may be able to leverage the ubiquity of mobile phones to stay engaged with citizens and vice versa. Policy initiatives should call for improved transportation and electricity infrastructure to complement communication technology access and thereby enable smooth movement of goods and services between locations. This may supplement and enhance the role of mobile phones in economic development and m-government initiatives. The same applies to supporting m-banking services, which may lead to financial inclusion for more citizens.

The potential for mobile technology to contribute to effective e-government through these kinds of initiatives needs to be further explored, to ensure that m-government addresses community needs and contributes to desirable and sustainable outcomes in PICs. Another possible area of research would be to take a political economy approach and analyze the political and economic power of private telecommunication companies in relation to the governments that might promote m-government but which also set the regulatory framework for the sector. Industry-focused research could examine ownership structures, pricing and regulatory frameworks.

In the 1950s and 1960s, communication scholars argued that the effects of mass media exposure would be a catalyst for positive socio-economic change. As Heeks (2009) has argued, participatory approaches to development initiatives are now considered more appropriate and more likely to lead to beneficial outcomes. As mobile phones allow for two-way communication, in contrast to the one-way flow of information from a television set or radio, it may be possible that mobile telephony can assist in addressing persistent development challenges. However, some of the examples cited in this chapter repeat approaches of the ICT4D 1.0 era, in that they have been designed without adequate community consultation and have been imposed on, or marketed to, potential users. Heeks argued for more than just a change of technology (from tele-centers to mobile phones). Instead, he sought to encourage development practitioners to engage with communities in an inclusive, participatory manner (ICT4D 2.0). Engagement with communities in developing these new government services brings its own challenges, but it may result in more effective m-government implementation in the Pacific.

References

Aker JC, Boumnijel R, McClelland A, Tierney N (2013) Payment mechanisms and anti-poverty programs: evidence from a mobile money cash transfer experiment in Niger. Unpublished working paper

Akter S, Ray P, D'ambra J (2013) Continuance of mHealth services at the bottom of the pyramid: the roles of service quality and trust. Electron Mark 23:29–47

Australian Government (2008) Pacific economic survey. [Online] http://www.pacificsurvey.org/site/. Accessed 7 May 2008

Bolitho S (2015) Tropical cyclone Pam: why the Vanuatu death toll was so low. http://www.abc.net.au/news/2015-04-01/explainer3a-why-was-the-vanuatu-death-toll-from-cyclone-pam-so/6363970. Accessed 17 May 2016

Cardi S (2014) Remarks. United Nations, New York. http://www.un.org/en/development/desa/news/sustainable/ict-and-e-government-for-sustainable-development-in-sids.html. Accessed 1 Jan 2016

Cave D (2012) Digital Islands: how the Pacific's ICT revolution is transforming the region. Lowy Institute for International Policy

Chan Mow IT (2014) Issues and challenges, strategies and recommendations, in the development of ICT in a small island developing state: the case of Samoa. Electronic J Inf Syst Dev Ctries 63:1–24

Chib A (2013) The promise and peril of mHealth in developing countries. Mobile Media Commun 1:69–75

Cobain G (2014) Digicel embarks on new service. The National, p 40

DIGICEL Group (2014) DIGICEL and Milne Bay Governor formalise agreement to establish maritime contact centre. Port Moresby, Papua New Guinea. https://www.facebook.com/notes/digicel/digicel-milne-bay-governor-formalise-agreement-to-establish-maritime-contact-cen/10152241260662655/. Accessed 18 May 2016

Donner J (2008) Research approaches to mobile use in the developing world: a review of the literature. The Information Society 24:140–159

Donner J (2015) After access: inclusion, development, and a more mobile internet. MIT Press, Cambridge, MA

Donovan K (2012) Mobile money for financial inclusion. Information and communication for development. World Bank, Washington, D.C.

Fijian Election Office (2014) How to cast your vote. http://www.electionsfiji.gov.fj. Accessed 28 Feb 2016

Flew T (2007) Understanding global media. Palgrave Macmillan, Hampshire

GSMA (2015) The mobile economy: Pacific Islands 2015. https://gsmaintelligence.com/research/?file=23485245295f02524925b2bd3aeec6de&download. Accessed 26 Aug 2015

Haley N, Zubrinich K (2015) Mobile phones and the 2014 Solomon Islands national elections. SSGM In Brief 2015/28. Australian National University, Canberra

Hau'ofa R (2014) Interview with Sundar Ramamurthy on FM100 radio talkback show, 25/11/2014. Papua New Guinea, Port Moresby

Heeks R (2009) The ICT4D 2.0 manifesto: Where next for ICTs and international development? Development Informatics Working Paper Series, Development Informatics Group, Manchester: Institute for Development Policy and Management

ITU (2011) The role of ICT in advancing growth in least developed countries: trends, challenges and opportunities. International Telecommunications Union, Geneva

Kaleebu N, Gee A, Jones R, Jauk M, Watson AHA (2013) SMS story impact assessment report. Voluntary Services Overseas, Papua New Guinea

Logan S (2012) Rausim! Digital politics in Papua New Guinea. SSGM discussion paper 2012/9. Australian National University, Canberra

Logan S, Gibbs P (2015) Using mobile phones to track anti-witchcraft violence in Papua New Guinea. SSGM In Brief 2015/19. Australian National University, Canberra

McChesney R (2010) The media system goes global. In: Thussu DK (ed) International communication: a reader. Routledge, London

Melkote SR (2010) Theories of development communication. In: Thussu DK (ed) International communication: a reader. Routledge, London

Minges M, Stork C (2015) Economic and social impact of ICT in the Pacific. Pacific Region Infrastructure Facility, Sydney, Australia

Miniwatts Marketing Group (2016) Oceania and South Pacific telecommunications reports. http://www.internetworldstats.com. Accessed 28 Feb 2016

Mutula SM, Olasina G (2014) E-government strategies in Sub-Saharan Africa: failures and successes. In: Mazebe P, Sebina M, Moahi KH, Bwalya KJ (eds) Digital access and e-government: perspectives from developing and emerging countries. IGI Global, London

Network Strategies (2010) Review of the Pacific region digital strategies network strategies report No 29029

Noske-Turner J, Tacchi J, Horst H, Papoutsaki E (2014) Locating disaster communication in changing communicative ecologies across the Pacific. Aust J Telecommun Digit Econ 2:66.1–16

Raja S, Melhem S, Cruse M, Goldstein J, Maher K, Minges M, Surya P (2012) Making government mobile. In: Maximizing mobile: information and communications for development. The World Bank, Washington, D.C.

Rogers EM (2003) Diffusion of innovations. Simon & Schuster, New York

Rosewell A, Ropa B, Randall H, Dagina R, Hurim S, Bieb S, Datta S, Ramamurthy S, Mola G, Zwi AB, Ray P, Macintyre CR (2013) Mobile phone-based syndromic surveillance system, Papua New Guinea. Emerg Infect Dis 19:1811–1818

Solomon Islands Government Information Communication and Technology Support Unit (ed) (2015) SIG ICTSU briefing note (for donors and SIG partner agencies): mobile data collection tools and GIS reporting tools. SIG ICTSU, Honiara, Solomon Islands
Suwamaru JK (2013) ICT initiatives in Papua New Guinea: impact of mobile phones on socio-economic development. Divine Word University, Madang, PNG
Suwamaru JK (2014) Impact of mobile phone usage in Papua New Guinea. SSGM In Brief 2014/41. Australian National University, Canberra
Suwamaru JK (2015) Aspects of mobile phone usage for socioeconomic development in Papua New Guinea. SSGM discussion paper 2015/11. Australian National University, Canberra
Taule T (2015) Open parliament project making progress. Port Moresby, Papua New Guinea: Loop PNG. http://www.looppng.com/content/open-parliament-project-making-progress-0. Accessed 18 May 2016
Tonga Cable Ltd. (2016). About us. Tonga. http://tongacable.to/?page_id=2. Accessed 10 Aug 2016
Trucano M (2013) 10 principles to consider when introducing ICTs into remote, low-income educational environments. EduTech. http://blogs.worldbank.org/edutech/10-principles-consider-when-introducing-icts-remote-low-income-educational-environments. Accessed 12 Mar 2014
United Nations (2014) E-government survey: E-government for the future we want. United Nations, New York
United Nations Development Programme (2012) Mobile technologies and empowerment: enhancing human development through participating and innovation. United Nations, New York
Va'a F, Unasa TV, Fonoti FL, Chan Mow I, Amosa D (2012) Aspects of economic development. In: Meleisea M, Meleisea PS, Meleisea E (eds) Samoa's journey 1962–2102: aspects of history. Victoria University Press, Wellington, New Zealand
Vanuatu Office of the Government Information Officer (ed) (2014) Mobile governance for small island developing states—Strategy knowledge base. Center for Electronic Governance, United Nations University IIST, Port Vila, Vanuatu
Vlies MVD, Watson AHA (2014) Can mobile phones help reduce teacher absenteeism in Papua New Guinea? In: Australian and New Zealand Communication Association annual conference, Swinburne University, Victoria, Australia
Watson AHA (2011) The mobile phone: the new communication drum of Papua New Guinea. PhD dissertation (unpublished). Queensland University of Technology, Brisbane
Watson AHA (2012a) Could mobile telephony be harnessed for development in Papua New Guinea? Contemp PNG Stud DWU Res J 17:46–52
Watson AHA (2012b) Tsunami alert: the mobile phone difference. Aust J Emerg Manag 27:46–50
Watson AHA (2013) Mobile phones and media use in Madang Province of Papua New Guinea. Pac Journal Rev 19:156–175
Watson AHA (2014) Mobile phones and development in Papua New Guinea: guiding principles. SSGM In Brief 2014/42 ed. Australian National University, Canberra
Watson AHA (2015a) Not apply in Melanesia. The interpreter. http://www.lowyinterpreter.org/post/2015/06/16/Not-appy-in-Melanesia.aspx?COLLCC=2317657416&. Accessed 16 Jun 2015
Watson AHA (2015b) Remote data collection in Papua New Guinea: an aid to policy deliberations. DevPolicy Blog. http://devpolicy.org/remote-data-collection-in-papua-new-guinea-an-aid-to-policy-deliberations-20150707/
Watson AHA, Duffield LR (2016) From garamut to mobile phone: communication change in rural Papua New Guinea. Mob Media Commun 4:270–287
Watson AHA, Morgan E (2014) Can mobile phones really be used for data collection? Results from a trial in Papua New Guinea. Contemp PNG Stud DWU Res J 21:127–138
Watson AHA, Morgan E (2015) M-government for courts systems: a data collection trial in Papua New Guinea. In: Robertson A (ed) Commonwealth governance handbook 2014/15. Nexus/Commonwealth Secretariat, Cambridge

Watson AHA, Poima R (2015) Healthy phoning: Papua New Guinea's first ever health call centre. Australas Psychiatry 23:48–51

Watson AHA, Sabumei G, Mola G, Iedema R (2015a) Maternal health phone line: Saving women in Papua New Guinea. J Personal Med 5:120–139

Watson AHA, Wingi T, Tiki T, Cvetanovska M, Balboa T, Erepan S, Jackson C (2015b) Phones against corruption: preliminary findings of user experience research. PNG Update. University of Papua New Guinea School of Business and the Development Policy Centre, Australian National University, Port Moresby, Papua New Guinea

World Bank (2011) Implementation and completion results report (IDA-37240) on Samoa Telecommunications and Postal Sector Reform Project

Yamo H (2013) Mobile phones in rural Papua New Guinea: a transformation in health communication and delivery services in Western Highlands Province. Master of Communication Studies. Auckland University of Technology, Auckland

Yamo H, Watson AHA (2014) How to maximize value when utilising a Closed User Group mobile phone service in the Papua New Guinea context. IBS J Bus Res 10:9–24

Part II
Sectors

Chapter 6
The Role of ICTs in Public Finance Management in Pacific SIDs: a Case for Good Governance

Rowena Cullen

Abstract The chapter examines the critical role of good governance and robust financial management in building stable democratic states in the Pacific Islands. It explores the role of ICTs in public financial management (PFM) as a means to this end and identifies some of the factors that lead to the success or failure of e-government initiatives in this core government activity. The chapter outlines the dominant PFM reform agendas in the Pacific, including those of the World Bank and the IMF, which have been endorsed by the Pacific Islands Forum Economic Ministers Meeting, and the role of PFM systems and effective audit systems in combating corruption. The need for capacity building and leadership in embedding public financial management reforms are discussed and the chapter concludes with a discussion of some issues emerging from the chapter: the value of centralized versus decentralized systems and the tension between development assistance and sovereignty. The chapter argues that ownership of the public finance reform agenda by Pacific governments and a commitment to good governance is essential for sustainable PFM reforms in the region and for development.

6.1 Introduction

Two core functions of government are revenue gathering to provide facilities and services to citizens that must be managed on behalf of all and accounting for the expenditure necessary to provide those services. This chapter focuses on the central public administration processes of managing revenue and expenditure or public financial management (PFM) and how ICTs have become an essential part of such processes. Good public financial management is regarded.

R. Cullen (✉)
School of Government, Victoria University of Wellington, Wellington, New Zealand
e-mail: rowena.cullen@vuw.ac.nz

R. Cullen and G. Hassall (eds.), *Achieving Sustainable E-Government in Pacific Island States*, Public Administration and Information Technology 27,
DOI 10.1007/978-3-319-50972-3_6

> as essential to the functioning of government. It covers planning, collection, spending, reporting and audit of public monies and involves systems and staff in every agency of government. Governments need a strong PFM to deliver macroeconomic stability vital for sustainable growth and to enable efficient and effective service delivery. (Davies et al. 2010, p. 1)

Despite the endorsement of the Pacific Islands Forum leaders and the PIF Forum Economic Ministers' Meeting, the 'institutionalization' of financial transparency and responsible governance is not easily achieved in the Pacific context (Joshi and Moore 2010). With the increasing availability of relatively inexpensive automated PFM systems, which bring standardized, centralized, transparent and auditable procedures, Pacific Island countries (PICs) have responded positively to considerable external pressure from international development agencies and partners to adopt such systems and to embrace the culture of shared information that they are configured to support (p. 11). This chapter records some of the policy reforms and systems that have resulted from this agenda. At the same time, as Joshi and Moore argue, PFM systems are being rapidly introduced into countries where fiscal control, transparency and accountability may be at odds with traditional ways of governance and where fragmented and isolated government agencies are unaccustomed to sharing information (Joshi and Moore 2010).

This chapter therefore explores some of these developments and their impact on the success or failure of ICTs in government. It begins with a discussion of the motives and activities of some of the leading international and regional organizations driving change in the region, the World Bank, the International Monetary Fund (IMF) and its regional office, the Pacific Financial Technical Assistance Centre (PFTAC). It analyses some of the PFM reform programs in PICs and the role of ICTs in combating corruption; it also outlines some of the initiatives and tools that have contributed positively to the development of Public Finance Management and Revenue Management systems in PICs. Boxed examples provide more detail about the systems being used. The role of government auditors in PFM and the need for capacity building are discussed and the chapter concludes with some comments on the key issues emerging from the chapter: the value of centralized versus decentralized systems, the tension between development assistance and sovereignty and several factors contributing to effective and sustainable public finance management reforms in the region.

6.2 ICTs in Public Administration in Developing Countries

Information technology has been used in public administration since computers were first introduced into the workplace. By 1990 ICT in the form of the desktop or personal computer, especially for handling financial data, was making a considerable impact on productivity in the public sector in developed countries (Lehr and Lichtenberg 1998, p. 2). Throughout the 1990s government agencies experimented with new ways of disseminating information through the Internet and the World

Wide Web (Eschenfelder et al. 1997). Thus, in the early '90s the focus of e-government began to change from these 'back-office' functions to the delivery of information and services through networks and a web interface, as the definitions in Chap. 1 show.

The use of ICTs by Pacific Island governments has lagged due to lack of financial resources, the availability of hardware and computing skills and lack of 'fit' with the actual administrative procedures to which ICTs could be applied. E-government in PICs is therefore still mainly centered on basic office systems, with specialized software and networked systems only now being introduced to support inter-agency information exchange. Access for citizens or businesses to government information and services via the World Wide Web is similarly underdeveloped, only starting to appear in certain key functions, such as the customs software ASYCUDA (which supports a web interface for clients; see below), while in some larger countries license forms are available online for download, or personal income tax is payable online. In general, e-government in Pacific SIDs is more accurately defined within the ICT-for-development framework "to improve the activities of public sector organizations" (eGovernment for Development 2015).

6.2.1 The Benefits of ICT-Related Change in Government Financial Management

The OECD identifies the benefits of e-government as: efficiency improvements in mass processing tasks and public administration operations; improvements in government services; a customer focus adding 'user value'; the sharing of information to contribute to policy outcomes; and the achievement of economic policy objectives. In addition, e-government can be a major contributor to reform, promoting open and accountable government and preventing corruption (OECD 2003). In Chap. 1 we argued that the history of ICT projects in Pacific governments is one of ad hoc interventions, project failures due to the inappropriate agendas of development partners (Heeks and Bhatnagar 1999) and failure to take into account the social and cultural environment (Budden 2005)—a history that has come at considerable cost. We also noted that most Pacific SIDs are micro-economies with tiny markets, making economies of scale unachievable. However, in the past decade new ICT initiatives in the core aspects of government administration have had considerable success in introducing more effective management practices and structures, increasing the efficiency of Pacific Island governments. In particular, broad-ranging public finance management reforms accompanied by changes in business processes to meet international standards have been instrumental in increasing revenue and reducing corruption, as some of the examples in this chapter will show. Furthermore, the cultural, physical and technological environment of Pacific SIDs means that efficiency gains will come more from standardized and timely processes and the ability to extract data for policy-making, rather than the mass processing of tasks envisaged by the OECD, although the value of these gains, as with gains from reducing corruption, may be easier to identify and proportionately more significant than in more advanced countries.

6.3 Regional and Global Interest in Public Finance Management in Pacific SIDs

Despite these recent gains, the pattern of development partners' mixed agendas continues and will probably remain as long as PICs are dependent on development aid. Economic Governance is one of the Pacific Islands Forum's four main activities; the Forum Economic Ministers' Meeting (FEMM), held annually, is a key advisor to Forum leaders on good governance and economic management and regularly recommits to the process of economic reform.[1] Its Eight Principles of Accountability, adopted by FEMM and endorsed by Forum Heads of Government in 1997 (PIFS 2000) is regularly reported on at FEMM [see for example the 2012 FEMM *Biennial Stocktake* (PIFS 2012)]. However, most recent initiatives introducing ICT to public finance management within a framework of PFM reforms, albeit with the strong support of the FEMM, have been led by international and regional organizations committed to global development. These include the World Bank (whose overarching mission is "a world free of poverty"); the Bank's Public Expenditure and Financial Accountability (PEFA) Program, aimed at assisting developing countries to improve their public finance management and build capacity in PFM; the International Monetary Fund's Pacific Financial Technical Advisory Centre (PFTAC) and the United Nations Development Program (UNDP).

Inevitably, the strong interest of these development partners (and the interest of the Commonwealth of Australia's Aid Programme and the New Zealand Aid Programme[2]) in public finance systems is both altruistic while at the same time protecting their own interests. (NZAID's mission in 2016, for example is to "support sustainable development in developing countries, in order to reduce poverty and to contribute to a more secure, equitable, and prosperous world" (NZAID 2015)). So, the interest in PFM systems and reform is not only because of the belief that good public finance management is essential for economic development, but also to close loopholes in international financial systems, ensure good border control in the Pacific, protect against money laundering, mitigate against corruption and ensure that substantial development funding can be accounted for.

The goals of their development partners have undeniable value for the Pacific region, but it is also true that the international agencies bring additional agendas,

---

[1]See the annual FEMM Action Plans, and meeting papers on the PIF Economic Governance web site. www.forumsec.org/pages.cfm/economic-governance/forum-economic-ministers-meeting-femm.

[2]The official name of the funding body is New Zealand Agency for International Development (NZAID).

which may not align with those of a country's own government. The agenda of the World Bank in public financial management systems, often in association with the IMF, is one of reform of public services and what it deems to be prudent accountable financial management. However, its reform model has not always been seen as appropriate. As the Public Sector Board of the World Bank's Poverty Reduction and Economic Management Network noted in 2000,

> The World Bank has had a mixed record in public sector reform to date... The bank has sometimes taken a rather narrow and 'technocratic' view of what is needed... it has sometimes relied on 'best practice' that has not been feasible in the particular country setting. (Gray 2000, p. iii)

> Moreover, the authors note, traditional applications of the bank's lending instruments, Structural Adjustment Loans (SAL) and Technical Assistance Loans... have not always allowed the long-term commitment and systemic viewpoint needed to achieve long-term results. (Gray 2000, p. iii)

The Bank's new approach, aimed at sustainable development, is to proceed with reform only when a country's leaders are "committed and in the driver's seat" (Gray 2000, p. iv), and it promises to work with partners to "understand and address the broad range of incentives and pressures—both inside and outside government that affect public sector performance." One of these pressures, and a prime motivator of the bank's focus on public sector reform as well as country "ownership and community involvement", is corruption. And at this point in its history the reform model is still one of decentralization, the central tenet of the Bank's *Three drivers of public sector reform*[3] (p. v). So despite its awareness of the need to take account of local context and country ownership, the Bank's agenda differs from what is happening in many of the systems and projects outlined below, where centralization is a key aspect to ensure accountability, counter corruption and secure the economic benefits of public sector reform and e-government.

In addition to the World Bank, an abundance of other regional development partners can also bring their own agendas when promoting PFM reforms in Pacific SIDs. The IMF's PFTAC, whose activities are discussed in more detail below, expended a total of NZ$8 million from 2002 to 2012 on tax reform projects in the region, working primarily with Australia and New Zealand, who together provided a further $47 million in both lead 'reform' roles and support or 'inline' roles, mainly through temporary staff placements, across most of the 15[4] core PICs (Murray et al. 2014). In addition, the Forum Economic Ministers' Meeting, the Pacific Community (SPC), the Oceania Customs Organisation (OCO), the Association of Pacific Island Public Auditors (APIPA) and the Pacific Organisation of Supreme Audit Institutions (PASAI) all develop action plans and provide technical advice and sometimes consultants to individual PICs. This influx of technical advisors and proliferation of policy frameworks, well-meaning as it is, results in different agendas that are not always easily reconciled and a certain amount of territoriality.

---

[3]The Three Drivers are: Rules and restraints, Competitive pressures, and Voice and partnerships.

[4]These are identified in Chap. 1, Sect. 6.2.

Policies are not always seen through to implementation and evaluation, and the work of a previous advisor might well be replaced by a new program. Although many of these organizations are focused on information sharing and cooperation [see for example the principles on which the Association of Financial Supervisors of Pacific Countries (AFSPC) was founded (Bucknall et al. 2004, p. xii)], in reality cooperation and the sharing of information between them is at the very least, challenging, and, in the views of some regional experts, rarely occurs (Joshi and Moore 2010).

6.4 The Impact of Good Governance on PFM Reform and ICT Use

Despite this intensification of effort, the two key factors in the effectiveness of PFM reforms in the region remain the quality of governance (which the reforms aim to enhance) and the nature of governance in the region. Most Pacific countries are recently formed independent democratic states. The majority of their populations still live in rural villages, following traditional life-styles based on community values and chiefly authority. In most PICs the Council of Chiefs still has some authority over traditional or 'customary' issues, including land titles and village justice systems. Apart from the larger towns, parliamentary electorates are based on local communities, dominated by tribal leaders. Political parties proliferate[5] and coalition governments made up of multiple parties are fluid and unstable; as a result, ministers are uncertain of their tenure in office. In addition, the dominant traditional culture is one of familial duty, obedience to chiefly authority, traditional obligations and responsibilities and reciprocal gift giving.

This model, often described as a system of relational ethics which provides acceptable rules for small communities to negotiate among themselves (Sen 1987; Thompson 2009, p. 20), is not easily transformed into a modern bureaucracy meeting the criteria of good governance; aspects of traditional society continue to influence the way the public service operates.

> The concept of a neutral society in which all are equally entitled to public services on objective criteria other than kinship, is a product of larger scale, more mobile societies. Even in the largest most modern nations the idea is never fully realized, but it remains the goal and efforts are made to achieve it. In the Pacific Islands, with smaller numbers, substantial subsistence [living], and limited formal social security, the traditional priority for kin and community more often conflicts with national interest and equal opportunity. (Crocombe 2001, p. 546)

[5]In Vanuatu's 2016 election, following the arrest and conviction of 14 MPs for corruption, 246 candidates representing 36 parties contested the elections for the 52 parliamentary seats available. Sixteen parties and four independents won seats.

In addition, the lack of a strong tradition of a neutral public service loyal only to the government of the day means that public servants can feel vulnerable to pressures which run counter to the management practices of development partners and which governments themselves aspire to. Balancing "indigenous cultures, global patterns and technological requirements" (Crocombe 2001, p. 547) continues to be a challenge for governments; and the nature of fragile multi-party coalition governments, supported by a public service which may have divided loyalties, can hamper the unity and leadership needed from both politicians and public servants to institute the major reforms of policy and practice necessary to successfully implement modern efficient ICT-based central financial and administration systems.

6.4.1 Corruption a Major Issue

The system of relational ethics outlined above leads to tensions and a number of practices which are frequently labeled 'corruption' and which highlight the conflict between relational ethics and normative ethics and the assumptions of 'mainstream economics' and rationality. Regardless of how this behavior is viewed, government revenue and development (or aid) funds redirected for personal gain or to ensure electoral success have major consequences. Corruption in the Pacific has been identified as a particular problem in the public sector and, as such, has a significant impact on the ability of PICs to achieve their development goals.

> In the Pacific, it is clearly evident that corruption hurts the poor indiscriminately, undermines state accountability and the capacity to provide equitable and responsive public services, and investments from infrastructure, institutions and social services. (UNDP 2012, p. 4)

The impact of corruption in PICs has been noted by a number of studies and was a frequent theme of Pacific economist Savenaca Siwatibau, former governor of Fiji's Reserve Bank and director of ESCAPs Subregional Office for the Pacific, after whom Transparency International's annual Siwatibau lecture is named. As Siwatibau observed in 1997, "Corruption, which poses unaffordable costs on the economy … is rampant … [and] … offers untold damage to our economies" (Siwatibau 1997). The sources of corruption, which flourished in the Pacific in the 1980s and 90s, and which Crocombe documents in detail, are attributed to significant rises in government revenue, facilitated by foreign and local operators offering bribes and "by decay in some checking mechanisms: accounting and audit, police and judiciary, ombudsmen and government legal services; as well as in the relaxed supervision of civil servants" (Crocombe 2001, pp. 512–514).

The problem remains. One of the most recent and high profile cases of bribery in PICs involves the arrest and subsequent conviction of 15 members of the Vanuatu Parliament in 2015, leading to its dissolution, fresh elections and a new prime minister focused on change (Connors and Bule 2016). Transparency International's Corruption Perceptions Index for 2015 lists only Papua New Guinea, which is ranked at only 139th of 168 countries (it was ranked 145th in 2014 and 144th in 2013). No other Pacific Island SIDs are included, possibly because the ranking, determined by a composite index that combines surveys and assessments of corruption in the public sector by a variety of reputable institutions (Transparency International 2016), is not feasible in most SIDs. In 2014 Samoa was also included (ranked 50th out of 175 countries) and the two scores for 2014 provide a credible representation of the range of Pacific countries, most others being closer to PNG. Apart from electoral fraud the most common forms of corruption (setting aside the cultural issues of the role of kinship and tribal or language affiliations in appointments) are bribes to speed up bureaucratic process and delivery of public services, administrative corruption—misappropriation, embezzlement, abuse of power and manipulations of budget processes to the benefit of vested interests (Larmour and Barcham 2006).

6.4.2 The Role of ICTs in Promoting Good Governance

Over recent years there have been various regional initiatives to remedy this situation and a number of international anti-corruption instruments developed. These include the United Nations Convention against Corruption, which was followed up by the joint UNODC-UNDP Pacific Regional Anti-Corruption Project (UN-PRAC) and the OECD's anti-bribery instruments, such as the *Principles for Enhancing Integrity in Public Procurement* (2009), the *Public Sector Integrity Framework for Assessment*, the *Bribery and Corruption Awareness Handbook for Tax Examiners and Tax Auditors* and the *Legal Instruments on Corruption Prevention*. The ADB/ OECD Anti-Corruption Initiative for Asia and the Pacific, formed in 1999, has also been very active focusing on legal and institutional reform and capacity building, especially between 2003 and 2010.

The United Nations has been actively working to assist PICs to ratify the Convention against Corruption (UNCAC) and meet their commitments under the Convention through its Pacific Regional Anti-Corruption Project (UN-PRAC). UN-PRAC, which included 13 Pacific Island states, was established in 2012 with the assistance of AusAid "to support Pacific Island countries (PICs) to strengthen their capacity to address corruption in order to provide better service delivery and development outcomes for their peoples". A second objective was to promote the ratification of the United Nations Convention against Corruption by the remaining five non-ratifying countries in the Pacific region[6] and "support the active participation of PICs in the UNCAC review process." The project is focused on helping

[6]This included Kiribati, Nauru, Samoa, Tonga and Tuvalu. Not all UN-PRAC signatories have yet competed the necessary legislation, but as a result of pressure arising from the adoption of the SDGs, all PICs are now engaged with the UNCAC, and have either agreed to sign in principle, or are considering such a move. See SDG Action 11973: https://sustainabledevelopment.un.org/partnership/?progress&id=231.

PICs develop the political will to establish strong policy and legal frameworks for implementing the Convention, strengthening the capacity of key national anti-corruption institutions and sharing knowledge (UNDP 2012). It is also focused on achieving a balance between the urgent need for action, the historical and cultural context, the resources and capacity to support the necessary reforms: institutional reform (of the core financial and administrative agencies), the application of international standards, the adoption of the necessary legal framework (including legislation to make bribery illegal) and the authority of the necessary accountability institutions. These latter include a national agency responsible for auditing government revenue and expenditure (known as a 'supreme audit institution'), a government official with the authority to deal with complaints (often bearing the title of ombudsman) and the judiciary.

Fiscal accountability and transparency in revenue collection and expenditure (which include resource allocation and procurement, standardized procedures for specifying required goods and services, selecting suppliers, negotiating prices, acknowledging receipt and approving and recording payment) are therefore major elements in the reforms that are the goal of all these initiatives. These are central reasons why financial management (and the financial sector generally) are so strongly supported in the Pacific by regional development partners and international institutions such as the IMF, as one of the many ADB/OECD Anti-Corruption Initiatives in Asia and the Pacific reports on bribery in the region asserts.

> Public procurement is now among the most dynamic areas of anti-corruption reform in Asia-Pacific. In the past years, many member countries of the Initiative have implemented significant reforms in order to counter corruption risk in their procurement frameworks. (ADB/OECD 2007)

6.4.3 The World Bank and IMF Focus on Good Governance Through PFM Reform

The World Bank's *World Development Report* for 1997 highlighted the significance of good governance as a major development issue and encouraged states to focus on core public activities that are crucial to development and to improving the state's capability by reinvigorating public institutions. The report put particular emphasis on "mechanisms that give public officials the incentive to do their jobs better and to be more flexible, but that also provide restraints to check arbitrary and corrupt behavior" (World Bank 1997). Public financial management and accountability became a central focus for the Bank's work, further developed by a paper issued in 2000 which outlined strategies for implementing the 1997 recommendations for institutional reform in a Comprehensive Development Framework. It focused on governance, the justice system (a topic in Chap. 7) and other 'pillars of public administration' (education, health etc., covered in later chapters). The new strategy, grounded in the understanding that the Bank's previous 'technocratic' approach had not succeeded, focused on the Bank's activities "in the core areas of public sector functioning—such as public expenditure analysis and management, tax administration, civil service reform, regulation, decentralization, and judicial reform" (World Bank 2000a, p. 3). A coordinated approach to assisting developing countries resulted in the founding of the Public Expenditure and Financial Accountability program (PEFA) in which the World Bank is a partner (and a trustee), along with the IMF, the European Commission (represented by Europe Aid) and other European aid agencies. As a consequence, Public Finance Management (PFM) reforms have increased across the Pacific, accompanied by "increased resourcing and policy attention" (Haque et al. 2015).

The Pacific Financial Technical Assistance Centre, established with the support of the UNDP in 1993 as one of the IMF's seven Regional Technical Assistance Centres, and partly funded by the Asian Development Bank (ADB), as well as Australia, the EU, Korea and New Zealand, has been a major driver of these PFM reforms. Its primary role is to provide technical assistance and training for 16 countries in the region (Timor-Leste as well as the Pacific SIDs) in their public financial management, revenue/customs administration, financial sector supervision (the role of central banks) and statistics. It works towards a framework of key outcomes for each country and plans its assistance each year dependent on their status in terms of their public financial management, tax base and the efficiency of their revenue systems, management of risk in the financial sector, timely and reliable economic statistics and macroeconomic analysis.

6.4.4 The Pacific Islands Forum Economic Ministers' Meeting Endorses the PFM Roadmap and the PEFA Framework

Acknowledging the need "to improve their public financial management to encourage growth and reduce poverty" the Pacific Islands 2010 FEMM unanimously adopted a *Public Financial Management Roadmap for Forum Island Countries* (Davies et al. 2010). In adopting the Roadmap, FEMM also acknowledged that PFM reforms would encourage "increased and better quality" development partner support (p. [1]). Progress of individual PICs and the region generally, in meeting the goals of the Roadmap are reported at the annual meeting of FEMM.

The Roadmap emphasizes the need for each country to have a basic well-functioning PFM system and endorses the PEFA framework as the most effective tool for monitoring progress, and recommends a PEFA assessment every three-to-four years. It also notes however, that PEFA is not comprehensive or infallible and that PFM reform is a long-term complex process that needs commitment from the government, should be tailored to each country's unique circumstance and should not to be driven by development partner's objectives. The

Roadmap acknowledges the demands on human resources capacity that PFM and PEFA assessments place on staff and that staff turnover and loss of experienced staff is a major problem. (Haque et al. (2015) go so far as to suggest that PEFA is beyond the capacity of most PICs.) The Roadmap recommends strategies to counter capacity issues, the role of PFTAC in building capacity and the importance of sharing local experience in the region (Davies et al. 2010).

In the public finance sector, the main instruments for assessment, strongly promoted by PFTAC, are PEFA reports and regular Public Expenditure Reports. PEFA was developed to be a key tool in promoting public financial management reform and identifying the adequacy of public expenditure, procurement and financial accountability systems in developing countries. The PEFA Framework employs seven ‘pillars’ in its assessment, all part of the budget cycle. They are: budget reliability, transparency of public finances, management of assets and liabilities, policy-based fiscal strategy and budgeting, predictability and control in budget execution, accounting and reporting, external scrutiny and audit. A World Bank report on Financial Management Information Systems (FMIS) published in 2011, which notes that PEFA is a “promising direction for developing robust performance indicators for FMIS projects” (Dener et al. 2011, p. 76), includes it in its initial assessment and design protocol (p. 88) and for monitoring progress in PFM functions and performance after FMIS installation (p. 131).

As the World Bank’s staff, led by Tobias Haque, note, “The PEFA framework has been enthusiastically embraced by Pacific development partners, regional organizations and governments” (Haque et al. 2015). Thirteen countries at that time had completed PEFAs, seven at least twice, and Haque et al. note that “in some cases donors may make aid resources conditional on improved PEFA scores or the existence of plans to address weaknesses identified in PEFA assessments” (p. 611). By 2015, the FEMM annual report on PFM reforms noted that Marshall Islands and Tonga had adopted PFM Reform Roadmaps and that drafts were under consideration in Fiji and Papua New Guinea (PIFS 2015).

6.5 Financial Management Information Systems an Essential Part of PFM Reforms

The PEFA assessment requirements and donor expectations for public financial management reforms thus highlight the demand for Financial Management Information Systems to provide adequate data and accountability. Donors also expect them, as evidenced by the OECD-DAC Task Force on Public Financial Management’s substantial manual that advises donors how best to make use of in-country FMIS, to encourage best practice in handling development funds, ensure accountability and facilitate “harmonization of donor and government procedures” (Sprietzer and Vargas 2011). PFM systems, essentially an annual accountable budget cycle, must be distinguished from Financial Management Information

Systems and the more sophisticated Integrated Financial Management Information Systems (IFMIS) that have been installed in most PICs. PFM reforms would be impossible without such systems, although, as Haque et al. (2015) note, both PFM reforms and the ICT systems that support them, raise major capacity issues for small island states. The World Bank has financed 87 FMIS projects in 51 countries as part of its PFM reforms program, although in the Pacific region its input has been based more on technical assistance alongside other partners (such as Australia and the ADB), which have funded the systems (ADB 2012).

The Dener report, prepared by World Bank staff involved in such projects, set out to analysis a number of critical success factors for FMIS projects. It reiterates that well-functioning accounting and financial systems underpin government capacity to allocate and use resources efficiently and effectively and should reflect the principles of comprehensiveness, legitimacy, flexibility, predictability, contestability, honesty, transparency and accountability that are needed for sound budget and financial management. The report defines an FMIS,

> as a set of automation solutions that enable governments to plan, execute, and monitor the budget by assisting in the prioritization, execution, and reporting of expenditures, as well as the custodianship and reporting of revenues. (Dener et al. 2011, p. 1)

When an FMIS and another PFM information system such as payroll share the same central database to record and report all daily financial transactions, "offering reliable consolidated results for decision support, performance monitoring, and Web publishing", the authors suggest, "they can be referred to as an 'integrated' FMIS, or IFMIS." But, they caution, "IFMIS solutions are rare in practice and, to avoid unrealistic expectations, the term should not be used as a synonym for core FMIS functionality" (Dener et al. 2011, p. 25).

Typically, the cost of installing an FMIS is only around 50% of the total PFM reform package, the rest being analysis, installation and training. The cost of the actual system installed varied (from $610,000 to around $12 million, calculated at 2011 values) depending on its functionality (i.e. from Treasury only to a full IFMIS), the number of agencies and locations connected and whether a commercial off-the-shelf system is purchased or the software is developed locally. (While costs per user tend to decrease when locally developed software is used, in small scale projects where FMIS users number less than 250 the total cost of FMIS solutions based on an off-the-shelf system and locally developed software are judged to be comparable.)

The Dener report notes that FMIS systems are benefitting from advances in technology (including hardware and networking capabilities) and are now designed "with better focus on the quality and security of systems to minimize the risk of corruption and improve the reliability of systems" (Dener et al. 2011, p. 78). Advances in FMIS technology that could bring significant benefits in terms of reliability, cost effectiveness and accountability, include use of electronic payments systems, use of digital signatures for financial transactions, electronic records management, dynamic web publishing abilities, interoperability with other systems/ software that supports electronic data exchange, the application of international standards for security and quality assurance and use of systems and applications based on Free/Libre Open Source Software. Many of these features are now available in FMIS systems used in Pacific countries although not all are used. They are essential tools for the next level of integrated financial management information systems that Pacific Island countries can afford, have the appropriate level of institutional capacity to manage and are moving towards (see for example Cullen and Hassall 2016a, p. 15). But, as noted below in the discussion of customs software, opting for free and/or open source software can carry substantial implementation and maintenance costs and may not be the appropriate choice for all countries, despite its appeal.

Not all projects are successful—Dener et al. identify several factors that contribute to the success or failure of projects and their sustainability. Key success factors are: a focus on capacity building and training; close World Bank supervision; a suitable political environment and committed leadership; flexible project management; a pre-existing enabling environment (ICT, HR, accounting capabilities and policies); adequate preparation and clarity of design; good project management and coordination; and the external environment (p. 42). In general, failures were due to the lack of these key success factors, in particular lack of human resource capacity, institutional and organizational resistance, complexity of project design and weak project preparation and planning (which, the authors suggest, should routinely include a review of PFM procedures, capacity and needs). The authors comment:

> From the project design stage (often even at the project development stage) to the project implementation and maintenance stages, a lack of interest and commitment from top leadership is often cited as one of the most critical impediments for project success and long-term sustainability. (Dener et al. 2011, p. 42)

Given that the report finds that a suitable FMS for Pacific SIDs is estimated to cost least $1 million and in light of their fragile economies and limited resources, failure is costly, not to mention the additional impacts—the opportunity costs, political costs and the loss of benefits—that the project would have brought (see Heeks 2003).

Box 6.1: Papua New Guinea's HR network and Payroll system benefits government and employees

The PNG Department of Personnel Management's HR COnNECT network when completed will provide access to the PNG government's Alesco HRMIS system to public servants in national and provincial agencies as well as to teachers and health workers in remote locations. The system is funded by the Papua New Guinea Economic and Public Sector Programme (EPSP) a long-term AusAid capacity development project focused on strengthening governance in selected central agencies. This program has recently been extended to cover whole-of-government service delivery, enhanced financial management, up-skilling the government workforce and gender equity and social inclusion throughout the public service. A strength of the program is the extensive relationships it has built up through institutional strengthening and partnerships, which the program's evaluators characterize as 'enablers of sustainable capacity development' and 'drivers of change'.

Getting HR COnNECT and the Alesco HR Payroll system out to all government agencies in the National Capital District and to the country's 22 provinces and the autonomous region of Bougainville is a challenge, with installation delays due to lack of local leadership, suitable buildings or lack of ICT equipment experienced in areas where the system has been piloted. But initial outcomes are very positive for PNG's government and its workforce. The Department of Personnel Management initially developed HR COnNECT to improve HR process in the government but, working with an EPSP HR Advisor, have been also able to use it to aid performance management in government agencies. A payroll data cleansing program to eliminate allowance overpayments, personnel cost overruns and eliminate 'ghosts' from payrolls is being enacted throughout the government with major savings (potentially in the millions of PNG Kina) resulting.

The system ensures prompt payment of government employees and allows them to access their pay slips online or via SMS. It has helped provincial administrators and hospitals deal with prolonged staff absences (by removing such staff from the payroll) and has improved staff attitudes as payroll issues are addressed promptly. The payroll data cleansing initiative is taking longer and is not yet fully accepted. It is meeting resistance in some sectors and provinces where local loyalties conflict with fiscal responsibility; and in some agencies personnel data does not match well with the HRMIS data held by the Alesco system. Strong political leadership is needed to ensure gains continue to be made. But overall, as the Economic and Public Sector Program transitions into the new PNG Governance Facility with its own training program at the PNG Institute of Public Administration, both the PNG and Australian governments are focused on the real achievements of EPSP program, on HR COnNECT and on the increasing ownership of the program by the PNG government. The ongoing role of Technical Advisors and long-established partnerships with government agencies will keep the program's initiatives moving forward.

Sources: Papua New Guinea Economic and Public sector program. Six month report Dec 2014. (http://dfat.gov.au/about-us/publications/Documents/png-epsp-6-month-report-july-dec-2014.PDF); Six month report Jan–Jun 2015. (http://dfat.gov.au/about-us/publications/Documents/png-epsp-6-month-report-jan-june-2015.PDF) Papua New Guinea Education news website.

6.6 Policy/Planning Link Essential to PFM Reforms

ADB and Australia's DFAT in a review of PFM reforms in Pacific Island countries. Aware that in the initial PEFA assessments external development partners, including PFTAC, tended to rely too heavily on consultants and involve a ministry staff primarily as "information providers and report recipients" (PFTAC 2015, p. 4) the authors readily admit that country ownership was missing in many PFM reforms and that plans were overly ambitious, "poorly attuned to political economy realities" (p. 4) and lacking any clear prioritization. This resulted in low PEFA ratings in several areas that countries were reluctant to address. From 2011 the policy has been more locally focused. PFTAC introduced a 'self-assessment' phase for PEFA reviews and PFM reform roadmaps that resulted in reviews being conducted with much "shoulder primary responsibility for authoring the reform plans." Moreover, the PFTAC reviewers noted that these changes revealed that ministry staff understood their country's PFM weaknesses very well, did not find the PFM criteria "inappropriate or too advanced" and that the obstacles were "mainly political". The review showed, however, that a lack of consensus remains about what the basic principles of PFM reform should include. In particular, the review identified a small set of key factors that were continuing to impact on the success of PFM reforms. These were: the need for policy-focused budgeting (linking plans and budgets); the need for medium-term budget planning; the lack of accrual accounting; commitment controls; and asset management in the PEFA framework, which the review considered were essential tools. The link between policy, planning and budgets is of particular concern.

Unfortunately, the practice in most PICs has been for the development of plans and budgets to be separate processes, involving different staff, following different timelines and being communicated through separate documents. The consequence has been little consistency between plans and budgets. Very often plans are too vague to be costed or used and there are too many of them—national strategic development plans, sector plans, corporate plans and annual plans for individual ministries.

> The PEFA framework is in fact silent on the specific form of plans. .. The important point is that they should be clear and specific enough to be costed, cover full implementation, and that the costing information is comprehensive... Merging planning content into budget documents with a medium-term orientation is one way of 'forcing' greater consistency between a government's policy plans and its budget. This is important for long-run accountability because it will improve the likelihood that promised service improvements will actually be implemented. (PFTAC 2015, p. 4)

The report to the FEMM 2015 meeting notes that the draft revised PEFA Framework, to be adopted in 2016, places more emphasis, among other things, on "the careful selection of projects based on a critical weighing of costs and benefits, performance measurement for significant public projects, and policy-focused budgeting" (PIFS 2015, p. 2).

The incorporation of medium-term budget planning in an FMIS is a way of ensuring that policy-decisions and funding for multi-year endeavors in areas such as education and health are included in annual budgets and that past and future spending commitments are factored into annual budgets. Similarly, accrual accounting, commitment controls and asset management, which are vital for good fiscal control and planning, are built into most modern FMIS systems but are not used in the way they should be in many PFTAC member countries. Asset management in particular and maintenance of an asset registry, enables "disciplined planning for financing asset repair/replacement and monitoring asset theft" (PFTAC 2015, p. 5); both have been identified as problems in the region. PFTAC workshops have been focused on these issues in 2015 and 2016 and technical assistance in PFTAC member countries focuses on supporting staff to use the full functionality of their FMIS for these essential procedures and for reporting.

6.7 PFM, FMIS and IFMIS Systems in Pacific Countries

Clearly, use of an FMIS is essential in meeting the PEFA assessment requirements and the demands of development partners, but the selection of an appropriate system is a difficult challenge even for larger government agencies, given the considerable range of systems available. Brisbane-based TechnologyOne's FinanceOne system is used in some Pacific countries, as is DataTorque's RMS. Companies specializing in tendering for projects in the South Pacific are often preferred because of their willingness to actively support and maintain their systems despite the isolated location of their clients and because of their knowledge of local systems and practices. Long-term relationships are important—for installing a system, training staff in new business practices as well as the system's functions and dealing with legacy systems that may have been significantly modified as part of a number of different aid projects can take many years. Papua New Guinea's installation of FinanceOne, which needed to be integrated with the PNG accounting system, PGAS, took over a decade and still is not fully functional (Papua New Guinea 2015). Vanuatu's implementation of SmartStream, a system extensively used in Caribbean SIDs, however, has been widely acknowledged as highly successful (Joshi and Moore 2010, p. 63) and is a model for sustainability, largely due to ownership by the Ministry of Finance, a positive long term onsite relationship with the developers and effective links with pre-existing 'feeder' PFM systems (e.g. HR, PayRoll and ASYCUDA).

A PFTAC report on the FMIS systems used in six PICs (Cook Islands, Kiribati, Tonga, Tuvalu, Vanuatu and the Marshall Islands) arose out of a regional workshop in 2008 at which several countries sought guidance on the key systems analysis issues[7] (Joshi and Moore 2010). The factors come under three headings:

- *Technical* (i.e. adequate network infrastructure to support a secure government data network, email, etc.), up-to-date government-wide ICT policies, good vendor relations and a contract for reliable affordable support);
- *Functional* (functional software that meets the needs of all stakeholders with suitable analysis and reporting tools); and
- *Management* (HR capacity, acknowledging the difficulties of recruiting and retaining staff, a consistent 'view' of the system to encourage consistency, stakeholder involvement, the incorporation of standards and best practice and legislative reform).

The Functional category also includes systems integration to allow automated data exchange between government PFM systems and revenue (tax and customs) systems, integration with the banking system to ensure automated payments and bank reconciliation, and the phasing out of cash and checks for both receipts and payments, including salaries. This last issue (abandoning the use of checks and cash) brings a major shift towards transparency and accountability and reduces the risk of fraud and corruption; but it presents huge challenges in countries where electronic banking is in its infancy (mobile banking, discussed in Chap. 5, is more widely used) and where much of the country remains a subsistence economy.

The contextual notes in the report acknowledge a number of inhibiting factors, including size and lack of human resources, as well as resistance to change in these small island countries. They also acknowledge a number of cultural factors that influence the effectiveness of centralized networked systems.

> Information sharing systems may sometimes be culturally at odds with traditional fragmented communities of information workers, such as government staff, who may not be comfortable with the idea of widely sharing information. In some organizations, there can also be traditional taboos regarding attributable information. In that case, people may be reluctant to write down information to be shared with others for fear of being held accountable. (Joshi and Moore Joshi 2010, p. 11)

This attitude to information, in particular to personal information, also affects HR systems and the willingness of citizens to share personal information with government. Users need to be assured of security of personal information and staff reassured about how information would be used; 'outreach and socialization efforts' are needed to ensure system users (whether government employees, clients, or citizens) understand the value and purpose of the system. In addition, the authors note "no information system can operate successfully without a well-trained and motivated support staff" (p. 11).

---

[7]The report refers to an IFMIS; but in line with the definitions of the Dener report cited above, the term FMIS is used here because the countries under review did not have a full IFMIS at this time of the review.

The review outlines good practice and advises on the effective management of a PMS ICT project, from putting a team together (to be led by a senior PFM staff member, not IT support staff), the importance of a steering committee of key stakeholders, through legislative and business process review and development of a business case and budget, to a conceptual design focused on what an FMIS is and what it should do, to a full project management process. Sustainability is acknowledged as a key issue, and from the inception of the project senior managers must understand that systems require ongoing maintenance, annual license fees, hardware and software upgrades and training. This aspect of sustainability is a significant issue in the Pacific where fragile economies and frequent changes of government put such funding at risk.

The section on technology selection weighs up the relative merits of off-the-shelf, Open Source (OSS) and customized or locally developed software, arguing that off-the-shelf and OSS 'packaged' systems are quicker to install and more sustainable as long as they are a good 'fit' with local legislation and practices. The systems and PFM processes reported on in the country reports in the Lessons Learned section comprise some valuable case studies of the status of the PFM systems in the six PICs during 2008/9, lessons which are still relevant. They highlight a myriad of success factors: good government ICT policy; integration of budget systems across government; documentation to ensure guidelines and encoding procedures are standardized; data quality assurance procedures; elimination of inconsistent redundant; and duplicate datasets. Impediments include: PFM silos and lack of the integration of financial data in a central repository; problems with multiple bank accounts (rather than a 'best practice' Single Treasury Account linked to existing line ministry accounts); incomplete reform of PFM and lack of alignment with system capacity; and problems where small systems that initially appeared more appropriate for small countries are acquired by larger systems over time and are no longer supported. Also important for success are: adequate back-up procedures (critically important in view of natural disasters and frequent power outages); secure systems and a climate controlled environment; a secure government network; harmonized laws, regulations and PFM procedures; analysis and reporting capacity in some systems; good technical support through well managed vendor relations; strong motivated teams; system ownership in the ministry of finance rather than in the IT unit.

Most of these issues need to be addressed by changing attitudes, by building capacity and commitment to government-wide perspectives on the use of ICTs and by understanding that the benefits of PFM reforms and effective use of FMIS systems are interdependent. But this, in turn, is highly dependent on secure, reliable, networked government operations (themselves dependent on a robust telecommunication industry and service) and a well-supported centralized system that allows full budget, HR/payroll and procurement functions to be carried out in agencies. When this can be achieved, as it has been shown for example, in Vanuatu (Cullen and Hassall 2016b), Papua New Guinea (Papua New Guinea 2015) and Solomon Islands (Cullen and Hassall 2016c) real gains are made in efficiency, productivity and accountability. The example from Samoa (see box) provides more detail on how some of these gains are made.

Box 6.2: Samoa's revenue increased through institutional strengthening and world-class software

The Samoan Ministry for Revenue (MfR) is responsible for collecting personal income tax and PAYE, company tax, capital gains tax, resident and non-resident withholding tax, business licenses, Value-Added GST, excise duties paid on products such as tobacco, alcohol and domestic vehicles and administrative charges. The Ministry is also responsible for customs and border control and facilitating trade. (Transport licensing is the responsibility of the Ministry of Works, Transport and Infrastructure.)

The Ministry undertook an Institutional Strengthening Project (ISP) in 2010, funded by the Government of Samoa's Public Sector Improvement Facility (established in 2002 with funding from Australia and New Zealand to support public sector reform). The MfR's ISP covered policy and legislation, staff development, business process, investment in IT and program management—all focused on better customer service, compliance and growing the country's revenue base. The ISP, in which the MfR brought in New Zealand Inland Revenue Department staff to work alongside MfR staff, resulted in considerable advances in corporate planning, organizational structure, compliance, risk management, an upgraded DataTorque Revenue Management System (RMS7) and enhanced public access. As a result of the upgrade and improved business process, MfR staff report that thousands of taxpayers pay their tax online, obtain a business license in as little as 5 min, and that revenue has increased by approximately $25 million in the first three years of operation.

The RMS7 system is linked to a new Public Finance Management System, FinanceOne, an integrated financial, supply-chain, payroll and HR management information system, which has greatly increased efficiency in the Ministry and all government agencies which access the system through a Local Area Network.

The latest and most powerful web-based version of the UNCTAD customs software ASYCUDA World was installed in 2013 as part of the second phase of the ISP, the Customs Modernization project. Because of the security and levels of access built into the system, the MfR is able to make full use of the external access modules of ASYCUDA, as well as the modules for Cashier, Declaration, Audit trail and transaction history. The system allows shipping agents to lodge manifests electronically and view their status and customs agents to lodge declarations and get a customs number online. Declarations are approved online and taxes assessed. The system allows external parties such as the Bureau of Statistics to access trade data and the Port Authority to access information on shipping.

Ministry staff are excited about the future benefits the system offers and recognize the value of restructuring, noting that they have developed a strong customer-focus and a clear perspective on the distinction between 'what is an IT problem and what is a business process problem'. The MfR is now in a position to look at workflow for gains in efficiency. In the view of the CEO, the ISP was very successful, "changing mindsets", which was an encouragement to other ministries to engage in similar development programs. In pursuit of greater accountability, the Ministry now strives to publish its financial reports on the web site as soon as possible. Making financial data publicly available is seen as critical to both national and international partnerships.

Sources: Interviews with MfR staff, 2014; Cullen and Hassall 2016a; Murray et al. 2014; DataTorque http://datatorque.com/case-studies/samoa-ministry-for-revenue/.

6.8 Revenue Systems Part of PFM Reforms

Revenue systems have also been transformed in most PICs by the use of dedicated software, which records the collection of a wide range of income and sales taxes, tariffs on imported goods, excise (on goods such as alcohol) and license fees. Many also handle border control and customs duties (thereby enhancing both trade and revenue gathering) and the collection of trade data collection for international reporting). As one report of these programs noted, "in few countries are tax and customs operations so limited that manual processing of returns and payments is adequate" (World Bank 2000b). Instead, embedded in broader public sector reforms and pursuing broader objectives, the report notes, "tax computerization can support fundamental changes in the vision, structure, and public perception of revenue administration." Such programs, which have significantly increased revenue in most PICs, have long been seen as an example and incentive to countries to proceed with further e-government projects, while the risk of sustainability of the system could be addressed by allowing a certain percentage of the additional revenue collected to be dedicated to maintenance of the system and retaining the IT skills necessary to its functioning.

Effective customs and border control systems are crucial in the region. Small isolated countries are at high risk of exploitation by unscrupulous entrepreneurs, and imports and exports comprise the major component of their Gross Domestic product. Customs duties account for up to 90% of tax revenue, as opposed to customs revenue in developed countries of around 30% of all tax revenue (WCO 2015). As with other PFM reforms, leadership and commitment to reform are critical to successful implementation; and a recent report on the use of the automated customs software ASYCUDA noted that successful implementation of such systems requires both strong political backing of the reform and modernization processes combined with a transparent and collaborative approach (UNCTAD 2011). Indeed, the management of border control is by its nature a collaborative undertaking and support in the reform process is also offered to PICs through the Oceania Customs Organisation (OCO) and its Secretariat, OCOSEC, whose goals are focused on improved law enforcement and border security, enhanced trade management and facilitation, institutional strengthening for small customs administrations and strengthened revenue management. Perhaps a symptom of the challenges of its work is that one of its key tools is a 'small crafts database', although tracking small craft across the vastness of the Pacific Ocean is no mean task. OCO provides a great deal of technical of assistance in the region; and the efforts of PFTAC and the OCO along with Australia and New Zealand have seen significant advances in customs administration in most PICs and progress against the PFTAC goal:

> To build simple and efficient revenue instruments and effective tax administrations capable of increasing tax compliance, and modernized customs procedures to secure revenue and facilitate trade – 17% countries reported goal achieved, 78% modest or substantial progress towards goal. (Murray et al. 2009)

The close association between these reforms and the installation of automated customs systems is different from other regions in that only two systems predominate here, ASYCUDA and PC/Trade (Bucknall et al. 2004). ASYCUDA (Automated System for Customs Data), developed by the United Nations Conference on Trade and Development (UNCTAD) in 1981 to assist developing countries modernize their customs systems, minimize costs and maximize revenue and produce reliable trade data is used by Fiji, Papua New Guinea, Samoa, Solomon Islands and Vanuatu. The ASYCUDA software is available at no cost, but depending on the size of the country using it and the complexity of the system being installed, the hardware, training, software customization and consultancy fees can cost anywhere from $1 m to over $30 m (costs in the Pacific region are in the lower range). The ASYCUDA software allows customs administrators and traders to handle most of their transactions themselves, from customs declarations to cargo manifests and transit documents, via a web-based interface. ASYCUDA promotes the use of the international UNCTAD Harmonized System for coding imports and exports and is consistent with the World Customs Organization's (WCO) Data model which streamlines government to business and government to government transactions. Through its in-built Integrated Customs Information System (ICIS) it produces comprehensive and reliable trade statistics and can include essential elements for 21st century trade such as bio-security data.

PC/Trade, a simpler system developed by Statistics New Zealand initially to provide reliable statistics for small islands states in the Pacific has been extended with the assistance of the New Zealand Ministry of Foreign Affairs and Trade to handle customs clearance procedures as well. It is used in more than a dozen countries in the region, (see box) and unless the more powerful ASYCUDA system is warranted because of the size of the country and its trading activity, PC/Trade is seen as a flexible, practical, sustainable solution that meets local needs. A few countries (Tonga for example) use the Customs Management System modules of their revenue management systems (such as the DataTorque RMS) and the Cook Islands, largely through its close relationship with New Zealand, has implemented the in-house system built by the New Zealand Customs Service for its own use.

Box 6.3: Statistics New Zealand's PC/Trade in use across the Pacific
PC/Trade was developed in the late 1980s as a joint venture between the Australian Customs and Border Protection Service and Statistics New Zealand and became the sole responsibility of Statistics NZ soon afterwards (funded in recent years by the NZ Ministry of Foreign Affairs and Trade). The aim was to develop a simple system to enable Pacific Island countries to manage their trade data through accurate customs documentation; the system includes core customs and border control functions and incorporates international standards such as the WCO system of harmonized codes for traded goods.

PC/Trade was initially developed as a database system using Access and SQL/Visual Basic (the latest version uses an SQL database and DotNet front-end) and supports a variety of functions: processing customs declarations and manifests, managing levies and tariffs, applying selected higher level WCO codes and retaining a record of transactions. Data reports are simply produced and can be exported to other systems. PC/Trade is a standalone system that supports manual input from forms as well as Excel-based electronic declarations that allow an automated load. For some countries, Statistics NZ intends to automate further Customs documentation (e.g. to include ship manifests) to meet a broader range of needs for automated customs management; in others, with only one ship arriving each month and several weeks taken to offload cargo, this is not a priority. The emphasis is on the quality of data, the ease of use of the system and the quality of the trade statistics it produces.

In addition to producing trade statistics and supporting effective border control, PC/Trade also supports the collection of excise and other duties. It can hold data on waste packaging of imported goods, (waste disposal being a critical issue in most Pacific SIDs) and its flexibility allows some smaller countries to use it for other purposes as well; for example, Kiribati uses it to handle its VAT, Tonga uses it to produce statistics drawn from its DataTorque Revenue and Customs Management Systems and some countries use it for collecting school fees.

Training and support is provided by Statistics NZ, but demand for that is ongoing because of frequent changes of personnel in island customs offices; and training often has to start with basic PC skills and standard customs and border control procedures. While the system is being installed it is customized to fit a mutually agreed business process which will best meet the country's needs. Its development has been driven by two factors—to reflect international best practice in border control and data management, while keeping the system simple enough for a small office with one PC to manage. Since PC/Trade serves a small fragile market there are few resources with which to proactively develop the system further, but it has shown that it is a trusted brand that meets the needs of many island states in the Pacific. Because that trust was built over two decades of building relationships and listening to clients the client focus remains strong. Even in Pacific countries where PC/Trade is no longer used Statistics NZ continues to assist with the management of trade data.

Source: Statistics NZ.

6.9 The Role of the Public Auditor

Because good governance and reducing corruption require effective audit systems, the quality of the public audit function has significantly improved across the Pacific since the late 1990s despite many challenges—including their "tight-knit social and cultural environments and significant political, professional and personal pressures" (ADB 2008, p. 1). One of the key features of an FMIS, therefore, is that it supports both internal and external controls for financial management. Internal controls regulate the cycle of recording, classifying, summarizing and analyzing financial information. The internal audit function helps management in assessing compliance with these controls. External control is exercised through an independent audit carried out by 'the supreme audit unit' i.e. the government auditor (Diamond and Khemani 2005).

The Pacific Association of Supreme Audit Institutions (PASAI), credited by the ADB (2008) for much of the improvement in the timeliness and quality of public auditing in the region, is one of the Regional Working Groups of the International Organization of Supreme Audit Institutions which sets the international standards for external government auditing. It aims to "improve government auditing worldwide and enhance professional capacities, standing and influence of member SAIs" (usually the Auditor General) in their respective countries (IOSAI 2015). PASAI's specific focus is on improving transparency and accountability among member nations and its primary activity is a Regional Institutional Strengthening Plan. In this endeavor, it works in partnership with PFTAC, holding regional workshops for government auditors (internal and external) to enhance capacity and foster reforms and the adoption of international accounting standards. It also provides technical assistance in audit procedures to individual member countries (PFTAC 2015). In some countries a module of the FMIS is adequate for carrying out the required audit function. In others, proprietary software such as TeamMate is also used. Effective auditing procedures are also of considerable interest to development partners, as the OECD DAC's Good Practice Guide for SAIs (published to assist development partners in their ODA choices) makes clear:

> Well-functioning SAIs can play an important role in confirming that controls are operating effectively, identifying waste and suggesting ways in which government organizations can operate better... By ensuring that money is well spent they can contribute to sustainable national development, reduce fraud and contribute to poverty reduction. (OECD 2011, p. 11)

6.10 Capacity Building and Leadership

Undoubtedly, many factors are putting increasing pressure on Pacific governments to ensure that their revenue, budgetary, reporting and audit systems meet international standards. In addition, a considerable number of international and regional organizations are helping them to get there. But among all the pressure that Pacific Island countries face to improve their public financial management and adopt the reforms recommended by the World Bank and the IMF, not the least of which are the demands of development partners and other independent donors, some real gains are being reported; and in plenty of instances highly functional systems, such as DataTorque, FinanceOne, ASYCUDA and even PC/Trade are helping countries achieve their goals for sound financial management. A strong culture of regional cooperation and recognition that country ownership is a significant success factor is helping build this momentum.

But capacity, which includes both individuals with the necessary skills and experience and adequate organizational capacity (i.e. the systems, resources, policies and legislative framework needed to support PFM) remains a critical issue, one that impacts on the sustainability of PFM reforms. As Haque (2013) notes "PFM systems in PIC contexts are often very different from the sophisticated and comprehensive systems operating in larger, wealthier countries," (p1) and PFM reforms thus harder to achieve, constrained by lack of leadership, capacity and resources. Lack of capacity is even more critical in line agencies than in central ministries of finance. With limited staff numbers and high turnover "PFM capacity should be prioritized to areas that matter most in achieving development outcomes, and reforms should be intended to address specific, identified, problems, rather than to achieve blueprint "good practice" (Haque 2013, p. 1). Haque argues that these constraints place such limits on standard strategies for capacity building (workshops, training programs etc.) that ongoing support should be externally accessed or outsourced, and policies of 'capacity supplementation' (delivered through technical advisors, regional institutions, internship schemes) and 'capacity substitution' (placing outside experts in line management roles) should be considered.

But this raises a major issue in relation to country ownership and sustainability. Technical advisors, usually on short term contracts funded by development partners in the region, have long been used in the Pacific, particularly in public financial management. Such contracts can often lead to stop-start progress and changing strategies unless very strong policy and planning frameworks are in place. Outside experts taking key roles in government or governance present an even greater risk of reducing local ownership and leadership. As Haque notes, on occasion "performance in some PFM areas will lag" (p. iv). Leadership, good communications, building capacity and local teams, as well as getting buy-in from "a range of PFM stakeholders" were also identified by Joshi and Moore as key factors in their analysis of lessons learned from PFTAC's extensive involvement in PFM reforms (Joshi and Moore 2010). More recently, lack of leadership was included as a major risk factor in the protracted history of PNG's IFMIS installation (Papua New Guinea 2015). In contrast, leadership was emphasized as a key success factor in the gains Vanuatu has made in both its PFM reforms and in its overall e-government program (Cullen and Hassall 2016b).

More enduring solutions may lie in identifying and fostering leadership within each country that encourages stable teams, securing champions among the political leadership for PFM and other administrative reforms and ensuring that external standards, such as the PEFA Assessment Framework, are realistic and achievable. The most recent report on the PEFA Framework and progress on the PFM Roadmap (PIFS 2015) comments PFTAC's approach to assisting with PFM reform Roadmaps "has placed more emphasis on country ownership through greater participation of PIC officials in their preparation" (PIFS 2015, p. 2).

Acknowledging the need for country ownership is perhaps the reason why PFM reforms and the implementation of new FMIS systems, when part of an institutional strengthening program, have been shown to be more successful, the reforms more sustainable and systems and policy better aligned (as noted in the example of Samoa in this chapter). Institutional strengthening programs, when used well as they have been in several government sectors in Samoa, ensure that capacity building is carried out with local leadership and ownership.

6.11 Centralization Versus Decentralization

The need to involve key stakeholders in PFM reforms raises one further critical issue, about which there is less consensus, the issue of centralization versus decentralization. Decentralization (administrative, fiscal, political, service delivery) is a policy still espoused, which argues that decentralization can "broaden participation in political, economic and social activities in developing countries" (World Bank 2015), while acknowledging that the issue is complex and carries risks. Decentralization of governance is also strongly supported by the United Nations (UNDP 2016). But research carried out on behalf of the IMF suggests that decentralization is effective only under certain conditions, which include an adequate political and institutional environment (Sow and Razafimahefa 2015). It is not necessarily the solution for PICs, partly due to lack of capacity, but largely because of their tiny size and still developing democratic processes and the fact that their central governments are generally not stable enough to avoid the risks of decentralization. As Crocombe argued in 2001, while "decentralization is popular with electors and aid donors" the checks and balances needed to provide the necessary controls in a decentralized system are not necessarily in place in agencies with offices in remote regions where governance systems are largely still traditional and the authority of the state weak (Crocombe 2001, p. 555). This is evident in Solomon Islands, where centralized HR and payroll systems have ensured that teachers and health workers are paid regularly, their attendance at work monitored and non-existent workers eliminated from the payroll, and in Vanuatu where a highly centralized FMIS combines government accounting with procurement, HR and payroll functions (Cullen and Hassall 2016a). In the small fragile economies of the Pacific Island states centralized accounting, revenue, procurement and HR systems are a major force against corruption. Centralized financial management has the best chance of ensuring the most efficient use of staff skills and enabling policy and planning to align with budgets and expenditure.

6.12 Conclusion

The role of information systems in effective public finance management in the Pacific Island countries cannot be overestimated. Without the robust financial and revenue systems now in place in most PICs the necessary PFM reforms which are finally having an impact on the misuse and mishandling of scarce government resources and development partner funds would not have occurred. Gains are also being seen in increases in revenue through taxes and trade (Kazumoto and Wilson 2008). The role of international and regional agencies in this achievement is also clearly very significant, although not always as beneficial as hoped. The tension between needed development assistance and sovereignty can be overlooked at times by agencies wanting to shorten the development process and ensure international standards are followed in all countries. This is not always productive unless development partners are aware of the necessary limits to their intervention.

> Policy-making is best carried out by the state. This includes setting the policy framework, and making decisions about functions or services. For example, while the state may contract out some aspects of budget design, it should not contract out decisions about budgetary allocations or priority setting. These are decisions that should be made by elected officials, accountable to the public. (Haque et al. 2015, p. 58)

But the international development community has learned from past experience. Development assistance in public financial management is now more frequently offered (and accepted) through regional institutions, with regional leaders playing a key role. The adoption by the Pacific Forum Economic Ministers' Meeting of the PFTAC PFM Roadmap is one such example, and many countries have gone on to develop their own roadmap following this initiative. Recent moves to use the PEFA Assessment tool more judiciously, to enable it to provide more useful information to governments and better reflect their information needs also reflects a changing attitude; less of a 'once size fits all' approach and the growing recognition of the need for country ownership is essential if PFM reforms and the ICT applications that enable them are going to be sustainable.

The sustainability of many of the reform processes outlined above is critical to the economic development of PICs. The increasingly successful introduction of what are sometimes sophisticated information systems into core aspects of public administration in PICs is due to several factors: the commitment of PICs to work towards good governance; the willingness of regional and international organizations and advisors to work alongside their colleagues in PICs, to seek to reconcile local practice with international standards, while understanding what is essential to good practice and good governance and what can be culturally determined. Understanding and working with the local information ecology, (see Chap. 1) is at the heart of understanding the factors that ensure the successful introduction of PFM reforms and help development partners (both international agencies and Pacific Island governments) find a path to sustainable change. It is also at the heart of understanding which elements of a Public Finance System should be prioritized, what data Pacific Island governments will need for their own planning and decision-making and what is essential for international compliance and good governance. The three narratives provided in this chapter, of Samoa's PFM Institutional Strengthening Program, Papua New Guinea's HR COnNECT system and Statistics NZ's PC/Trade software, show that where projects take into account the local context (in terms of what information is valued, what technology is available and affordable and what systems it must interact with, what existing processes can be built on, what objectives and values the government and its agencies hold, what capabilities there are in terms of line staff and management systems and what additional resources are available), success and economic gains are more likely.

While international development partners will continue to play a major role in assisting PICs to realize their goal of good governance, in the end, good governance requires ownership and leadership from Pacific governments in the core area of public financial management. That ownership and leadership is starting to emerge and the next decade should see major advances in the economies of the Pacific Island countries and their ability to benefit from these far-reaching e-government initiatives.

References

ADB/OECD Anti-corruption Initiative for Asia and the Pacific (2007) Fighting bribery in public procurement in Asia and the Pacific. Asian Development Bank, Manila. http://www.oecd.org/site/adboecdanti-corruptioninitiative/40838411.pdf. Accessed 5 Mar 2016

Asian Development Bank (ADB) (2008) Strengthening governance and accountability in Pacific Island countries (Phase 2). ADB technical assistance report. Asian Development Bank, Manila.

http://www.adb.org/sites/default/files/project-document/67647/42454-reg-tar.pdf. Accessed 6 Jun 2016

Asian Development Bank (ADB) (2012) Nauru: public financial management reform program. Project number 45032-01, January. ADB concept paper. Asian Development Bank, Manila. http://www.adb.org/sites/default/files/project-document/60115/45032-001-nau-cp.pdf. Accessed 10 Mar 2016

Bucknall J, Allen P, Vaai K (2004) Evaluation of Pacific financial technical assistance centre. International Monetary Fund, Washington. http://www.imf.org/external/np/pp/eng/2004/093004.htm. Accessed 20 Feb 2016

Budden J (2005) E-government in the Pacific Island countries: opportunities and impediments—with an emphasis on the governance of ICTs. University of the South Pacific, Suva

Connors A, Bule H (2016) Vanuatu's new parliament elects Charlot Salwai nation's 11th prime minister. ABC, Sydney. http://www.abc.net.au/news/2016-02-11/vanuatu-new-parliament-elects-charlot-salwai-pm/7160362. Accessed 21 Feb 16

Crocombe RG (2001) The South Pacific. University of the South Pacific, Suva

Cullen R, Hassall G (2016a) E-government in the Pacific island states: ICT policy and implementation in small island developing states. Samoa country report. http://pippr.victoria.ac.nz/handle/123456789/31. Accessed 2 May 2016

Cullen R, Hassall G (2016b) E-government in the Pacific Island states: ICT policy and implementation in small island developing states. Vanuatu country report. http://pippr.victoria.ac.nz/handle/123456789/30. Accessed 10 Mar 16

Cullen R, Hassall G (2016c) E-government in the Pacific Island states: ICT policy and implementation in small island developing states. Solomon Islands country report. http://pippr.victoria.ac.nz/handle/123456789/35. Accessed 10 Mar 16

Davies M, Joshi S, Naidu S (2010) A public financial management roadmap for forum island countries. Pacific Financial Technical Assistance Centre, Suva

Dener C, Watkins J, Leslie W (2011) Financial management information systems: 25 years of World Bank experience on what works and what doesn't. World Bank, Washington

Diamond J, Khemani P (2005) Introducing financial information management systems in developing countries. IMF working paper WP/05/196. International Monetary Fund, Washington

eGovernment for Development (2015) What is eGovernment? [Website]. http://www.egov4dev.org/success/definitions.shtml, 10 Mar 2016

Eschenfelder KR, Beachborder JC, McClure CR, Wyman SK (1997) Assessing federal US web sites. Gov Inf Q 14(2):173–189

Gray C (2000) Reforming public institutions and strengthening governance: a World Bank strategy. World Bank, Washington. http://www1.worldbank.org/publicsector/civilservice/Strategy.pdf. Accessed 5 Mar 2016

Haque T (2013) Planning public financial management reforms in Pacific Island countries: guidance note. World Bank, Washington. http://documents.worldbank.org/curated/en/2013/01/17886478/planning-public-financial-management-reforms-pacific-island-countries-guidance-note. Accessed 10 Mar 2016

Haque TA, Knight D, Jayasuriya D (2015) Capacity constraints and public financial management in small Pacific Island countries. Asia Pac Policy Stud 2(3):609–622

Heeks R (2003) Most e-government-for-development projects fail: how can risks be reduced? Institute for Development Policy and Management, Manchester. http://unpan1.un.org/intradoc/groups/public/documents/cafrad/unpan011226.pdf. Accessed 12 Feb 2016

Heeks R, Bhatnagar S (1999) Understanding success and failure in reform. In: Heeks R (ed) Reinventing government in the information age: international practice in IT-enabled reform. Routledge, London, pp 49–71

International Organization of Supreme Audit Institutions (IOSAI) (2015) http://www.intosai.org/about-us.html. Accessed 3 Mar 2016

Joshi S, Moore J (2010) IFMIS systems in Pacific Island countries—lessons learned. Pacific Financial Technical Assistance Centre, Suva. http://www.mfed.gov.ki/sites/default/files/IFMIS%20Lessons%20Learned%20in%20PICs.pdf. Accessed 3 Mar 2016

Kazutomo A, Wilson JS (2008) Governance, corruption, and trade in the Asia Pacific region. Policy research working paper no. 4731. World Bank, Washington. https://www.openknowledge.worldbank.org/handle/10986/6961. Accessed 29 Feb 2016

Larmour P, Barcham M (2006) National integrity system in small Pacific states. Publ Adm Dev 26:173–184

Lehr W, Lichtenberg FR (1998) Computer use and productivity growth in US federal government agencies, 1987–92. J Ind Econ 46(2):257–279

Murray B, Abrams R, Vaai K (2009) Independent evaluation of the Pacific Financial Technical Assistance Centre: final report. PFTAC, Suva, Fiji. www.pftac.org/filemanager/files/FinalPFTACEvaluationReport_PhaseIV.pdf

Murray K, Oliver R, Wyatt S (2014) Evaluation of taxation reform in the Pacific. Ministry of Foreign Affairs and Trade, Wellington

New Zealand Aid Programme (NZAID) (2015) New Zealand aid programme strategic plan 2015–2019. Ministry of Foreign Affairs, Wellington

OECD (2003) The e-Government imperative. OECD, Paris

OECD (2011) Good practices in supporting supreme audit institutions OECD-DAC. OECD Publishing, Paris

Pacific Financial Technical Assistance Centre (PFTAC) (2015) Quarterly report, December 2015

Pacific Islands Forum Secretariat (PIFS) Forum Economic Ministers' Meeting (2000) Progress report on the implementation of the forum's eight principles of accountability. Pacific Islands Forum Secretariat, Suva, Fiji

Pacific Islands Forum Secretariat (PIFS) Forum Economic Ministers' Meeting (2012) FEMM biennial stocktake 2012. Pacific Islands Forum Secretariat, Suva, Fiji. http://www.forumsec.org/resources/uploads/attachments/documents/FEMK.05%20-%20Biennial%20Stocktake%202012.pdf

Pacific Islands Forum Secretariat (PIFS) Forum Economic Ministers' Meeting (2015) Public financial management reforms. Pacific Islands Forum Secretariat, Suva, Fiji. http://www.forumsec.org/resources/uploads/attachments/documents/2015FEMM.10_Public_PFM_Reforms.pdf. Accessed 16 June 2016

Papua New Guinea (2015) Review of the integrated financial management system (IFMS) project. Office of the Auditor General, Waigani. http://www.ago.gov.pg/images/Report_No5_IFMS.pdf. Accessed 10 Mar 2016

Sen AK (1987) On ethics and economics. Basil Blackwell, Oxford

Siwatibau S (1997) Pacific islands economic management: challenges for the 21st century. In: Kumar M (ed) Islands in the Pacific century. VIII pacific science inter-congress, July, 1997. University of the South Pacific, Suva, Fiji, pp 48–59

Sow M, Razafimahefa IF (2015) Fiscal decentralization and the efficiency of public service delivery. IMF working paper WP/15/59. International Monetary Fund, Washington. https://www.imf.org/external/pubs/ft/wp/2015/wp1559.pdf

Sprietzer D, Vargas M (2011) Using country public finance management systems: a practitioner's guide. Prepared for the 4th high level forum on aid effectiveness, Busan, Korea, 29 Nov–1 Dec 2011. OECD, Paris

Thompson LJ (2009) The moral compass: leadership for a free world. IAP, Charlotte, NC

Transparency International (2016) Corruption perceptions index 2015. http://www.transparency.org/cpi2015. Accessed 16 Feb 2016

UNCTAD (2011) Financial implications of the introduction of the ETIR international system: informal input paper of UNCTAD ASYCUDA program to the 18th session of the UNECE informal ad-hoc expert group on conceptual and technical aspects of computerization of the TIR procedure. Informal document GE.1 No. 2. Economic Commission for Europe, Brussels. https://www.unece.org/fileadmin/DAM/trans/bcf/adhoc/conc_tech/documents/ECE-TRANS-WP30-GE1-2011-inf02e.pdf. Accessed 25 Feb 2016

UNDP (2012) UN Pacific regional anti-corruption (UN-PRAC). Project (2012–2016). Project document. dfat.gov.au/about-us/publications/Documents/project-document-unodc-undp-un-pacific-regional-anti-corruption-project-un-prac.pdf. Accessed 19 Feb 2016

UNDP (2016) Decentralized governance for development: a combined practice note on decentralization, local governance and urban/rural development. United Nations Development Program, New York. http://www.undp.org/content/undp/en/home/librarypage/democratic-governance/local_governance/decentralised-governance-for-development-a-combined-practice-note-on-decentralisation-local-governance-and-urban-rural-development.html. Accessed 11 Mar 2016

World Bank (1997) World development report 1997: the state in a changing world. https://openknowledge.worldbank.org/handle/10986/5980

World Bank (2000a) Reforming public institutions and strengthening governance. A World Bank strategy. http://www1.worldbank.org/publicsector/Reforming.pdf

World Bank (2000b) Computerizing tax and customs administration. PREM notes, public sector 21600, Oct 2000, no 44. http://www1.worldbank.org/prem/PREMNotes/premnote44.pdf

World Bank (2015) Decentralization topics: what, why and where. http://www1.worldbank.org/publicsector/decentralization/what.htm. Accessed 4 Mar 2016

World Customs Organization (WCO) (2015) Annual report 2014–15. http://www.wcoomd.org/en/topics/nomenclature/overview/~/~/media/WCO/Public/Global/PDF/Topics/Research/Annual%20reports/WCO_AnnRep_EN_2014_2015.ashx. Accessed 25 Feb 16

Chapter 7
E-Government in Support of the Institutions of Democracy and Justice in Pacific Island States

Rowena Cullen

Abstract A critical aspect of e-government is how countries use ICT to support the institutions of democracy and the rule of law to facilitate citizens' access to information about these institutions. This chapter discusses how Pacific Island countries (PICs) are applying ICT to further this goal, with the support of regional organizations and development partners. It outlines how PICs are using e-government applications to improve their parliamentary process and provide citizens with information about parliament and the legislative process and how Pacific parliaments are using the web to communicate this information. The chapter discusses the use of ICT in the electoral process and the benefits and challenges these new technologies present. The contribution of justice information systems and the critical role of PacLII in providing access to information about legislation and court records in Pacific Island states are also discussed. Finally, the importance of government record-keeping and how ICT is used to ensure access to historic and current records is considered. The chapter explores some of the organizations and strategies helping to promote the use of ICT in democratic processes and concludes that the collaborative, knowledge sharing model that pervades this sector is an effective and sustainable model of e-government for development in the Pacific.

7.1 Introduction

The Pacific Islands Forum leaders' Auckland Declaration of 2004 articulated a vision which included the statement, "We seek a Pacific region that is respected for the quality of its governance, the sustainable management of its resources, the full observance of democratic values and for its defense and promotion of human rights" (PIFS 2004). In Chap. 6 we considered the role of sound financial management as the predominant element of good governance and the use of information and communication technologies (ICTs) to improve transparency and the effective

R. Cullen (✉)
Victoria University, Melbourne, Australia
e-mail: Rowena.cullen@vuw.ac.nz

R. Cullen and G. Hassall (eds.), *Achieving Sustainable E-Government in Pacific Island States*, Public Administration and Information Technology 27,
DOI 10.1007/978-3-319-50972-3_7

management of resources in the region. In this chapter, we turn to the observance of democratic values and the role of ICTs in ensuring these, including the democratic principle of the rule of law, also embraced by the Forum under its Human Rights and Security policies and in the Framework for Pacific Regionalism adopted in 2014 (PIFS 2014).

The three pillars of democracy are often depicted as the legislature, the executive and the judiciary, with the media forming a fourth pillar as 'the fourth estate'. These pillars are especially relevant to this chapter which explores not only the use of ICT to support democratic processes and the work of the judiciary, but also the use of ICT to disseminate information about these activities, with or without the assistance of the media. The right to information, which UNESCO defines as the right to access information held by public bodies, "is an integral part of the fundamental right of freedom of expression" recognized in Resolution 59 of the UN General Assembly adopted in 1946. UNESCO also notes:

> The free flow of information and ideas lies at the heart of the very notion of democracy and is crucial to effective respect for human rights. Democracy demands that individuals are able to participate effectively in decision making and assess the performance of their government. … Freedom of information thus contributes to government openness and accountability, and represents an important instrument to prevent and combat corruption. It can also help increase government efficiency and responsiveness, along with civic trust. … [Freedom of information] can be curtailed by burdensome mechanisms for information access and weak enforcement, the arbitrary use of exceptions or reference to other legislation to deny public information access, the bad state of record-keeping and archive management systems, and poor monitoring of the law′s implementation. (UNESCO 2016)

This chapter will therefore discuss the use of ICT in parliamentary and electoral processes, especially in disseminating information about parliamentary and legislative processes via the Internet and the use of ICT in the justice system and in government record-keeping. It will explore how ICT can play a major role in improving these crucial aspects of governance and look at some of the strategies and organizations helping to promote the use of ICT in democratic processes.

7.2 Parliamentary Processes Informed and Supported by ICTs

The parliaments of the Pacific Island countries (PICs) were established by constitution with the approval of both their departing colonial governments and the newly-installed political leadership at the time of independence. They mirror, consequently, either the Westminster or the republican model of governance with which the departing colonial powers were familiar, and incorporate the strengths and weaknesses inherent in both. They were imposed with minimal consultation, replacing the limited powers of the existing legislative assemblies of the pre-colonial period (Larmour 2002). As Hassall (2012) notes, because these parliamentary systems were imposed quite rapidly rather than developing locally and over a longer time,

> some of their characteristics are defined by law rather than convention so as to ensure the existence of practices which might otherwise take a much longer time period to settle on … [T]he roles of 'the government', 'the opposition', and political parties have in some places been legislated rather than explored through practice. Provisions about the offices of speaker and attorney-general, as well as the major parliamentary committees, are also detailed in some constitutions. (p. 215)

In many cases these 'provisions' mandated independent funding to ensure the roles were adequately supported.

The current functioning of Pacific parliaments is influenced by their history and by other factors: the extremely limited resources available which affects the availability of accommodation and support services for ordinary members; the 'archipelagic' nature of the territories to be governed (including the diversity of cultures within these islands, the remote location of many of them and the fact that daily living is based on traditional forms of governance and gender specific roles); and the lack of any tradition of political participation, of political parties or the development and implementation of policy (Hassall 2012). PICs therefore struggle to deal with the inevitable conflicts between traditional forms of governance and authority and introduced democratically elected positions of authority such as ministers, prime ministers—and even parliamentarians—as well as senior public servants, especially permanent Heads of Department and CEOs.

Serious problems persist concerning the functioning of Pacific Island parliaments. This is a matter of concern to the United Nations Development Program (UNDP). It runs substantial UNDP regional and national education programs from the UNDP Pacific Office for Pacific parliamentarians and parliamentary officers to encourage a better understanding of the parliamentary process that PICs have inherited and (in recent years) to encourage the inclusion of more women in parliament (UNDP Pacific Office 2016). In this context, and in "the absence of strong civil society, public interest advocates, and mass media" (Hassall 2012, p. 222), access to information about parliament, its procedures and its decisions, becomes an essential part of the parliamentary process and must be provided by government itself. Given the distance of many Pacific citizens from their parliament, dissemination of this information via the Internet is the only real option, even though this leaves such access dependent on the limited bandwidth of telecommunications in most PICs.

While ICT has a vital role to play in the effectiveness of parliamentary democracy in the Pacific, its adoption in this aspect of government has been piecemeal and driven by a variety of factors, such as resources (funding and skills capacity), the views of parliamentarians, the priorities of development partnerships and technological infrastructure. Even the use of standard technologies such as office software, Internet access and email cannot be taken for granted and many parliamentarians resist pressure to use ICT in their daily work.[1] In addition, support for technology in parliaments is limited. Although the most recent World e-Parliament Report (2012) indicated that "parliaments in the lowest income level are closing the technology gap" (p. viii), the report acknowledged that "many parliaments still face substantial obstacles in their efforts to enhance the state of ICT to support the work and purposes of the institution" (Global Centre for ICT in Parliament 2012, p. viii). The provision of personal computers to members, Internet access, ICT support, web publishing and network operations are lacking in many of the parliaments studied in the survey, limiting their ability to provide timely information and documents to members and to disseminate information to citizens. Lack of resources and skilled staff were identified as key reasons for these deficiencies, along with a 'technology gap' and a 'knowledge gap' between parliament and citizens.

Despite the fact that no Pacific Island country participated in the survey, the situation in Pacific parliaments is very similar. To remedy this a range of targeted projects, largely funded by the UNDP, and New Zealand, have seen the introduction of systems for electronic drafting of Bills, Acts and Regulatiocns (or Resolutions), the recording of parliamentary sessions for Hansard (the official verbatim record of the proceedings of parliament)[2] and the proceedings of committees, transcription software to create the published record, parliamentary intranets and Wi-Fi to support the sharing of knowledge and online access in the House to official documents (such as Standing Orders and Bills), access to Parliamentary Library resources and online interaction with citizens. In Solomon Islands, for example, electronic submissions to select committees is part of the e-Parliament strategy (Cullen and Hassall 2016a), a key focus of the National Parliament of Solomon Islands Strategic Plan (Solomon Islands 2012). But whereas Solomon Islands, with considerable international and regional support and a large number of ICT staff, is introducing most of the technologies above, resource constraints and lack of human capacity limit what other countries in the region can achieve. Vanuatu's sole Parliamentary ICT staff member manages the Parliamentary website and Parliament's ICT systems, wireless Internet, technical support and training for MPs and staff and the tablets provided for MPs. Producing Hansard in Vanuatu therefore remains a manual process (although like all Vanuatu official documents it is published in English, French and the local pidgin Bislama) and only sessions up to 2012 are available online. Negotiations began in 2015 to acquire and install a Legislative Management Information System (LMIS) to make Bills, Acts, Order papers and other parliamentary papers available online to members via their laptops or smart phones and ensure that members have read and considered Bills that come before the House (Isno 2015). Cook Islands, by contrast has made a deliberate

---

[1]It should be acknowledged that this would have been the case in parliaments of many industrialized nations 25 years ago.

[2]Hansard is the traditional name for transcripts of the British Parliament; the name is widely adopted by countries in the Commonwealth of Nations (which a majority of independent Pacific Island states belong to), and commonly used in the Pacific region.

decision not to make Bills, Acts, Regulations and other parliamentary papers available online at this time, being reluctant to forgo the small income stream it gets from selling printed copies in the country and internationally. Its population of 12,000 is mainly found on the island of Rarotonga, and published Acts, Bills etc. are readily available for reading at Parliament House.

7.2.1 Parliamentary Websites

The well-established systems for ensuring access to information about parliament and its business in industrialized nations, previously through the deposit of 'parliamentary papers' in public and university libraries and now through online access to documents published on a parliamentary website, has not been easy to duplicate in PICs and has led in the past to serious gaps in citizens' access to information. Equally important for functional democracy and the accountability of parliaments is the issue of timeliness. Unless information is made available when decisions are being made and those making decisions have access to all necessary information, democracy is not well served. But the lack of structures through which to make information available (the 'technology gap' and 'knowledge gap' identified by the 2012 World e-Parliament Report) has hindered the development of participatory democracy in the region and slowed the progress of these emerging democracies.

> The foundation of democracy is an informed constituency that is able to thoughtfully choose its representatives on the basis of the strength of their record and one that is able to hold their government accountable for the policies and decisions they promulgate … Democracy is enhanced where people meaningfully engage with their institutions of governance and form their judgments on the basis of facts and evidence, rather than empty promises and meaningless political slogans. Where people do not have access to information, as has so often been the case in the Pacific Islands states, voters will often fall back on ethnic, religious or geographic affiliations when choosing a candidate. (Mistry and Rodrigues 2007, pp. 93–94)

Given the lack of a tradition of libraries and civic engagement through print and the remoteness of the majority of many Pacific Island populations from the seat of government, the Internet has the potential to transform citizens' access to information about the parliamentary process and the laws and policies being enacted in their name. In particular, the use of websites for disseminating information about such matters, and websites designed for access by mobile technologies (smart phones and small tablets) is essential to the delivery of participative democracy in the Pacific Island countries in this study.

A global survey undertaken by the Inter-Parliamentary Union (IPU) in 2000 indicated that only 29% of the Pacific parliaments had websites, a figure which had risen to approximately 40% by the end of 2006, or nine of the current 22 members of the Pacific Community (which includes sovereign states, dependencies and territories). In an independent study of the content of these Pacific parliamentary websites, Hassall (2007) noted the need to weigh the considerable cost of setting up and maintaining a parliamentary website compared with the cost of providing access to printed copies of Bills, Acts and Hansard, or when balanced against the need for ICT support for members and staff. In these small jurisdictions such choices must often be made.

At the time of going to press not all PICs and territories have what could be described as parliamentary websites. The larger and more developed independent states (Papua New Guinea, Fiji, Solomon Islands, Vanuatu, Samoa, Kiribati, Tonga and the Federated States of Micronesia) all have some form of parliamentary website. So do some of the larger territories (French Polynesia, New Caledonia, the Northern Mariana Islands and Guam). But the smaller independent states struggle to maintain a web presence for their government, let alone their legislative assembly (which may be called a Parliament, Fono, Congress or carry a name based on the country's language, e.g. the Maneaba Ni Maungatabu in Kiribati or the Nitijela in the Marshall Islands). Among the smaller countries with populations of less than 20,000, Cook Islands and Nauru manage a Parliamentary web presence, even if irregularly updated, but tiny nations such as Niue and Tokelau (with around 2000 and 1000 inhabitants respectively)[3] do not attempt to maintain websites for their legislatures. (The Tokelau government website refers to the General Fono which shares power with the Council for Ongoing Government (the main administrative department) which lists Ministers and their portfolios, and notes Decisions taken by the General Fono on its government home page.) The website of Palau's National Congress (which comprises the Senate and the House of Delegates) is currently under construction.

The current state of the websites of the independent Pacific Island states are surveyed in Table 7.1, using abbreviated criteria based on the International Parliamentary Union (IPU)'s *Guidelines for Parliamentary Websites*, which include:

- information on access to parliament including visitor information and educational tours, seating arrangements in the debating chamber and a virtual tour;
- history and role of the parliament, responsibilities of the legislature and the text of the constitution or other founding documents;
- functions, composition and activities, events, sitting days, reports, statistics and press releases; elected leaders (Speaker and Deputy), powers, biodata and photograph;
- list of parliamentary committees, commissions and other non-plenary bodies, responsibilities, membership, report on activities;
- list of members of Parliament, biodata and photo, constituency and political party affiliation along with roles, committee membership etc., contact information and email address, activities of current members and lists of previous members;

[3]Tokelau is a non-self-governing territory of New Zealand, with a resident NZ Administrator, but defines itself as "largely self-governing with a system of government based on traditional village leadership".

- political parties represented in parliament and links to their websites; elections and electoral system including detailed results of recent and previous elections; administration of parliament, description of key roles and vacancies;
- publications, documents and information services, including where to obtain documents if not online, information about parliamentary library and services;
- general links to other relevant websites, ministries, other government agencies, Constitutional and Supreme court, etc. and the IPU website (IPU 2009).

Not all of these are relevant to all PICs—some do not have a system of political parties; others do not yet have websites for courts. In other PICs the technology available to citizens does not make the use of virtual online tours a high priority for the country. Financial resources, staffing levels and the technology available present challenges in keeping information up-to-date online, as will be seen. Hence a more conservative, but essential set of criteria are applied in Table 7.1.

This table shows that only 12 of the independent member states of the Pacific Community (SPC) have what could be described as a functional Parliamentary website and that few cover all the expected core elements. Many are several years out of date. In particular, information about sitting dates and the business to be considered on those days (the 'Order Paper') is rarely available and direct contacts for MPs, committee reports, other parliamentary papers and online access to the ParliamentaryLibrary (which might fill some of these gaps) is very limited. Hansard (which might be expected to be available at least for the previous year) is also rarely available online. But the creation of Hansard and keeping a journal record of the proceedings of parliament is a labor intensive skilled task; in the past it required high-level shorthand and editing skills. When technology is used to record and transcribe debates specialized software is needed to create an appropriate record and a considerable amount of training is involved. Instead, many countries, while committed long term to the Hansard concept, have adoptedlive-streaming audio/video in the interim, which is accessible via smart phone or tablet. Where such video recordings are archived and remain available they can be useful in providing access to the business of the house, but they are no substitute for the permanence and searchability of a Hansard report. The lack of information available in most jurisdictions and the lack of timeliness of the information available in most of the others, while due in large part to lack of resources, is still a major brake on good democratic process in Pacific Island states.

To the extent that they can, Pacific parliamentary libraries play a key role in allowing citizens access to parliamentary papers, even if they are unable to support a web presence. Thirteen of the 15 SPC independent Pacific member states are members of the Association of Parliamentary Librarians of Asia and the Pacific, which is focused on the role of parliamentary libraries in support of democracy and which is engaged in training and knowledge sharing in the region (APLAP 2016). Their role, in both providing research services to MPs and parliamentary staff and in providing access to parliamentary publications to citizens is a crucial one; they, along with the Pacific Legal Information Institute (PacLII), which makes all Pacific

Table 7.1 Summary of content of Pacific Island Countries' parliamentary websites, using IPU criteria

Country[a] (population)	MP's use of ICTs	Parliament structure and processes, constitution, standing orders, tours etc.	Information on MPs, affiliations, contacts	Order papers, journals and sitting days notified	Legislation (acts, bills, regs.) online	Hansard or other record online	Committees and reports	Other parliamentary papers and reports	Parliamentary library
Cook Islands (14,300) http://www.parliament.gov.ck/index.html	MPs have MS Office, Internet and Wi-Fi in the House	Brief information about Parliament given. Constitution and Standing orders online. Speaker not named, but Clerk's name, address and email address given. No tours mentioned	Name and party affiliation. No individual contacts	Not found	Acts, bills and regulations listed, with price. Not accessible online	Described and available in print at Parliament House	Committees described and membership listed but no reports available	Not found	Not found
Fed States Micronesia (FSM) (102,800) http://www.fsmcongress.fm	Not known	No information currently available ('History' tab not functioning); Speaker and Vice Speaker's names and photos provided	Name, electorate and photograph only (14 members only)	Sitting days online for 2014 only	Acts, bills and resolutions online line up to August 2016	Not available	Up-to-date online for 2016. Committee members listed in reports	Not found	Not found
Fiji (867,000)	MPs and staff have MS Office, Internet	Full information about history and	Name, party and photograph only	Sitting days notified and		Hansard, journals and	Committee membership and		Research and library

(continued)

Table 7.1 (continued)

Country[a] (population)	MP's use of ICTs	Parliament structure and processes, constitution, standing orders, tours etc.	Information on MPs, affiliations, contacts	Order papers, journals and sitting days notified	Legislation (acts, bills, regs.) online	Hansard or other record online	Committees and reports	Other parliamentary papers and reports	Parliamentary library
http://www.parliament.gov.fj	and access to GovNet through Parliamentary web site	function of Parliament online, with seating plan. Tours, education program, constitution, standing orders available online	(ministerial roles noted). No contact details	Order papers up-to-date online	Acts, bills and resolutions up-to-date online	voting records of members available and up-to-date online. Live streaming video on sitting days	reports online, but currently unavailable	Range of annual reports available	services noted on website, email link available for MPs to make requests
Kiribati (113,400) Website not available but cached site at http://www.maneaba.net/constitution.html indicates it usually provides the items noted	Standard ICTs and knowledge sharing platform provided by UNDP	About Parliament pages available, content not known. Constitution available on cached site. Link to standing orders broken	Members pages available, content not known	Not known	Bills/acts pages available, content not known	Hansard pages available, content not known	Committees pages available, content not known	Not found	Not found
Marshall Islands (54,900) http://rmiparliament.org/cms/	Standard ICTs and knowledge sharing platform provided by UNDP Parliamentary staff also receive ICT training from SPC	Brief history of the Nitijela (parliament) and roles of Speaker, Vice Speaker, Clerk and Legislative (Parliamentary) Counsel, and incumbents noted. Constitution	MPs listed, some link to personal page which lists electorate and brief bio. Ministerial roles noted. Party affiliations not noted	Session calendar and order papers to mid-year 2016 available under 'Library' tab	Acts and Resolutions available online	Live streaming on sitting days, recorded	Committees and membership listed but no reports available	Not found	Not found. 'Library' is tab heading for Parliamentary papers (Order papers and journals), legislation etc.

(continued)

Table 7.1 (continued)

Country[a] (population)	MP's use of ICTs	Parliament structure and processes, constitution, standing orders, tours etc.	Information on MPs, affiliations, contacts	Order papers, journals and sitting days notified	Legislation (acts, bills, regs.) online	Hansard or other record online	Committees and reports	Other parliamentary papers and reports	Parliamentary library
		available through search facility							
Republic of Nauru (10,800) http://www.naurugov.nr/parliament-of-nauru.aspx	Standard ICTs in use in MPs and staff offices, with Internet access (through UNDP strengthening project 2009–)	Summary information, constitution and standing orders online	MPs listed, with electorate and brief bio. Ministerial roles noted. Members of previous parliament listed	'Notice' papers, Votes and proceedings available online up until June 2013	Bills available online up to June 2013 through RonLaw database. Acts up to 2016 available through RonLaw database	Votes and proceedings available online up to 2013. Hansard being digitised. Online up to 2007	Committees and members listed, as at 2013, but no reports available	Only Annual report of the PCO—up to 2013	Parliament has a Library Committee for the Library and Research and Information Service
Niue (1500) http://www.gov.nu/wb/pages/parliament.php (In free association with New Zealand since 1974)	Standard ICTs available. Parliamentary staff also receive ICT training from SPC	Web page for Parliament (Fono Ekepule) on government web site. Brief description of political system (Constitution online on the NZ Legislation web site. Standing orders available as part of Niue Consolidated Laws)	Cabinet, responsibilities of Ministers and MPs listed, with constituency. No contact details, but ministry email contacts online. This includes Director of Cabinet and parliamentary services	Not available	Consolidated and sessional Acts and regulations available online up to 2014. Bills before parliament sometimes noted in government news	Not available	Not available	Not available	Not available

(continued)

Table 7.1 (continued)

Country[a] (population)	MP's use of ICTs	Parliament structure and processes, constitution, standing orders, tours etc.	Information on MPs, affiliations, contacts	Order papers, journals and sitting days notified	Legislation (acts, bills, regs.) online	Hansard or other record online	Committees and reports	Other parliamentary papers and reports	Parliamentary library
Palau: republic with elected president (21,000) http://www.palauoek.com	Not known	History and composition of the National Congress (parties not part of system) Mission statement	Photo, email, brief bio, role and phone number of each member	Not available. Records of sitting days, bills considered and voting online to early 2014	Bills and Senate Joint Resolutions online to 2014. (Sessional legislation up to end of previous year on PacLII)	Out-dated 'Archived Posts' (news) available online	Committees and membership listed	Not available	Not available
Papua New Guinea (7,744,600) http://www.parliament.gov.pg	Basic ICT services for members and staff, internet access, and Intranet for document and knowledge management	Brief description and history of Parliament. Sitting dates notified, Constitution link broken, Standing Orders not available. Tours advertised online	Members, parties and ministerial roles noted, bios and email contact. SMS messaging invited to find out 'what your MP is doing' Links to party websites inactive	Order paper (Notice papers) and Questions for written answers avail online on the day. 'Minutes' for previous year available online	Acts available online up to end of previous year. Regulations and current Bills, not available online	Hansard up-to-date online	Committees and membership listed. Meetings dates and minutes not available	Other Parliamentary papers (referred to in Notice papers) not available online	Online link to Information Support Division (Library ICT support and telephones)
Samoa (187,300) http://www.palemene.ws	ICT services available to MPs in office and parliamentary library online access to library resources and documents	Info sheets on aspects of parliament available online, roles of Speaker and Clerk outlined.	Members listed, with constituency and initial of party. One general phone number given. Members of earlier	Order papers available (in Samoan and English) up to end of previous year available online, as well as Procedural Digest, 'Daily	Acts and regulations up-to-date online. Bills available up to end of previous year	Hansard available online for previous sessions, in Samoan, and in English at a later date. Audio	Committees and membership can be found on search (no tab). Committee reports tabled available up to end of previous session	Annual reports for ministries and SOEs available online	No information found although information management services listed in organizational

(continued)

Table 7.1 (continued)

Country[a] (population)	MP's use of ICTs	Parliament structure and processes, constitution, standing orders, tours etc.	Information on MPs, affiliations, contacts	Order papers, journals and sitting days notified	Legislation (acts, bills, regs.) online	Hansard or other record online	Committees and reports	Other parliamentary papers and reports	Parliamentary library
		Standing orders cited but not available online	parliaments listed. Parties tab has link to ne party's website	progress' and Journals. Sitting days not available		recordings available for current session while Hansard is being prepared			structure chart and references to the library can be found
Solomon Islands www.parliament.sb (641,900)	Well-funded ICT Unit, Intranet/Internet access for MPs and staff, training provided. Tours and civic education noted	Brief description and history of Parliament. Law-making process, role of committees, and Officers Constitution and Standing orders available online (with updates). Session dates not noted	Current members, parties and ministerial roles noted, bios and phone numbers	Current Order papers (and business papers) avail online, Minutes up to end of previous year	Acts avail online to end of previous year. Bills avail but not current, committee schedule out of date	Hansard available online up to end of previous session	Committees listed, with membership. Bills under consideration and committee reports online	Press releases and news	Online link to library, role, and access to parliamentary publications including committee reports (not up-to-date). Access for MPs to library online catalogue
Tonga (103,300) Constitutional monarchy Parliament includes appointed nobles and	Standard office software available	Description of how parliament functions, history, accountability measures, visits and education programme.	Photo, constituency and brief biographical information provided by member; contact details if member provides them	Sitting days noted online. Order papers available up to end of previous year. Journal online but not up-to-date. News	Parliament website describes legislative process, link to Tonga legislation website (corrupted and inactive).	Online and up-to-date in Tongan and English. Video of specific events available.	Standing and select Committees listed, no reports available	Available in Parliamentary library	Library services described, research and resources for members, and public access. Aiming to have all

(continued)

Table 7.1 (continued)

Country[a] (population)	MP's use of ICTs	Parliament structure and processes, constitution, standing orders, tours etc.	Information on MPs, affiliations, contacts	Order papers, journals and sitting days notified	Legislation (acts, bills, regs.) online	Hansard or other record online	Committees and reports	Other parliamentary papers and reports	Parliamentary library
elected members www. parliament.sb		Constitution and standing orders available online Sitting days noted online	(No party system)	(including information about sitting days) is current	Cur-rent legislation, consolidated and sessional avail on Tonga Crown Law website				parliamentary papers avail online in the long term
Vanuatu (277, 500) https:// parliament. gov.vu	PCs and standard office software, and training available for all MPs. Wi-Fi in debating chamber gives access to web site only. Tablets supplied to MPs	Political system and its history described. (English only). Role of Clerk and Speaker, and legislative process, outlined. Constitution and Standing orders available on PacLII (link is on website) but this is not stated	Members of current and past parliaments listed, with electorate and party noted, and role (ministry or committee membership). Approx 2/3 link to personal page with photo, bio and phone contact	'News' announces start of a session. No schedule for sitting days available. Order papers not available	Acts (sessional and consolidated) in English and French available on PacLII website. Titles only of Bills available on website up to 2012	Hansard (usually both English and French) available to download to end 2011. Live streaming on session days	Committees and membership provided. No reports available	Website notes limited range of documents available for public to read in library	Library's role and collections described. Email address for librarian given

[a]Only independent Pacific Island states are included here. NB: Cook Islands and Niue are self-governing states in free association with New Zealand which has responsibility for their defense and foreign affairs. FSM is in free association with the USA which is responsible for its defense. Other Pacific Island territories are not included although most do have either a parliament or a governing council. (Population data sourced from SPC website.) NB: Not all countries use the party system

legislation (Acts, Regulations and court information) it can obtain available on its website, to some extent can fill the gap in essential information available to citizens in many countries. But access to Bills before the house, sitting days and order papers are essential tools of democracy and it is disappointing to see them not more widely available on Pacific parliamentary websites.

7.2.2 Barriers to the Use of ICTs in Pacific Parliaments

As Table 7.1 shows, the use of parliamentary websites to disseminate core information recommended by the Inter-Parliamentary Union's *Guidelines* is patchy, and the lack of currency in the information provided is a signal of the difficulties PICs experience in maintaining them once launched. Effective use of technology in parliaments and the sustainability of technological innovations are all affected by the same issues. The lack of financial resources and qualified personnel, the cost of maintaining software, the need to train parliamentary staff and members, the need for local champions,[4] and the difficulties of maintaining hardware in the tropical environment of most Pacific SIDs are all challenges to be overcome. Vendor support for ICT hardware is a major issue, while having hardware which can be supported by local firms can be a factor in success even though it may come at a higher cost (Lomp 2013). Staff turnover is high (and sometimes changes dramatically with the election of a new government), positions remain vacant for lengthy periods and retraining is constantly needed (Hassall and Talifu 2015). Where technical assistance is provided through development partnerships (whether for reforming parliamentary procedures, the production of Hansard, library systems or staff in the Clerk's or Speaker's office), projects often stall when that technical assistance comes to an end, unless a commitment is made to ongoing staffing (Hassall and Talifu 2015). Funding for software or technical support can run out, so to save licensing fees open source software may be used, as in Vanuatu where the parliamentary ICT staff member uses Joomla to create a fairly comprehensive parliamentary website [which, since it is using Joomla, can be read in over 100 languages, although substantive content is only in English and French (Cullen and Hassall 2016b)].[5] Political factors may also come into play. For example, information on Nauru's website was frozen after the election of a new government in 2013 following which many expatriate staff and advisors, had their visas withdrawn or their contracts not renewed (Johnson 2015; Garrett 2016). It has since been

[4]Even where there is a champion, maintaining momentum in making information available is a challenge. The Speaker of Papua New Guinea's House of Representatives has launched a program of Reformation, Restoration and Modernization, and has made a commitment to greater community access to information about the activities of the parliament (which has had considerable UNDP support). But despite the fact the PNG is the largest and wealthiest of the Pacific Island countries in our study there is still significant information missing from its website (such as current Bills).

[5]The Vanuatu lingua franca Bislama, is not supported by Joomla.

updated and the RonLaw database continued to hold current legislation throughout.[6]

Several reports make clear that the introduction of ICT involves considerable change in the way the business of the parliament is conducted and is most effective where it is accompanied by a substantial review of procedures, workshops on parliamentary process and democracy and a well-managed reform or strengthening program (Lomp 2013). Such ongoing training has been a major feature of some development partnerships in the region. For example, the program in Samoa began with a Legislative Needs Analysis and had a strong training component based on the identification of lack of basic ICT skills among members of the Samoan parliament. Training in ICT skills for members and parliamentary staff is a key part of ensuring that members are committed to their ongoing use, enhancing sustainability. But even with such training and support, members may opt not to make use of systems to provide them with Bills and other crucial papers (Lomp 2013). But one further element of democracy is also crucial: an informed citizenry. A commitment to democracy and the rule of law requires that citizens are vigilant and actively seek information to hold leaders accountable. Hassall's questions in his 2007 review of Pacific Islands remain as relevant today. Who drives these innovations? Is demand coming from MPs reaching out to their electors from administrators, citizens or development partners? "Website development will only be encouraged if sufficient demand for this is expressed by one or other of these actors in the democratic process, and if sufficient resources continue to be supplied for this purpose" (Hassall 2007, p. 127).

7.2.3 Regional and Global Support for ICTs in Pacific Parliaments

A number of organizations are helping PICs develop their parliamentary processes and introduce technology to increase their effectiveness, including the Inter-Parliamentary Union (IPU), the Commonwealth Parliamentary Association (CPA) and the UNDP. These organizations, as well as the Pacific Parliaments Network, are engaged in regional networking, workshops and in-country development projects, all focused on strengthening Pacific parliaments and, inevitably, making more effective use of ICTs. Several countries (Fiji, the Federated State of Micronesia, Papua New Guinea and Samoa) have joined the IPU which is a major sponsor of the *Global Centre for ICT in Parliament.* The Centre, which describes itself as a "hub for information, research and technical assistance to parliaments in the area of ICT", is a strong advocate for the use of ICTs to progress its goal of strengthening representative institutions. It holds the biennial World e-Parliament Conference, produces the regular *World e-Parliament Report,* provides technical

[6] http://ronlaw.gov.nr/nauru_lpms/.

assistance to some IPU members and has issued the most recent edition of the *Guidelines for Parliamentary Websites* for the IPU (2009).

The UNDP has also played a major role in the Pacific, working to strengthen the legislatures in Fiji, Solomon Islands, Samoa, Tonga, Vanuatu, Palau, Kiribati, PNG and the Republic of Marshall Islands as part of its Responsible and Accountable Institutions and its Parliamentary Development programs, often referred to as Strengthening Programs. It has a number of regional bases in Asia and the Pacific—the UNDP Regional Hub (Bangkok), the UNDP Pacific Centre (Suva) and eight in-country offices. It is also a joint partner with the Australian federal and state parliaments in a twinning program with the 12 Pacific members of the CPA, the Pacific Parliamentary Partnerships (PPP) program.[7] As part of this program the UNDP supports the Victorian State Parliament as it assists the Tuvalu Parliament develop a strategic plan, human resource strategy and assess its information technology needs (equipment is donated through the PPP Fund). The Fund has supported many IT projects in the past including the Samoan Legislative Needs Assessment and an ongoing Parliamentary Strengthening Project in Solomon Islands that began in 2001. The Solomon Islands project, which has supplied computing equipment in the past and funded Internet access, funded salaries for an additional 14 (of 71) Parliamentary staff (including an ICT officer) and supported training and an ICT Policy for the Parliament. These staff have been transferred to Solomon Islands Parliament budget as the project winds down, although the past history of such projects suggests that ongoing funding is not assured at this point. The use of ICT to produce Hansard via a fully automated system was also a focus of the UNDP Solomon Islands program. UNDP is currently actively working in Papua New Guinea and its semi-autonomous state Bougainville (which has a separate legislature), building on an earlier EU supported Open Parliament project. The project involves an ICT-assessment to assist both legislatures develop their ICT strategies and determine future ICT needs, which will then be implemented.

But in many ways, local partnerships and the commitment of Pacific neighbors to work side by side produce the most sustainable change. Under the PPP program, primarily funded by the Australian Commonwealth government, the parliament of each Australian state and territory (including the ACT[8] but not the federal parliament) has a development relationship with one or more Pacific Island parliaments,[9] bringing staff in to observe, and making in-country visits. (The New Zealand and Australian national parliaments have an overall coordinating role.) Individual

[7]The Association of Parliamentary Libraries of Australasia also participates in this, working together in the same paired relationships (http://apla.org.au/twinning-program/).

[8]The Australian Capital Territory (ACT), a self-governing enclave similar to Washington, DC, is the home of the federal government, but has its own Legislative Assembly.

[9]The Pacific Parliamentary Partnerships program links ACT with Kiribati; NSW with Bougainville and Solomon Islands; NT with Niue; Queensland with PNG and Vanuatu; SA with Tonga and Marshall Islands; Tasmania with Samoa; Victoria with Niue and Tuvalu; WA with Cook Islands.

programs cover all aspects of governance through effective parliamentary systems and focus on capacity-building among parliamentary staff (e.g. the Clerk, Parliamentary Counsel, Hansard staff and committee secretaries) as well as elected members, such as the Speaker. The Speaker has a key role in reform programs and in the effective use of ICTs in the debating chamber and in the overall management of parliamentary services (CPA 2016).

These partnerships often include a focus on the use of ICT in creating better information management processes and dissemination. Although the PPP program is currently under review and its future uncertain, program participants and parliamentary staff (Surtees 2014) confirm that working alongside parliamentary colleagues in the region and having Pacific colleagues observe and work in Australia is the preferred mode of knowledge sharing, which "provides a sustainable model for democratic development as it is helping to build enduring relationships" (p. 4), one that better fits with the cultural values of Pacific people. And this relationship is not one way—Australian and New Zealand parliamentary staff acknowledge the value of learning from Pacific neighbors, not just how to target aid and development partnerships, but how to work with other sovereign states in the region and understand their perspective (Surtees 2014). Commitment over time is also seen as a particular virtue of the UNDP Strengthening Programs in the region, although because of the high levels of staff turnover and unstable funding throughout some of these projects, not all are sustainable after their conclusion. Australia's commitment to its Pacific neighbors and to effective governance in the region is likely to see the program continued even with any reduction in funding.

Cooperative arrangements between the New Zealand Parliament and Pacific parliaments are less formal, but equally productive. As well as its shared oversight over the PPP programs, the New Zealand Parliament has arrangements with its own dependencies, Cook Islands, Tokelau and Niue. Much of New Zealand's support is focused on legislative drafting and sharing the knowledge of the Parliamentary Counsel Office and Hansard staff, with the costs of these activities being funded by the NZAID program. The Parliamentary Counsel Office also assists in law drafting for Fiji as part of a UNDP program (see box 7.1) and has been providing law-drafting assistance to other Pacific countries, an activity regarded as "effective, efficient and sustainable" in the most recent evaluation (Schmich et al. 2015, p. 7) (NZ.PCO 2014, sect. 3.5). The New Zealand Parliamentary Counsel Office also contributes to PILON (Pacific Islands Law Officers' Network) meetings in relation to legislative drafting as well as drafting and training forums organized by the Pacific Islands Forum Secretariat. Hansard staff from the New Zealand Parliament, whose own Hansard Production System has recently been upgraded to an integrated web-based collaborative document management and publishing resulting in faster, more reliable publication of Hansard, will share this experience with Pacific countries and seek funds to assist them to deploy equivalent software.

Box 7.1: Fiji Parliament's new state-of-the-art ICT system supports the return to democracy

When the new Fijian Parliament opened in October 2014 after a lapse of over eight years Members of Parliament and the Parliament Secretariat staff were equipped with the latest technology, making it one of the most modern parliaments in the region. Parliament met under a new Constitution which required parliament to give due consideration to each Bill before the House, obliged Ministers to provide Parliament with 'full and regular reports concerning matters for which for which they are responsible' (sec 90(4)) and Parliament to facilitate public participation in its processes (sec 72). These provisions in support of more open democracy are enhanced by the use of ICT in the new Parliament.

The United Nations Development Program (UNDP) worked throughout 2014 to provide support to the Parliament of Fiji to achieve the goals of the new constitution. The work undertaken included a needs assessment of the original parliament building, abandoned after the coup of 1987, which was to be refurbished and brought back into use to make parliament more accessible to the people and provide offices for MPs; technical assistance for the development of revised Standing Orders and laws relevant to the functioning of parliament; the development of training programs for secretariat staff and MPs; and the installation of state-of-the-art technology to support the work of parliament. Other nations in the region (New Zealand, Japan and Australia) and the European Union contributed funding and expertise; staff from the Indian Parliament provided Hansard training.

ICT equipment worth over US$893,000 (FJD$1.6m) donated by UNDP included a delegate conferencing system that enables MPs to vote electronically, a portable delegate conferencing system to enable the various Committees to engage with citizens during the public consultation process, the latest Hansard software and hardware to ensure accurate recording of the proceedings in Parliament, iPads to access order papers and documents related to each day's sitting online, desktops and laptops for Parliamentary Secretariat staff. The parliament now has an active and current website, a FaceBook page and live-streaming of the Parliament when in session.

The New Zealand Parliament and the Victorian State Parliament in Australia have both been longstanding supporters of the UNDP Parliamentary Strengthening Program, which is seen as an important cornerstone in Fiji's return to democratic rule. Since the new parliament was formed in 2014 they have been lending the expertise of their parliamentary librarians to help the Fiji Parliament develop its library, to ensure that legislators have relevant historical and legal resources to hand and are supported by a strong research team. A recent donation by UNDP, of FJD$20,000 worth of books and online resources to strengthen the library's ability to serve the interests of democracy, is enhanced by the training provided by New Zealand and Victoria to ensure the parliamentary library remains a "crucial foundation of a well-functioning parliament."

Sources: The Interpreter (Lowry Institute), 9 October 2014, 12:10; FijiLive March 03, 2014; RadioNZ; Fiji Parliament website.

7.3 ICTs in Electoral Processes in the Pacific Islands

A sound electoral process is as important as good governance in promoting democracy, the rule of law and reducing corruption. The Pacific Islands Forum plays a major role in providing expert teams to monitor elections in the region and ensure the principles of the *Bitikawa Declaration* (PIFS 2000) are upheld. Regional monitoring by peers and colleagues is a considerable incentive for free, fair and democratic elections (PIFS 2015).

ICT has a significant role to play and has been adopted more rapidly than in many neighboring industrialized countries. Its use is mainly in the creation of electoral rolls, including the use of biometric technology as part of Biometric Voter Registration (BVR) systems and the use of SMS messaging, which suits the 2G phone systems used by the majority of the voters that candidates wish to reach in the remote rural electorates of most Pacific Island states. BVR systems store thumb prints and photographs of citizens registering and issue secure voter ID cards which must be presented at polling booths. BVR was first used in Samoa's General Election in 2006 and again in 2011 and 2016; it was used in the Fiji elections of 2014, the Solomon Islands election of 2015 and was planned for Vanuatu's 2016 election (Ligo 2013) and for Papua New Guinea's 2017 election (RadioNZ 2012). In Vanuatu the system was not ready when a snap election was called and the use of BVR is now deferred until the 2020 elections (Cullwick 2016). BVR systems play a valuable role in eliminating deceased persons (particularly in regions where registering a death is not common practice) and eliminating the problem of multiple identities in communities where one individual legitimately or otherwise bears many names. In practice, where electoral fraud is entrenched, they do not resolve all forms of electoral fraud and questions have been raised about whether the cost is justified (Haley and Zubrinich 2015a).

An investigation conducted by the European Commission—UNDP Joint Task Force on Electoral Assistance on ICT Solutions in Electoral Processes (EU-UNDP 2010) raises concerns that the introduction of ICTs into electoral processes can carry risks and often suffers from unrealistic expectations; it noted that in 2009 the UN Secretary-General had expressed concern that "… some of the poorest countries in the world have chosen some of the most expensive electoral processes and technology" (UN Secretary General 2009). The report notes that ICTs have already dramatically changed the way elections are conducted in many developing countries and that the process, driven by demand from developing countries and pressure from the suppliers of ICT solutions (regardless of the level of preparedness to introduce such applications and despite significant cost implications) is likely to affect more emerging democracies. Trust is also a major issue: "while elections are technical processes, they are fundamentally political events. However, the true measure of an election is whether it engenders broad public confidence in the process and trust in the outcome." The report also notes the value of biometric systems in voter registration and voting operations, even considering them "the definitive solution to voter impersonation and multiple voting practices" (p. 64), but cautions that such systems must be backed up by efficient and timely matching of data from different regions, a process that adds costs and which must be supported by sanctions. Even so, the report notes, the environment in which such systems operate present additional risks (power failure, connectivity, challenging tropical environments, procurement risks, data security and adequate technical support).

By and large, Pacific countries that have implemented biometric voter registration and identification seem to have been satisfied that it reduces multiple voting and enables them to eliminate non-existent voters (PIFS 2015), a view endorsed by Transparency International's Steve Julax (ABC 2014). Lack of trust in electoral systems in the Pacific islands is based on allegations and substantial evidence of corruption (vote buying and financial rewards for political support when coalition governments are being formed), intimidation and candidates of uncertain quality (RadioNZ 2012b, 2014, 2016). Because of the cost of BVR systems development-partner funding is usually necessary to implement them (RadioNZ 2016). As is the case with the conduct of parliament, the development of democracy and the election of members of parliament according to rigorous international standards (a challenge shared at times with the most industrialized nations) is compounded by the co-existence of traditional forms of governance, relational ethics and the obligations of traditional leaders.

Box 7.2: Solomon Islands' new voter registration system

Parliamentary elections are managed by the independent Solomon Islands Electoral Commission (SIEC), chaired by the Speaker of the Parliament (who need not be a sitting MP); two other members are appointed by the Governor-General. An electoral database was launched before the parliamentary election held in November 2014. This used a biometric voter registration system commissioned by the Minister of Home Affairs, and supplied by Canadian software company, Electoral Services International. The BVR system stores a photograph and both thumbprints of all citizens registering and issues secure voter ID cards. The registration process required eligible citizens to register at a 'roving' registration center, which became the location where they were to vote. The process was regarded as highly successful, reducing the number of voter registrations by more than one third. A text message campaign was used to inform the public about the voter registration process, about candidates, voting procedures and locations of polling stations.

The SIEC reports that approximately 85% of all eligible voters registered to vote and that a record 89.93% of those registered voted. This compares with a figure of 52.4% in the 2010 elections, although this figure is considered to be distorted by the number of multiple registrations and deceased persons on the roll before the introduction of Biometric Voter Registration. The percentage of votes rejected during counting, 0.63%, was low by international standards, an outcome attributed to the clarity of information given to voters.

However, there were several reports that not all eligible citizens were able to register during the registration period and some voters experienced difficulty in locating the polling station where they were registered. Despite these problems SIEC remains confident that the recent elections were a 'significant milestone in the management of clean and fair elections' in Solomon Islands.

Sources: SIEC (2014), Haley and Zubrinich (2015a).

With the massive increase in the use of mobile phones and limited impact of newspapers and other media, mobile telephony plays an increasing role in the electoral process in PICs. Since most people do not have access to more than 2G networks (see Chap. 5), SMS messaging is mainly used to inform voters about how to register, where to vote and who the candidates are. It is also used to send messages from candidates to voters, to notify them of meetings, solicit their vote or spread negative information about other candidates. There is some evidence that SMS messaging has also been used to intimidate voters, facilitate vote buying, distract polling officials and facilitate "block voting, where voters queuing to vote receive text messages containing a photograph of the candidate's symbol they were to mark" (Haley and Zubrinich 2015b). Mobile phones, the authors conclude, are having a profound effect on elections in the region, an effect which has some unforeseen benefits apart from the dissemination of information to voters. Both Haley and Zubrinich (2015b) and Logan (2014) report incidents of voters and official observers using smart phones to record irregularities. Logan, in particular, notes the use across Melanesia of both 'bounded' crowd-sourcing (ICT-enabled monitoring by trained observers, usually members of civil society organizations with a standard reporting mechanism) and 'unbounded' crowd-sourcing (the same communications transmitted by untrained, usually anonymous, individuals; this information can be shared with others, with traditional media outlets or transmitted to a website). This spontaneous form of electoral monitoring, Logan suggests, may bring the benefits of added scrutiny and voter engagement. Equally, she suggests, such reports can produce uneven and unregulated election observation in ways that may "reframe the conduct of the election in public discourse" and lead to increased tension in these already fraught contexts (Logan 2014).

7.4 The Use of ICT in the Justice Sector in the Pacific Islands

Access to information about the constitution of a country, its laws and the operation of its courts is an essential element of a modern democratic state. The second and third principles of the World Justice Project's definition of the Rule of Law state:

> 2. The laws are clear, publicized, stable, and just; are applied evenly; and protect fundamental rights, including the security of persons and property.
> 3. The process by which the laws are enacted, administered, and enforced is accessible, fair, and efficient.[10] (World Justice Project 2016)

Ready access to information is clearly critical to the achievement of these principles, and technology has a clear role to play in facilitating this access.

However, law in the Pacific islands is a complex matter that can create barriers to access to legal and judicial information. The national jurisprudence comprising the customary, equity and common law of each Pacific country comes from many sources, as Jennifer Corrin explains.

> These different kinds of laws originate from different sources: the constitutions, legislation, and subsidiary legislation enacted from outside the country reflect the views of governments of former controlling countries [colonial powers]: the constitutions, legislation and subsidiary legislation enacted within the country are based on the policies of governments of the country; the principles of common law and equity derive from the practices of English people and English judges, and the unwritten customs and customary law derive from the usages and practices of communities within the country. (Corrin and Paterson 2011, p. 5)

The constitutions of several Pacific Island countries include reference to culture and tradition, as do judgments in some cases (Corrin and Paterson 2011, p. 5, 6) and there have been calls across the Pacific for laws to reflect indigenous values rather than English or French law (p. 6). Moreover, where a matter is not covered by the law of the country, English or French common law to which citizens may not have access, may apply. To complicate matters further 'custom law', which is recognized in all countries in the region except Tonga (p. 41), is not well defined and there is considerable dispute about which actions are customarily sanctioned and should have the force of law in the country. Customary land title is one area where advances are being made in creating information systems, but in other areas where customary law may be applied, only court records and law reports provide any access to information about how customary law has been interpreted. Some of these issues are discussed below as we explore the application of ICTs to justice and court systems in PICS.

---

[10]Principles 1 and 4 are: The government and its officials and agents as well as individuals and private entities are accountable under the law. Justice is delivered timely by competent, ethical, and independent representatives and neutrals who are of sufficient number, have adequate resources, and reflect the makeup of the communities they serve.

Table 7.1 indicates whether each country's parliamentary website provides access to the constitution and enacted legislation (Acts and Regulations). This same information is also provided where possible by the Pacific Legal Information Institute (PacLII), funded by the Australian government (see box 7.2); several parliamentary websites link to PacLII (although usually without any explanation as to what is on the site). PacLII has made strenuous attempts to collect as much information about each of the Pacific Island jurisdictions as it can, including substantial historical material dating back to the colonial period. For 20 PICs it collects, digitizes and publishes online as much information as it can in the form of the constitution, law reports, court decisions, decisions of tribunals and quasi-judicial bodies where available, sessional, subsidiary and consolidated legislation and any information about parliamentary process (see box 7.3). Where the information is available, PacLII also reports on legislation and court decisions related to customary law. PacLII performs an essential and critically important service in providing access to legal information in the Pacific that is otherwise inaccessible to most.

In addition, technology makes available the extensive resources and online learning materials of the Pacific Judicial Development Programme (PJDP), funded by the New Zealand Ministry of Foreign Affairs and Trade and hosted on the website of the Federal Court of Australia;[11] its goal is "strengthening governance and rule of law in PICs through enhanced access to justice and professional judicial officers who act independently according to legal principles". The PJDP conducts a considerable range of judicial and court development activities in partner courts across the Pacific focused on capacity-building, backed up by its online Toolkit, reports, templates for surveys and decisions and judges' bench books. The Toolkit includes a module on public information that outlines the responsibility of the judiciary to enable access to justice, the duty of courts to ensure that the people understand what the court can and cannot do for them and how they can use the court to protect their rights. "Public Information must be provided in an easily accessible form which the public can readily understand."

Box 7.3: The crucial role of the Pacific Legal Information Institute (PacLII)

Access to legal information to promote justice and the rule of law (legislation, constitutional information, court decisions and reports of quasi-judicial bodies such as tribunals and ombudsman's offices),[12] and the right of legal information institutes to publish legal information is affirmed in the Montreal Declaration on Free Access to Law (WorldLII) 2007). Although Pacific Island countries are gradually placing legislation on their parliamentary websites, this is a slow process and the principles of democracy require

[11]http://www.fedcourt.gov.au/pjdp/home.

[12]Cook Islands, Palau, Papua New Guinea, Samoa, Solomon Islands, Tonga, Tuvalu, and Vanuatu all have Ombudsman Offices. The role is otherwise usually filled by the Auditor General's office or the state Audit Office.

access to much more than just legislation. Other forms of legal information are a great deal harder to find. The Pacific Legal Information Institute (PacLII), supported by the Australasian Legal Information Institute (AustLII) plays a critical role in the region to ensure that, subject to the information being provided and the format in which it is provided, up-to-date legal information is made available for 20 of the 22 Pacific Island countries.[13] This is done via a reliable comprehensive website at www.paclii.org.

Although founded by the University of the South Pacific (USP)'s School of Law based in Port Vila, Vanuatu, to provide access to their own legislation and court decisions for students at the campus and across USP's eleven other regional campuses, PacLII has since broadened the scope of its publications. It now covers decisions of various tribunals, panels, ombudsman's office reports or secondary information such as court rules or bench books and is seeking to expand its collections to include law related materials such as subject specific papers and reports. It has also created an Intellectual Property Library and a Maritime Law Library. The functionality of the website, which utilizes the SINO search engine and automated 'markup' software to insert hypertext links (both developed by AustLII), enables users to search across the published material on a regional basis and by database type without conducting separate searches for each country. PacLII also includes primary sources (including the *Journal of South Pacific Law* and the *Melanesian Law Journal*) and manages the Pacific Legal Gateway, a website that organizes legal content and research resources that are relevant for the Pacific Island countries. As such the PacLII website has developed from its initial educational role to one that has become a critical pillar in the support of the Law and Justice Sector in the Pacific as a whole. The PacLII website also holds electronic versions of historic legislation from the pre-independence era[14] such as legislation related to Vanuatu from 1884 to 2014.

Source: PacLII website, Senakuraciri et al. (2015).

7.4.1 *Justice Information Systems*

In addition to the legislative and online searchable indexed databases of legislative and judicial information for each country available on PacLII's website, ICTs can be used to improve the administration of justice, covering everything from warrants

[13]French Polynesia and Wallis and Futuna are not included.

[14]This includes New Hebrides Joint legislation from 1909-1973, and Western Pacific Legislation (made by the Western Pacific High Commission) as it relates to Vanuatu.

of arrest and charge sheets, prosecutions and filing of civil cases to the scheduling of cases, court case records (decisions) and law reports (significant decisions including judges' opinions, usually from a higher court). Integrating all this into one system, usually referred to as an Integrated Justice Information Management System (IJIMS, sometimes IJIS) has been a challenge even for much wealthier industrialized countries because it involves pulling together information from separate government agencies and sub-units, often described as 'silos' of information, using different types and definitions of data which must be 'harmonized' for an integrated system (Rottman et al. 2006). Expert knowledge, an excellent system architecture, consultation with potential users, (judges, clerks, lawyers and their clients) and good telecommunications systems (most IJIMS are web-based) are all critical factors in the success of these new JIMS (Rosa et al. 2013).

The role of judges, especially, can be crucial where e-court systems are being developed (Lupo and Bailey 2014). In spite of a history of judges on secondment from countries such as Australia and New Zealand (Baird 2013) bringing their own 'home-grown' systems and high expectations of technological support with them which drive such innovations (Cullen and Hassall 2016a, c), small standalone systems can only provide temporary solutions and hinder the implementation of cross-sector integrated systems. Moreover, it is important to ensure that any introduced system enhances access to justice, particularly in contexts where paper-based alternatives are not readily available and where justice can often be a protracted process.

The essential components of a Justice Information Management System usually include: a Court (or Case) Management System which includes a case registry with details of each case; a scheduling and court room management module; a document management system; an electronic filing service (for filing charges, applications and supporting documents, including affidavits); and an email notification system. Functionalities must include searchability and reporting and audit trails; electronic systems to support the judiciary (multi-media court rooms, Internet access, secure communications and collaborative workspace); a publicly searchable record of judgments and court decisions; web access to case schedules, templates and forms. A fully integrated JIMS is less easily defined, since different jurisdictions will align information systems that they have determined to be the most useful to their administration. However, often they include sentencing databases, to ensure consistency in the application of justice. In some cases, a Courts Management System will be linked to the police and the corrections/prison systems to track individuals and cases through the entire system (see box 7.4). In another example, the Solomon Islands' Courts Management System is linked to the Police Information Management System (which includes biometric data) and is scheduled to link to the Transport Management System which includes traffic offences and vehicle and driver registration. This level of integration has been possible only by the extensive centralization of systems in the country's ICT Support Unit (Cullen and Hassall 2016a).

Box 7.4: Vanuatu's Justice Information Management System

In January 2016, the Vanuatu Chief Justice Vincent Lunabek opened the 2016 judicial year with a report on 2015 in which Vanuatu had implemented a new Court Management System that he expected would enable judges and magistrates to run their caseloads more effectively, and "increase the transparency of where each and every case is up to". The project, part of the Australian funded Vanuatu Law and Justice Partnership Stretem Rod Blong Jastis Program is focused on a high quality and sustainable Justice and Police sector to support Vanuatu's development objectives.

The Vanuatu system, based on the Lexis VisualFiles system, will be used for the Supreme Courts and Magistrates' Courts of Vanuatu replacing most existing manual processes. It will be used for assigning and scheduling cases and producing orders and judgments; it includes a document management repository where users can draft, view and save documents and emails; it delivers a wide number of reports, including cases and financial information and it will eventually include Arrest Warrants and Notices of Conference. Currently covering Courts and the State Law Office, it is being extended to include the State Prosecutor's Office and the Public Prosecutor's and Ombudsman's Offices.

A Police Information Management System installed by Brisbane-based IMA partially replaces the earlier Saperion based CRIMs system already linked to the Government's HR and Finance systems and will support new functions: capturing incidents, managing investigations, recording criminal history. The system will be available to all Vanuatu police throughout country and is regarded as "a major step for the sector as a whole". It will also be linked to the Offender Management System being developed locally (using Saperion) and Corrections' existing Document Management System and Financial system. This will allow tracking of convicted criminals, holding basic data on each individual, when they are due for parole, the terms of parole etc.* As Correctional Services develops more community sentencing and rehabilitation programs the system will include data on such initiatives, which will form the basis of a full reporting system. The three systems, when integrated into an IJIMS, will provide a new standard of reporting and accountability and enable better use of resources through the linking of case data and management with budgeting across the sector.

*Decisions of the Vanuatu Community Parole Board are available on the PacLII website from 2012, to enable the Board "to record its decisions in writing, maintain a register of its decisions and to make its decisions available as a matter of public record" as required by the Correctional Services Act 2006 (Sect. 58(g)).

Sources: Vanuatu Law and Justice Partnership Stretem Rod Blong Jastis Program (http://www.mjcs.gov.vu/images/stretem_rod/SRBJ_Progress_Report_Jan-June_2015_Final.pdf), Vanuatu Daily Post, Ministry of Justice and Community Services Newsletter, Lexis/Nexis, Cullen and Hassall (2016b).

Despite many of these initiatives, it is still not easy to access justice information in most PICs. Information about the Fijian Justice system is available on the Judiciary Fiji website (information is provided about each court and scheduled hearings); sitting dates and cases being heard by the Cook Islands High Court (Land Division) are prominent on the website of the Cook Islands Ministry of Justice (disputes over land titles are a major issue in Cook Islands, as noted below) but little else is available online. This highlights the role of the invaluable PacLII website and its extensive collection of Pacific court and tribunal decisions and its "awareness, advocacy and training role" (PacLII website).

7.5 Land Information

In countries where the majority of inhabitants have little wealth beyond the land owned by their village or kinship group, reliable, accessible information about land tenure is crucial and essential to their basic human rights. It is also a critical development issue. The potential conflict between customary land title in the Pacific and introduced concepts of land ownership, primarily freehold ownership, has been exacerbated over the years by considerable uncertainty over customary land tenure and collective ownership (Corrin and Paterson 2011, pp. 272–77). Customary boundaries and ownership are dependent on local landmarks and oral histories and frequently contested (p. 273). While customary ownership is still recognized and customary authorities may be used to resolve land disputes, some PICs have attempted to deal with land rights through the introduced court system. This is made more complex by some of the decisions concerning the registration of land during the colonial era made by administrators with little understanding of customary land tenure and succession (Corrin and Paterson 2011, p. 273; Crocombe 1987). In the post-colonial era registration of land and procedures for challenging past decisions have become more urgent. All PICs now have legislation outlining the procedures for determining land ownership and land rights, and increasingly are requiring agencies to maintain records of 'land titles', leases and ownership. In many countries, therefore, considerable effort has been made in recent years to create reliable databases with this information—a timely endeavor given that such land ownership information can now be tied to a geographical information system (GIS) that uniquely identifies the land in question. It is also timely, given the opening up of many Pacific countries to business interests which seek to purchase or lease land for business premises or to exploit local resources (such as forestry).

Electronic land registers are now underway or in use in Samoa, Solomon Islands, Papua New Guinea, Fiji and Vanuatu; an example of the value of such initiatives is indicated by Kwapena's report on how Papua New Guinea's Revised Incorporated Land Groups (ILG) Registration System and the New Voluntary Customary Land Registration (VCLR) System are empowering customary landowners to give economic value to their land (Kwapena 2014). In many other countries, GIS systems showing land use (but not title) are being developed; the PALARIS national geographic information system in Palau (Palau 2010) and the Cook Islands' use of its GIS which records data about geographical environments, roads, rivers, water quality, land use and population data (Cullen and Hassall 2016c, p. 28) show the value of combining GIS with other data. Cook Islands is also considering digitizing its land titles database, but a barrier to doing so is the high level of concern about the quality of the data that the process would expose and the potential for disputes, due to the inappropriate method used by the Cook Islands Land Court in assigning title in the early 20th century. Samoa's SOLA system and the value derived from it is another example (see box 7.5).

Box 7.5: Samoa's Land Registration and land use system

Samoa was the first country to trial the United Nations new software *Solutions for Open Land Administration* (SOLA), open source software developed by the UN Food and Agriculture Organization (FAO) to make computerized cadastral mapping and registration systems more affordable and sustainable for developing countries. The Samoa Ministry of Natural Resources and Environment (MNRE) has been working with Computer Services Ltd and the FAO (as donor) to customize and install the SOLA system. Although more than 80% of the land is held in customary ownership outside the land title registration system, transactions affecting the remaining 20% of freehold and government land have to date been managed using two separate computerized systems based on commercial software and dependent on overseas software support. SOLA integrates both title registration and cadastral mapping functions in an open source solution.

The Land Registration System (LRS) held by the Samoan Ministry of Justice and Courts Administration is also being migrated to an open source environment as part of the same project; it includes the digitization of fragile historic records of the Samoa Land and Titles Court which contain a wealth of information about family genealogy, matai[15] title holders and ownership of customary land. The new system will preserve all Land and Titles Court records in fully indexed electronic form and enable backup copies to be created in the event that the images on paper fade or documents themselves are destroyed. As well as improving access by the public to these important heritage documents, the combined databases will provide MNRE staff with title registration linked to cadastral information, including land ownership, tenure, precise location, dimensions and area and the value of individual parcels of land, via a GIS viewer. Additional benefits from this project include not only a digital archive of scanned registration documents, but also the development of local expertise capable of supporting and enhancing the

[15]Matai are Samoan chiefs responsible for the administration of the extended family and village and the maintenance of traditional customs. The Samoan matai title system includes many levels and roles, such as orator, and these traditional titles are often used before a holder's given name and surname.

open source system. The introduction of publicly available information on land registration is one of the key ways to reduce corrupt and non-transparent land management practices and provide better access to justice: "the SOLA project promotes affordable IT-systems that enable improvements in transparency and equity of governance" (McLaren 2013).

Sources: Samoa Law and Justice Sector (2013), FAO (2012).

7.6 Government Record-Keeping Strengthens Democracy

An equally important part of the democratic process, and essential for government accountability, is government record-keeping and archiving of government records, a function managed or overseen by appointed government archivists in nearly all Pacific countries. Most also have a national archive where such records are transferred; the majority have some form of Archives Act ensuring the safe-keeping of government records and assigning authority in that matter to the Government Archivist. ICTs have an important role to play in helping to preserve, index and make such records available in a context where the very existence of government records is threatened. Archives[16] "by their very nature are unique both as individual documents and as document[s] in context. Lost archives are irreplaceable, any loss is final, and in most cases reconstruction is impossible", noted Joan Van Albada, Secretary General of the International Council on Archives in 2001 (Teygeler 2001). Government records do not signify only constitutional documents and parliamentary papers (although government archives may hold these along with other institutional records) but rather the unique records created by the functions of all government agencies, as well as Cabinet papers, records from the Prime Minister's or Premier's department and the papers of individual ministers and members of parliament.

Many barriers in the Pacific islands prevent a national archive from collecting all such papers. There is no long standing convention of preserving written records in the oral traditions of the Pacific. History is passed down in song, narrative and other art forms. Government archivists can therefore struggle to persuade government agencies not only to maintain adequate records of their activities and house them securely, but appraise and select records that should be passed to a national archive for long term preservation. In addition, for much of their recent recorded histories, Pacific Island countries were subject to colonial rule and do not 'own' the government archives of that time, although many are acquiring copies, sometimes in

[16]The ISO defines records as "information created, received, and maintained as evidence and information by an organization or person, in pursuance of legal obligations or in the transaction of business." Archives refers to both records and materials that are appraised to have archival value in addition to the physical place where archival materials and records are stored (United Nations 2016).

electronic format (see, for example, information about digitization of historic archives of Samoa and Solomon Islands below). In addition, given the threat to their indigenous cultures and the availability of new technologies capable of creating a record of these, the modern Pacific Island states are especially motivated to create and retain records of their traditional culture, often requiring their national archival institutions to record traditional customs and culture (using audio-visual technologies). It is common to find the national archives part of the same institution as the agency responsible for cultural heritage and their work closely aligned with the cultural preservation mission.

PICs archivists face several additional challenges. Where government resources are very limited it is hard to get adequate budgets allocated to record-keeping and archiving. The environment, most obviously, the moist tropical climate and inadequate housing of records (both the buildings and the storage materials used), is hostile to the preservation of paper records, which can be affected by heat, dirt, fungi, insects and even small animals and birds (Teygeler 2001). Lack of trained staff and other aspects of the island environment (cyclones, earthquakes and tsunamis, for example) also threaten both records and archives. In the contemporary era, the collection, care and retention of electronic records (produced as a normal part of the daily business of government agencies), are even harder to manage; statutory authorization is usually required to give legal recognition to them and validate electronic contracts and signatures, and to provide for data security through encryption. Agencies (and politicians responsible for them), unfamiliar with standard professional practice in the management of records and archives, may have concerns about security and the safe transfer of electronic records. When the management of electronic records becomes too difficult many agencies, as happens still in some industrialized nations, simply print them and treat them as paper records. And, as in industrialized nations, increasing use of email and social media for government communications creates an even greater problem for record-keeping and archiving.

However, despite these challenges, government records and archives in Pacific Island countries are starting to get the attention they urgently need. Archives in many of the larger PICs are housed in purpose-built facilities, often funded by aid agencies as part of a larger archives project (Cullen and Hassall 2016b), although even in these new premises air conditioning can often be used for only part of the day and humidity control is rarely affordable. Papua New Guinea has a large collection of over 100,000 items including pre-independence records and private papers of former leaders, and is digitizing the Department of Agriculture and Livestock land use records to allow the originals to be transferred to the National Archive (Papua New Guinea 2016). Archives Acts in most PICs date from the 1980s or early 1990s (Tonga and Federated States of Micronesia lack such Acts); this legislation has sometimes been updated in the past decade, usually to cover electronic records and strengthen the authority of the institution. A survey carried out in 2003 by the Pacific Branch of the International Council on Archives (PARBICA) as part of a training needs assessment (Millar 2003) provides some information about the legal status of government archives and records management in PICs, as does the website. The survey data still gives a reasonably accurate picture of the situation in most PICs and often shows a range of other cultural activities and artifacts that a 'national archives' might be expected to collect and preserve.

The Vanuatu National Archives, for example, are housed in a purpose-built National Library and Archives building within the Vanuatu Cultural Centre Complex (funded by AusAid and completed in 2013). The National Archives of the Cook Islands (NACI) was established in 1974 and has been a division of the Ministry of Cultural Development (MOCD) since 1991. The Public Records Act of 1984 which conferred statutory responsibilities on NACI and empowers the Archivist to collect and store government archives also requires NACI to collect and preserve cultural artifacts and oral traditions of the country. In the Northern Mariana Islands, the national archives are held in the country's tertiary institution, the Northern Marianas College, and contain parliamentary records. In Tonga, which is a hereditary constitutional monarchy, the Tonga Traditions Committee has responsibility for the Palace records, cultural artifacts and buildings and some archives, but not government activities. Tonga does not have a national archives repository per se, although there is a strong system of government record-keeping and archiving and management of electronic government records (PARBICA 2014).

7.6.1 Regional Support Facilitates Better Government Record-Keeping

The Pacific Regional Branch of the International Council on Archives (PARBICA) plays a major role in supporting the professional care of government records and archives to the Pacific; much of the training it offers makes use of new communications technologies as well as providing advice on the use of ICTs in records management. The PARBICA website has a strong focus on knowledge sharing, the promotion of regional workshops and conferences in which participants are encouraged to start applying the PARBICA Record-keeping for Good Governance Toolkit specifically aimed at Pacific countries, and the Digitisation, Disaster Preparedness and Conservation resources on the PARBICA website. The Toolkit includes needs assessment, the development of a model record-keeping policy and plans for most record-keeping activities, including categories of files and titles, appraisal and disposal, digital record-keeping strategy, the basics of digital record-keeping, managing email and scanning paper records to a digital file. The Model Record-keeping policy provides a template for Pacific countries to use when drafting policy that was developed with local input from Pacific archivists and records managers and recognizes some of their challenges. The section on tailoring the template to local needs, which was developed by groups of Pacific archivists (PARBICA, n.d.) reflects local conditions with a strong focus on sustainable strategies. Virtually all Pacific Island archivists are involved in PARBICA in some way and have benefited from the training PARBICA offers through its annual conferences and regional workshops, as well as the online training resources, although progress towards full implementation of the recommended policies and practices is limited by the resources available.

In addition to this activity, digitization projects, supported by PARBICA and donors (Australia and New Zealand and sometimes former colonial powers) are underway in many Pacific Islands' archives; these projects included digitization and retrospective cataloguing of paper records that have been sent to the archives for storage and the digitization of colonial records held in-country or elsewhere. Samoa, for example, is funded by the Federal Republic of Germany to digitize records of the German administration of Samoa from 1900 to 1914, assisted by Archives New Zealand (Cullen and Hassall 2016d) and Solomon Islands is self-funding the digitization of 8000 records in the archives of the British Solomon Islands Protectorate, also with the assistance of Archives New Zealand, creating an electronic Finding Aid while doing so (Cullen and Hassall 2016a).

7.7 Conclusion

Although the adoption of ICT in the core institutions of democracy in Pacific Island states is piecemeal and sometimes uncoordinated, the degree of regional cooperation ensures that progress towards the overall objective—to make institutions more effective and efficient and to make information more readily available to citizens—is being made. Full implementation of the technologies used by industrialized Pacific nations to support the rule of law, democracy and manage the parliamentary, electoral and justice systems is beyond their resources, but in most PICs the benefits of new ICT applications and better access to information about parliament, legislation, the electoral process and the justice system is evident. Further progress is dependent on budgets being stretched to include the ongoing maintenance of ICT systems and ICT support, on funding from development partners and regional programs based on collaboration and knowledge sharing. New communications technologies are assisting with training and access to knowledge, while the availability of news media on the Internet strengthens the role of the 'Fourth estate' in drawing political concerns to the attention of the people. The role of the Pacific Islands Forum in asserting the importance of good governance and democratic values, the committed relationships with Australian and New Zealand parliamentary staff, the role of organizations such as PacLII and PARBICA demonstrate what can be achieved by working towards a common shared goal. The Pacific aid model of mutual understanding, focused on sustainable development and based on regional resources and collaboration, has been described as 'high value, low cost' (MFAT 2015). Where resources are limited, it is the only way forward.

Admittedly, the current state of 'e-democracy' in the Pacific Islands represents what could be called a 'minimalist', traditional, representative version of ICT-enabled democracy, which is a long way from the vision articulated by

Chadwick and May (2003) of a 'renewal of democracy' through e-government and citizen participation, or the 'participatory democracy' that Clift (2004) argued ICT's would bring. But it is a significant development, even though the eventual form of e-democracy that will emerge in the region is not yet clear. Models of democratic parliamentary practice and open and accountable government that can incorporate customary forms of governance in Pacific Island states are slowly emerging, and despite their short histories as modern democratic states and the lack of resources to fund new systems, the examples above show that the use of ICTs in support of democracy and justice in Pacific Island countries is having an impact. Where ICTs are well used, democracy is well served.

A program of ongoing regional and international support, along with regular regional meetings, relationships and reporting, helps maintain commitment and knowledge; and it motivates individuals and countries to keep the core mission of enhancing democracy and citizens' access to information through new technologies to the fore. Moreover, the use of ICTs to inform people has clear benefits in countries with dispersed and rural populations. Although the greatest early gains are to some extent confined to the urban centers of Pacific countries, where 3G and 4G phone networks are available, and the flow of information is limited, over time access to information helps build demand. Through SMS messaging, news is more rapidly passed to more remote areas, spread through NGOs and civil society. The opportunity for participation in national politics, to know and understand what is going on, also helps build demand for information, creating what could be described as a virtuous circle. Building public participation over the long term may answer the questions put by Hassall and lead to a more robust form of 'e-democracy'. Demand from the people will encourage the further development of parliamentary websites, access to information about their MPs and the legislative process, access to land data and to the justice system and encourage more openness among politicians. The free flow of information is at the heart of parliamentary democracy and the rule of law. The collaborative, knowledge sharing model that reflects this principle and which pervades this aspect of e-government in the Pacific, is an effective and sustainable model for development.

References

APLAP (2016) http://asiapacificparllibs.org

Australian Broadcasting Corporation (2014) http://www.abc.net.au/radionational/programs/lawreport/solomon-islands-hc-judge-speaks-out-v2/5912956#transcript

Baird N (2013) Judges as cultural outsider: exploring the expatriate model of judging in the Pacific. Canterb Law Rev 2013(4):80–96

Chadwick A, May C (2003) Interaction between states and citizens in the age of the internet: "E-government" in the United States, Britain and the European Union. Governance 16(2): 271–300

Clift S (2004) E-government and democracy. Representation and citizen engagement in the information age. Publicus, Minneapolis, MN. http://stevenclift.com/e-government-and-democracy-report-to-the-united-nations-by-steven-clift-2004/. Accessed 23 June 2016

Commonwealth Parliamentary Association. Australian Region (2016) Programs. http://www.cpaaus.org/programs.html http://www.cpaaus.org/programs.html. Accessed 7 Apr 2016

Corrin J, Paterson D (2011) Introduction to South Pacific law, 3rd edn. Palgrave Macmillan, Melbourne

Crocombe R (1987) Land tenure in the Pacific. 3rd ed. University of the South Pacific, Suva

Cullen R, Hassall G (2016a) E-government in the Pacific island states: ICT policy and implementation in small island developing states. Solomon Islands country report. http://pippr.victoria.ac.nz/handle/123456789/32. Accessed 2 May 2016

Cullen R, Hassall G (2016b) E-government in the Pacific island states: ICT policy and implementation in small island developing states. Vanuatu country report. http://pippr.victoria.ac.nz/handle/123456789/33. Accessed 2 May 2016

Cullen R, Hassall G (2016c) E-government in the Pacific island states: ICT policy and implementation in small island developing states. Cook Islands country report. http://pippr.victoria.ac.nz/handle/123456789/30. Accessed 2 May 2016

Cullen R, Hassall G (2016d) E-government in the Pacific island states: ICT policy and implementation in small island developing states. Samoa country report. http://pippr.victoria.ac.nz/handle/123456789/31. Accessed 2 May 2016

Cullwick J (2016) Biometric voting system confirmed for 2020 election. Vanuatu Daily Post, 1 Feb 2016. http://dailypost.vu/news/biometric-voting-system-confirmed-for-election/article_f1936890-61a2-5578-ab69-d827b8498a42.html?mode=print. Accessed 7 Apr 2016

EC-UNDP Joint task force on electoral assistance (2010) Procurement aspects of introducing ICT solutions in electoral processes. The specific case of voter registration. Operational paper. United Nations Development Programme, European Commission and International IDEA. EC-UNDP joint task force on electoral assistance, Brussels. http://www.ec-undp-electoralassistance.org/index.php?option=com_content&view=article&id=147&Itemid=103&lang=en. Accessed 19 Apr 2016

EC-UNDP Joint task force on electoral assistance (2013) Working together in electoral assistance: formulating and implementing electoral assistance projects in the context of the EC-UNDP partnership. Operational Paper II. EC-UNDP joint task force on electoral assistance, Brussels

FAO (2012) Sub-regional office for the Pacific. News. Samoa to pilot open source land administration software, 23 Sep 2012. http://www.fao.org/asiapacific/sap/home/news-room/detail/en/?news_uid=157790

Garrett J (2016) Nauru rule of law concerns remain unaddressed, New Zealand foreign minister says. Pacific Beat. ABC, Sydney. www.abc.net.au/news/2016-03-15/Nauru-rule-of-law-concerns-remain-unaddressed-says-nz-fm/72499416. Accessed 2 Apr 2016

Global Centre for ICT in Parliament (2012) World e-parliament report 2012. http://www.ictparliament.org/WePReport2012.html. Accessed 4 May 2016

Haley N, Zubrinich K (2015a) Is biometric voter registration the way forward for Melanesia? State, Society and Governance in Melanesia, Canberra. Australian National University. In Brief 2015/26. http://ips.cap.anu.edu.au/publications/biometric-voter-registration-way-forward-melanesia

Haley N, Zubrinich K (2015b) Mobile phones and the 2014 Solomon Islands national elections. State, Society and Governance in Melanesia, Canberra. Australian National University. In Brief 2015/28. http://ips.cap.anu.edu.au/publications/mobile-phones-and-2014-solomon-islands-national-elections

Hassall G (2007) Pacific islands parliamentary websites: a preliminary survey. In: Prasad BC, Roy DC (eds) Development problems in Pacific Islands States. Nova Science Publishers, New York

Hassall G (2012). Pacific Island parliaments: Developmental aspirations and political realities. Australasian Parliamentary review, 27 (1):213–37.

Hassall G, Talifu C (2015) Performance and reform in the national parliament of Solomon Islands. Paper presented at the annual conference of the Australasian Study of Parliament Group, Parliament House, Wellington, New Zealand, Sept–Oct 2015. http://www.parliament.nz/resource/en-nz/00CLOOCaspg11/3c697bed3c2195256a31adebf4bc883c8a3bd53c. Accessed 3 Apr 2016

IPU (2009) Guidelines for parliamentary websites. IPU, Geneva. http://www.ipu.org/PDF/publications/web-e.pdf. Accessed 22 June 2016

Isno V (2015) Negotiations are underway for MPs to access bills online. Parliament of the Republic of Vanuatu, Port Vila. https://parliament.gov.vu/index.php/documents/news/162-negotiations-are-underway-for-mps-to-access-bills-online. Accessed 2 Apr 2016

Johnson G (2015) Nauru needs to meet its human rights obligations. Pacific institute for public policy, Port Vila. http://pacificpolicy.org/2015/06/nauru-needs-to-meet-its-human-rights-obligations/. Accessed 2 Apr 2016

Kwapena F (2014) Empowering customary landowners to give value to their land. National Research Institute, Port Moresby. http://www.nri.org.pg/index.php/22-news/180-empowering-customary-landowners? Accessed 5 May 2016

Larmour P (2002). Westminster Constitutions in The South Pacific: A policy transfer approach. Asian Journal Political Science, 10(1):39–S4

Ligo G (2013) Biometric voter registration eyed for Vanuatu in 2016. Electoral officer says Vanuatu may use system like Solomons. Vanuatu Daily Post, Port Vila, East-West Center. http://www.pireport.org/articles/2013/12/10/biometric-voter-registration-eyed-vanuatu-2016. Accessed 9 Dec 2013.

Logan S (2014) Electoral integrity, citizens and ICT in Melanesia. In Brief 2014/9. State Society and Government in Melanesia Program(SSGM), Canberra. Australian National University. http://ips.cap.anu.edu.au/sites/default/files/IB-2014-9-Logan-ONLINE.pdf

Lomp A (2013) Final evaluation report on the UNDP Palau parliament support project, Dec 2013. Inter-Parliamentary Union, Geneva. https://erc.undp.org/evaluation/documents/download/8436. Accessed 2 Apr 2016

Lupo G, Bailey J (2104) Designing and implementing e-justice systems: Some lessons learned form EU and Canadian examples. Laws, 3:353–387

McLaren R (2013) Technology to promote transparency around land acquisitions. Evidence on Demand, London. doi:10.12774/eod_hd031.feb2013.mclaren

Millar L (2003) Education and training for records and archives management in Pacific island nations: a needs assessment and report prepared for PARBICA. PARBICA, Canberra. http://www.parbica.org/content/DistanceEducationProjectReport_tcm50-74752.pdf. Accessed 1 May 2016

Mistry IJ, Rodrigues C (2007) E-Governance in the Pacific islands: entrenching good governance and sustainable development by promoting ICT strategies based on the right to information. In: Prasad BC, Roy DC (eds) Development problems in Pacific islands states. Nova Science Publishers, New York

New Zealand. Ministry of Foreign Affairs and Trade (2015) New Zealand aid programme. Strategic plan 2015–2019. MFAT, Wellington. https://www.mfat.govt.nz/assets/_securedfiles/Aid-Prog-docs/New-Zealand-Aid-Programme-Strategic-Plan-2015-19.pdf. Accessed 7 May 2016

New Zealand. Parliamentary Counsel Office (2014) Briefing notes for the attorney general on the role and operations of the parliamentary counsel office. PCO, Wellington. http://www.pco.parliament.govt.nz/bim2014. Accessed 8 Apr 2016

Palau. Ministry of Resources and Development. Office of the Palau automated land and resource information system (2010) A National Geographic information system, PALARIS. http://palau.chm-cbd.net/palaris-land-and-resources-gis/about-office/office-palau-automated-land-and-resource. Accessed 4 May 2016

Papua New Guinea (2016) Introducing national archives and public records services of Papua New Guinea. http://www.parbica.org/content/naprs-png_tcm50-76140.pdf. Accessed 22 June16

PARBICA (2014) Institutional, state and country reports. Tonga. PARBICA, Canberra. http://www.parbica.org/conferences-and-events/conferences/parbica14/reports/tonga.aspx. Accessed 1 May 2016

PARBICA (n.d.) The recordkeeping for good governance toolkit. PARBICA, Canberra. http://www.parbica.org/sharing/publications/recordkeeping-for-good-governance/index.aspx. Accessed 1.5.16

PIFS (2000) Bitikawa declaration. http://www.forumsec.org/resources/uploads/attachments/documents/BIKETAWA%20Declaration.pdf. Accessed 21 June 2016

PIFS (2004) Leaders vision in the Auckland declaration, Apr 2004. PIFS, Suva. Available at http://www.forumsec.org/resources/uploads/attachments/documents/AucklandDeclaration1.pdf. Accessed 31 Mar 2016

PIFS (2014) The framework for pacific regionalism. PIFS, Suva. http://www.forumsec.org/resources/uploads/embeds/file/Framework%20for%20Pacific%20Regionalism_booklet.pdf. Accessed 31 Mar 2016

PIFS (2015) Solomon Islands national elections 2014: report of the Pacific islands forum election observer team. PIFS, Suva. Available at http://www.forumsec.org/resources/uploads/attachments/documents/2014_Solomon_Is_Elections_PIF_Observer_Team%20_Report_FINAL.pdf. Accessed 20 Apr 2016

RadioNZ (2012a) Electronic registration for voters in PNG planned for next election. Wellington. RNZ. http://www.radionz.co.nz/international/pacific-news/208149/electronic-registration-for-voters-in-png-planned-for-next-election. Accessed 5 Apr 2016

RadioNZ (2012b) http://www.radionz.co.nz/international/pacific-news/205461/electoral-official-in-png-talks-of-the-dangers-posed-by-vote-buying.

RadioNZ (2014) http://www.radionz.co.nz/international/programmes/datelinepacific/audio/20143166/push-in-solomons-to-stop-vote-buying

RadioNZ (2016) Money urgently needed for PNG elections. 12 Jan 2016. http://www.radionz.co.nz/international/pacific-news/293941/money-urgently-needed-for-png-elections. Accessed 13 May 2016

Rosa J, Teixeira C, Sousa Pinto J (2013) Risk factors in e-justice information systems. Government information quarterly, 30: 241–256

Rottman JW, Smith LD, Long DA, Crofts C (2006) Implementing judicial management systems within an Integrated Justice Information Framework: A case study on Information Systems development in the public sector. Am Rev Publ Adm, 37(4):436–456.

Samoa Law and Justice Sector (2013) Project update: Digitisation of land and titles court records. Tuvao: Samoa law and justice sector news, 1(2):6 http://www.samoaljs.ws/english/images/stories/uploads/tuvao_sljs_newsletter_july_2013.pdf

Schmich L, Pierce PV, Allen M (2015) Parliamentray Counsel Office legislative drafting assistance to Pacific nations evaluation report. MFAT, Wellington https://www.mfat.govt.nz/assets/_securedfiles/Aid-Prog-docs/Evaluations/2016/March-2016/Parliamentary-Counsel-Office-Legislative-Drafting-Assistance-to-Pacific-Nations_Evaluation-Management-Response2.pdf. Accessed 2 May 2016

Senakuraciri R, Hamilton L, Freriks K (2015) Reclaiming historical legal material in the Pacific. Pacific legal Information Institute (PacLII), Port Vila http://www.lvi2015.org/programme/papers/LVI2015_Resina_PacLII_paper_on_Reclaiming%20_Historical_Leg.pdf. Accessed 2 May 2016

Solomon Islands Electoral Commission (SIEC) (2014) Media release Thursday 26th June, 2014. The SIEC provisional figure is 287,438 registered voters. SIEC, Honiara http://www.siec.gov.sb/index.php/resources/media-releases. Accessed 2 May 2016

Solomon Islands (2012). National parliament of solomon islands strategic plan. National parliament of solomon islands, Honiara. Available at: http://www.parliament.gov.sb/files/Parliament%20Strategy/National%20Parliament%20Strategic%20Plan%202012-2016.pdf. Accessed 2 May 2016

Surtees C (2014). Co-ordination of assistance and support to other parliaments. Presented at the Inter-Parliamentary Union, Geneva session October 2014. www.asgp.co/sites/default/files/Surtees%20EN.docx. Accessed 2 May 2016

Teygeler R (2001) Preservation of archives in tropical climates. An annotated bibliography. International Council on Archives, Paris

UNDP Pacific Office (2016) Democratic governance. About Freedom of Information (FOI). UNDP, Paris. http://www.pacific.undp.org/content/pacific/en/home/operations/projects/democratic_governance.html. Accessed 5 May 2016

UNESCO (2016) About freedom of information. UNESCO, Paris. http://www.unesco.org/new/en/communication-and-information/freedom-of-expression/freedom-of-information/about/. Accessed 31 Mar 2016

United Nations. Archives and Records Management Section (2016) Glossary of recordkeeping terms. United Nations, New York. https://archives.un.org/content/glossary-recordkeeping-terms. Accessed 27 Apr 2016

United Nations Secretary General (2009) Strengthening the role of the United Nations in enhancing the effectiveness of the principle of periodic and genuine elections and the promotion of democratization. Report of the Secretary-General to the General Assembly, 14 Aug 2009. http://reliefweb.int/sites/reliefweb.int/files/resources/338606C6819560214925765C00235387-Full_Report.pdf. Accessed 5.5.16

World Justice Project (2016) What is the rule of law? World Justice project, Washington, DC. http://worldjusticeproject.org/what-rule-law

WorldLII (2007) Declaration on free access to law. WorldLII, Montreal. http://www.worldlii.org/worldlii/declaration/montreal_en.html

Chapter 8
Official Statistics and Information and Communication Technologies in Development in the Pacific

Len Cook and Rowena Cullen

Abstract The collection, management and dissemination of official statistics is essential for good governance and planning, but is a formidable task for most Pacific Island countries due to lack of resources, capacity and their dispersed rural populations. Information and communication technologies (ICTs), especially in the context of regional cooperation and leadership, have the potential to significantly improve the collection of data and the dissemination and analysis of statistics in formats that will ensure its value is maximized. This chapter discusses the role of statistics in development and their role in monitoring the Millennium Development Goals (2000–2015), as well as the recently adopted Sustainable Development Goals (2016–2030) and how ICT can be used to manage this process. It examines technologies used in PICs and the regional initiatives that are contributing to capacity building and more effective statistics management through the use of emerging technologies. It considers some of the issues that must be taken into account to maximize the sustainable use of ICT in statistics, the value of statistics for decision-making in PICs and highlights the importance of leadership in regional initiatives as well as in individual countries.

8.1 Introduction

Official statistics are an essential instrument for the modern state and for good governance and planning. Indeed, the gathering of official statistics is one of the earliest recorded forms of government activity.[1] Official statistics today encompass government revenue, expenditure, economic indicators, data on the industrial and

[1]The oldest census recorded was undertaken by the Babylonians in 3800 BC.

L. Cook (✉)
Superu, Wellington, New Zealand
e-mail: len_cook@xtra.co.nz

R. Cullen
Victoria University of Wellington, Wellington, New Zealand
e-mail: rowena.cullen@vuw.ac.nz

R. Cullen and G. Hassall (eds.), *Achieving Sustainable E-Government in Pacific Island States*, Public Administration and Information Technology 27,
DOI 10.1007/978-3-319-50972-3_8

services sectors, education and training, health, labor and income statistics and any other measures a country considers of value in planning and decision-making. Much of this activity involves information and communication technologies (ICTs) —the collection and storage of large datasets and analysis using powerful application software. Managing official statistics in the fragile economies and dispersed rural populations of the Pacific's small island developing states (SIDS) presents opportunities and challenges in collecting, disseminating and analyzing data in formats that will ensure their value is maximized. International and national demand for such information is increasing, but the use of ICTs to meet this demand requires the capacity to both afford and exploit such powerful tools.

As earlier chapters establish,[2] the populations, economies and cultures of Pacific SIDs vary considerably. In size, they range from some 1200 citizens to several hundred thousand (with the exception of Papua New Guinea which has a population of over seven million). The information required to govern them differs from that required by governments of larger industrialized countries. This chapter will focus on the impact of size and the extent to which a regional approach is helping exploit the opportunities that ICTs bring and provide solutions to some of the challenges the use of ICTs present. The chapter will discuss the role of statistics in development, in monitoring the Millennium Development Goals (2000–2015) and the Sustainable Development Goals (2016–2030), the potential for ICT to transform the management of statistics in Pacific Island countries and territories and the role of the Pacific Community (SPC) in developing regional initiatives. It concludes with a discussion of the main challenges presented by the adoption of ICT to manage statistics in Pacific Island countries (PICs) and the need for sustainable solutions and leadership in this endeavor.

Small Pacific Island states are limited in their ability to gather social and economic data, especially using international tools and applying standards for official statistics which have their origins in early industrial societies and which are designed for considerably larger populations. Contemporary statistical systems generally place little value on approaches to measurement developed outside industrial societies in the last century, but which may still be relevant in PICs. Differences between industrial and Pacific Island societies include the following.

- Population in some tribal societies is traditionally monitored by genealogical knowledge of past generations rather than by demographic analysis.
- Measures of the health and wealth of traditional societies was historically (and sometimes still is) directly related to the continuity of food supply.
- Valuing work outside of the job market or valuing resources in the seabed has not been of high interest to the industrial economies that have mainly determined international standards.

As a consequence, the methods, tools and legal frameworks that PICs have been encouraged to adopt for official statistics have brought challenges in their

[2]See Chaps. 1 and 5.

implementation that have tested the capacity of Pacific SIDS. Meeting these challenges has given impetus to national and regional statistical leadership and encouraged the adoption of pragmatic, common sense solutions more attuned to the way that small countries operate. It has also encouraged PICs to seek changes in the components of official statistics used in the region and to design measures that more closely relate to their needs.

However, PICs do have some advantages over larger countries in producing official statistics. Response rates in most PICs are generally high by world standards for household surveys. The development and unit operational costs of an appropriate survey frame (i.e. the target population) for population surveys, address registers and business registers increases with the size of country, which favors small countries. New technologies come at much lower costs than in the past, and Pacific states have shown a willingness to adopt global systems such as Google for innovative solutions. Collaboration between National Statistical Offices (NSOs) in PICs, especially through the Ten Year Pacific Statistics Strategy, 2010–2020, (see Secretariat of the Pacific Community 2010) has been growing in recent years, driven by a number of factors. These include the unaffordably high cost of standalone systems, the lack of expertise available to larger countries and the fact that Pacific governments are usually not large enough to function in isolation, particularly with regard to analytical and statistical capability. Problems which require very different solutions to those applicable in their Pacific neighbors and development partners, Australia and New Zealand, can be met by solutions shared among Pacific states. For these reasons, a strong history of 'south-south' collaboration which is being extended year by year, has built up in the management of official statistics in the Pacific.

8.2 Statistics in Development

On 20 October 2015, the United Nations celebrated the second World Statistics Day,[3] which was dedicated to 'Better Data. Better Lives'; and "the critical role of high-quality official statistical information in analysis and informed policy decision-making in support of sustainable development" (United Nations 2015). The provision of country data to the United Nations is virtually a requirement of membership, supported by a series of resolutions of the UN's Economic and Social Committee, ECOSOC. Civil registration and vital statistics (CRVS) is now regarded as an essential development and human rights issue, included in the Convention of the Rights of a Child. The Convention, which includes in Article 7 the statement, "The child shall be registered immediately after birth and has the right to a name and nationality…", was signed by all countries of the Pacific (United Nations 1989).

[3]The first was in 2010 and in 2015 the General Assembly decided to celebrate World Statistics Day every 5 years.

Over the past century technological advances have transformed the collection, storage, analysis and dissemination of data, such that today ICTs are an essential part of statistics at national, regional and global levels, influencing the scope, methods and scale of statistical systems. Official statistics is now a globally networked activity with strong common elements, particularly classifications, standards and frameworks, including the system of national accounts required for reporting. Common processes are expected in the gathering and reporting of information and in the ways it is made available to users. Technology has both enabled these processes and increased their complexity. It has even been argued that the prevalence of these tools and the generation of digital information have caused a fundamental change to the nature of data itself (United Nations Statistics Division 2015).

In the Pacific, the initiatives to advance the scope, quality and analytical dimensions of official statistics are as much driven by official statisticians as they are from domestic policy makers and ministers. Although Samoa has led the way in political leadership with its Samoa Plan which highlights the significance of statistics to measure progress against Samoa's development goals (Samoa 2012), in most other PICs there is still a critical need for evidence on which to base policy. This was noted at a recent meeting of PARIS21,[4] where Gerald Haberkorn, Director of the Pacific Community (SPC)'s Statistics for Development Division, launched a multimedia presentation featuring Pacific Island policy-makers and government statisticians, emphasizing the importance of accurate and timely statistics to assist evidence-informed policy development, planning and decision-making. In addition to the benefits of statistical information for policy, Haberkorn also acknowledged the significance to aid partners of the regular monitoring of development progress, stressing that with billions of dollars likely to be invested in post-2015 development efforts worldwide, it was essential to convince political leaders of the need to consider statistical development as a development goal in its own right, in order to provide countries and their development partners with evidence that money invested is well spent, making an impact and achieving desired development outcomes.

8.2.1 *From the Millennium Development Goals to the Sustainable Development Goals*

The United Nations' Economic and Social Council (ECOSOC) has taken a global lead in the collection and publication of official statistics since its establishment in

[4]PARIS21, the Partnership in Statistics for Development in the 21st Century, was founded in November 1999 as a global network of statisticians, analysts, policy-makers and development practitioners committed to evidence-based decision making. With the objective 'to achieve national and international development goals and reduce poverty in low and middle income countries, the Partnership facilitates statistical capacity development, advocates for the integration of reliable data in decision making, and coordinates donor support to statistics (PARIS21 Secretariat 2016).

1945 as one of the six main organs of the United Nations. Data has been an essential part of measuring progress on the Millennium Development Goals (MDGs) adopted in 2000 and the subsequent Sustainable Development Goals (SDGs) adopted in 2015. The monitoring requirement of the MDGs is widely credited with driving significant changes in data production at the national level in developing countries and enhancing "national statistics capabilities, measured by statistical methodology, data sources and data periodicity" (Chen et al. 2013). At the same time, the authors note, the demand for statistics that will demonstrate progress against the MDGs stretched the resources of the very countries they are intended to assist and required an increased investment in statistical capacity development.

ECOSOC more recently renewed its commitment to global, regional and national statistics, proclaiming "statistics are essential for sustainable economic, environmental and social development" and that "high quality international statistics, accessible for all, are a fundamental element of global information systems" (UN ECOSOC 2013). A set of *Fundamental Principles of Official Statistics* was adopted in a resolution passed on 24 July 2013, the resolution noted

> the critical role of high-quality official statistical information in analysis and informed policy decision-making in support of sustainable development, peace and security, as well as for mutual knowledge and trade among the States and peoples of an increasingly connected world.

The ten principles, which start by stating "Official statistics provide an indispensable element in the information system of a democratic society…", cover the need for professional practice and ethics, the use scientific statistical principles, confidentiality, public access to empowering laws and regulations and the use of international standards and definitions (UN ECOSOC 2013). The final principle states "Bilateral and multilateral cooperation in statistics contributes to the improvement of systems of official statistics in all countries."

8.2.2 Capacity a Major Constraint

With their limited financial and technical resources and lack of human capacity (either because staff lack the necessary skills or, if they have them, are 'poached' by other government agencies) statistical offices in Pacific Island countries usually struggle to meet all the demands placed on them. On the other hand, technology is bringing new opportunities—for efficiencies in the collection, storage, analysis and dissemination of data—and also strengthening opportunities for greater collaboration in the region. The MDGs were initiated with little involvement of developing nations and belated consultation with official statisticians. The scale of demands they placed on national statistical offices led to the recognition that capacity had become a major constraint in the use made of statistics for development. Capacity building, led by the PARIS21 initiative, was therefore aimed at developing the national statistical system of each country as a whole and embedding the role of statistics in national policy through the development of National Statistics Development Strategies (NSDS), to be funded by PARIS21. Samoa, Tonga, the Cook Islands and Vanuatu have completed or have an NSDS in train. Fiji, Papua New Guinea and Solomon Islands are in the process of finalizing theirs, with other countries having also expressed interest to develop an NSDS in the coming years.

Unlike the intense MDG focus on developing countries, the SDGs place demands on all UN countries, and it has been argued that only a small minority will be in a position to implement the SDGs fully in the near future (Igoe 2015; Kroll 2015). The capacity to implement and report on the SDGs, which quadruple the number of indicators compared with the MDGs, will place further pressure on developing states. The small island states of the Pacific, which have struggled to cope with the indicators of the MDGs will face serious challenges reporting on the SDGs as most indicators, or their underlying statistics, have thus far never been collected and lie outside the current business scope of Pacific Island statistical agencies (Haberkorn 2016). Although Pacific countries (Fiji and Samoa) and the SPC have been involved as part of working groups in specifying the Inter-Agency Expert Group's (IAEG-SDGs) indicators to monitor the SDGs, the demands of the SDG indicators are such that the resources of PICs, or their development partners, are unlikely to meet country needs if solutions have to be found on a country by country basis. The burden becomes even greater for those independent countries with populations of around 100,000 or less (Federated States of Micronesia, Kiribati, Marshall Islands, Nauru, Niue, Northern Mariana Islands, Palau, Tokelau, Tonga, Tuvalu and Wallis and Futuna). Regional solutions and collaboration have therefore been identified as critical to fill this capacity 'gap'. Section 8.3 discusses the current situation and Sect. 8.4 some of the regional solutions that are emerging.

8.3 The Scope and Quality of Pacific Statistics

The core set of fifteen independent Pacific Island states that are members of the SPC[5] each have a National Statistics Office or statistics unit, which is expected to regularly provide a set of essential statistics to the United Nations and its technical agencies. Much of this essential data is transferred to the United Nations Statistics Office through the SPC's Statistics for Development Division on behalf of its member countries. Along with all other members of the United Nations, PICs are expected to report demographic, economic and social statistics to the UN at regular intervals. These include data derived from national population and housing

[5]This does not include the seven 'territories' which are also members of the SPC or the four remaining 'founder' members, Australia, New Zealand, France and the United States. See Chap. 1 for discussion of this point.

censuses (to be carried out every 5 or 10 years), Household Income and Expenditure Surveys (ideally to be carried out every five years) and Demographic and Health Surveys, which cover population health comprising maternal and child health, fertility, nutrition and diseases and are also expected to be carried out every five years). This is not the same as the core set of statistics reported annually to the WHO which is included in the National Minimum Development Indicators dataset discussed below (the collection of health data and other applications of ICTs in health services in PICs is covered in more detail in Chap. 12). Few Pacific Island countries manage to collect all the required data, although in the past decade all twenty-two SPC member countries and territories have completed a population and housing census, some more than once (primarily those countries which are protectorates of larger 'metropolitan' countries such as the United States, France and New Zealand). For the fifteen independent Pacific Island countries, changes to population are estimated biannually in the years in between censuses and medium-term population estimates up to the year 2050 are made.

The difficulty of estimating population growth has been a major cause for concern in the past decade, but is expected to improve rapidly due to the implementation of the Pacific Vital Statistics Action Plan in which the SPC Statistics Division is a lead partner as a member of the Brisbane Accord Group (BAG). The Pacific Vital Statistics Action Plan (PVSAP) 2011–2014 was developed by BAG partners to improve vital statistics in the region and improve coordination between development partners in response to the realization that "no one agency is responsible for civil registration and the vital statistics in the Pacific region" (Pacific CRVS 2014). The goal of the plan was to ensure that all Pacific countries have reliable civil registration systems to improve the quality of population data in the region. The CRVS systems being put in place in each country will register all births and deaths, issue birth and death certificates and compile and disseminate vital statistics, including cause of death information. The "paucity of reliable birth and death data across the region, as well as quality and timeliness issues" has hindered accurate reporting against the MDGs, reduces the reliability of health and mortality data (including cause of death) and is a significant barrier to cost effective planning and resource allocation. Moreover, CRVS systems will, in contrast, reduce the reliance on "costly household surveys undertaken on an ad hoc basis (and often guided by availability of funding rather than local/national demand for such information)" (p. 1). The impact of such systems is shown in the example form Vanuatu (see Box 8.1).

Box 8.1: Vanuatu's mobile birth registration system

In Vanuatu, a mobile birth registration system piloted in 2010 resulted in registration of close to 17,000 children in Tafea Province, some 87% of all children 0–18 years, a significant advance on earlier rates of registration, which were as low as 28% in 2008, according to a UNICEF report (UNICEF 2014).

Historically, with the majority of births taking place outside health facilities, families have been required to travel to rural centers to register a birth (with penalties for late registration). The inefficient and costly transport of paper records to regional centers and to the capital, along with lack of understanding in 'kastom' villages of the importance of registration has led to a situation where not only were the rights of children denied, but totally inaccurate birth data impacts on their access to health and education services and the value of national statistics for government planning, international reporting and the monitoring of progress on MDGs.

The new Vanuatu birth registration system simplifies the process by allowing birth registration information about a newborn baby to be entered into a pre-programmed mobile phone which sends the data to the Civil Registry database. This pilot is now being expanded across the country. In addition, mothers who give birth in the main 'referral' hospital in Port Vila can register their baby's birth and receive a birth certificate and photo while still in hospital, a major draw card since this is often the only photo of mother and child available to poor rural women. This is possible because of the permanent placing of a civil registry office at the hospital, so that essential (and overloaded) maternity services are not interrupted.

This is part of what has become known as an 'opportunistic registration' policy, providing services at key points in the community such as health and education facilities. A decentralization policy for the collection of birth data and the use of an 'innovative' e-database, utilizing open-source software and specifically tailored to suit the information needs of the government and key agencies "demonstrates the importance of tailoring a resource to suit a specific purpose and skills set." Examples such as this show what affordable technology, culturally compatible systems and incentives and regional and national partnerships can achieve.

Sources: UNHCR (2013); UNICEF (2014).

Other statistics that are required of all Pacific Island member countries include the SPC's own National Minimum Development Indicators (NMDIs)[6] which were originally developed in 2011–12 to assist PICs with regional and international monitoring and reporting required by the Pacific Islands Forum Secretariat's Pacific Plan, the MDGs and several international treaties and conventions such as CEDAW[7] and CRC[8]. Economic statistics required for the NMDIs include historical

[6]www.spc.int/nmdi. The NMDI database is currently being expanded to also include the three French and three US Pacific territories.

[7]The Convention on the Elimination of all Forms of Discrimination Against Women (CEDAW) is an international treaty adopted in 1979 by the United Nations General Assembly.

[8]The United Nations Convention of the Rights of the Child, adopted in 1989 (United Nations 1989).

and current analytical tables showing gross domestic product (GDP), prices, international merchandise trade, balance of payments, tourism arrivals, tourism earnings and employment. Also part of the NMDI dataset are: human development statistics on gender, youth and culture (possibly unique to the SPC, but an expression of the concern of most Pacific Island countries that globalization and economic development is threatening local culture and cultural practices); data on specific industries (agriculture, forestry, fisheries, aqua-culture, communications and infrastructure); and public health data (vital statistics, vector-borne diseases (such as malaria and dengue), communicable diseases (e.g. tuberculosis), sexual health (HIV/AIDS), maternal health, child health, non-communicable diseases, environmental health and health systems—all WHO requirements).

Most of this data is collected from administrative systems in-country (these may be either electronic or based on manual surveys, entered into a spreadsheet for in-country use and transferred to the SPC to be entered into the regional database). Most countries have at least a paper-based national Education Management Information System that records the necessary data; the SPC's Statistics for Development Division's Strengthening Education Management Information Systems (EMIS) initiative, funded by the Commonwealth of Australia, is helping to put in place appropriate systems to enhance the collection and dissemination of information on the region's education systems. Centralized financial and HR systems that are routinely being installed in most countries will provide more reliable employee data in both the health and education sectors, both the largest employee groups in most countries, including the most remote and inaccessible of employed staff.

The NMDIs that all fifteen Pacific Island countries are expected to report on require a breakdown of some of the economic data into more analytical statistics including GDP per employed person, real per capita growth, annual change in CPI, trade balance-GDP ratio and annual tourists. Labor statistics required include the labor force participation rate, unemployment, the employment to population ratio and 'own account family workers' (these last two reflect the particular nature of the agricultural economy that supports the majority of the population in some Pacific countries). Statistics on social services provision includes government expenditure on education and health expenditure per capita. Not all countries are able to provide up-to-date data on all these indicators, as the underlying data sources are, more often than not, population census and household surveys rather than drawn from administrative databases which would provide more 'real-time' data. Gaps in data and currency of data are noted in the SPC's summary tables, and in some cases data is either non-existent or over a decade out of date. Much of the data is gathered manually, often on foot (some villages where information is collected are several days' walk from the nearest road), and is carried across considerable distances by sea to provincial and national capitals where it is entered into spreadsheets. New technologies such as tablets and smart phones have much to offer in terms of currency and accuracy in this process; some examples of how mobile technology is changing data collection are offered in this chapter.

Even with these initiatives, it is clear from this picture that, overall, most Pacific Island states are too small to maintain the capacity to operate and develop official statistics of the mix needed in a small economy either to meet their international obligations or to provide timely and appropriate data for government decision-making. The level of government funding may never be stable enough or even sufficient to maintain ongoing programs in the form originally developed and operated and to international standards. The investment in statistical co-ordination and standards in these small countries would be a prohibitive share of the statistics budget, were it carried out to the level required by the various UN standards and other international standards such as the International Monetary Fund's General Data Dissemination System (GDDS). The ongoing preparation of statistics for IMF's GDDS goes far beyond each country's domestic needs for budget planning, management of the tax base and other domestic decisions, let alone the demands of metadata and documentation that they require. Furthermore, the level of technology applied to administrative processes across Pacific Island governments (apart from financial, revenue and HR systems) is poor and has significantly jeopardized how far administrative sources can underpin both social and economic statistics. Even the requirements of the SPC's NMDIs place significant demands on national statistics offices or units in Pacific Island countries and depend on statistical infrastructures usually found in larger countries.

A number of other factors also affect the ability of these small island developing states to produce reliable statistics. For example, although sampling frames exist, the sampling fractions needed for reliable survey estimates are disproportionately large in small countries, although advances in methodology could reduce some of this particular burden. However, despite their size, several Pacific countries have made effective use of available administrative records, most significantly Samoa and Cook Islands, which are now able to regularly produce a very comprehensive and timely set of national accounts; and many smaller countries are making progress revamping their civil registrations systems. The sharing of this sort of experience plays a major part in raising expectations of what can be achieved within available resources.

Local economy factors must also be taken into account. Pacific Island countries tend to have two tier economic systems with a mix of a subsistence economy for the majority of their population and globally competitive economic sectors (e.g. export commodities such timber, processed fish, pearls and tourism). This makes it difficult to apply standard economic measures to the overall economy. Furthermore, with highly mobile populations, their economies are very dependent on overseas remittances either from their citizens working in seasonal (mainly agricultural) industries in Australia and New Zealand or from those who have gained residency and employment in either of these countries but who retain citizenship. This exemplifies the mismatch between the statistical data of most value to economic development in the Pacific and the statistics that are required internationally;

statistics which inform natural resource management are probably of greater value for Pacific Island governments than others that are an obligatory element of international economic accounts.

8.4 The Value of Regional Collaboration

Regional leadership and collaboration have long been identified as critical factors enhancing statistics in the region.

> The future impetus for change among the statistical offices of the Pacific Island region will come from two directions. The first is the intensity of purpose that users bring to the use and usability of official statistics. The second will be the capacity for regional decision-making on common infrastructures, systems and tools, and the statistical infrastructures of harmonization, in particular classifications, standards and methods, and in the instruments used in the collection and presentation of statistics. (Cook and Paunga 2010, p. 5)

However, the complex history of relationships among Pacific countries means that development partners have not always had a commitment to common regional ICT-based solutions (SPC 2010). But a paper prepared for the 2015 PARIS21 Cross Regional Forum argues that regional statistics cooperation among developing countries has enabled the contributions of development partners to be distributed in an organized manner and statistical operations to be planned in such a way that country investments can be maintained and built up (PARIS21 Secretariat 2015, p. 4).

A lot of the groundwork for securing investment for such regional solutions is already in place. The Pacific Islands Forum's *Pacific Plan* and subsequent *Framework for Pacific Regionalism* (PIFS 2005, 2014) are effectively a blue-print for regional cooperation and investment. These frameworks enable the SPC, the inter-government organization responsible for regional statistics, to share resources and any investment in the statistical tools, information systems and skill sets routinely available to more developed countries. The SPC's Statistics for Development Division plays a key support role in the region and its various technical and working groups lead most initiatives at regional and national level. In addition, its web-based data portal PRISM that connects to Pacific statistical agencies' websites and its National Minimum Development Indicator database coordinate systems that are used to disseminate and compare data from countries in a region or with similar contexts; these systems add significant value to official statistics.

The Statistics for Development Division has for many years been the central element of the Pacific Statistics System as it has been emerging and its roles have reflected the unique needs of Pacific Island states and what is needed to complement the capability provided by the states themselves. The SPC hosts the three-yearly Regional Conference of Pacific Heads of Planning and Heads of Statistics (HOPS) group whose Ten Year Pacific Statistics Strategy (2010–2020) (TYPSS), drives regional development in the sector. Collective leadership between meetings has been provided since 2010 by the Pacific Statistics Steering Committee (PSSC), which has responsibility for overseeing the implementation of the TYPSS. PSSC comprises the heads of six Pacific Island national statistical agencies elected by HOPS for a three-year term and various technical and financial partners involved in statistical development across the region.[9]

The Ten Year Pacific Statistics Strategy grew from awareness of the need for a stronger regional approach to national statistics at the 2009 Pacific Islands Forum Economic Ministers Meeting (FEMM), the FEMM-initiated Regional Statistical Benchmarking study in 2008–2009 and the Ministers' recognition of the importance of statistics to inform evidence-based policy. It was intended to provide stronger regional leadership for statistical developments and operations in the Pacific region, mobilize technical experts to provide technical assistance and training, supplement capacity in National Statistical Offices where necessary and strengthen governance of regional statistics through more control by member countries. The proposal was endorsed by the SPC Committee of Representatives of Governments and Administrations (CRGA). Subsequently, an implementation plan entitled *A Pacific Island Region Plan for the Implementation of Initiatives for Strengthening Statistical Services through Regional Approaches, 2010–2020* (Cook and Paunga 2010) was tabled by SPC at the 3rd Regional Conference of HOPS in July, 2010. The implementation plan built on the recognition of the original FEMM initiative that although the pressures for improvement are mostly at a country level, many of the solutions would be regional and ICT-based. "In this plan we have identified the key areas where a strengthened regional leadership of the Pacific Island statistical system could increase the impact of country and development partner initiatives" (p. 3).

Whereas the plan's authors considered that, when adequately resourced, most Pacific Island countries have managed statistical collections in a timely manner with high response rates and possibly at lower cost than developed countries would manage, it also noted the limited usability of statistical collections and their inability to provide the value that users expect from them. They concluded:

> …to add considerably to the value that users get from statistics in the Pacific Islands, the statistical systems of Pacific Island countries need to use the information and communications technologies that are now common elsewhere. Access to a personal computer is simply not sufficient (Cook and Paunga. 2010, p. 3)

---

[9]The 2013–2016 PSSC comprises the following partners: Australian Bureau of Statistics, Australian Department of Foreign Affairs and Trade, Asian Development Bank, Pacific Community, Pacific Financial Technical Assistance Centre, Paris21, Statistics New Zealand, UNDP Pacific Regional Centre, University of the South Pacific, World Bank.

8.5 The Potential for ICTs in the Transformation of Pacific Statistics

New tools and technologies certainly have the potential to resolve many of the problems faced by national statistical offices and units in the Pacific. As the design for Phase 2 (2015–2017) of the Ten Year Pacific Statistics Strategy noted:

> Official statistics are highly dependent on information technology, which is a significant driver of innovation. A further key principle for Phase 2 is the expectation of Pacific-wide innovation, particularly those sourced from the application of technology, adoption of systems or analytical practices that have worked in one of the Pacific countries. This principle is about development partners resourcing regional leadership to promote adoption and sharing of innovation. (Kelly et al. 2014, p. 27)

Pacific countries thus face strong imperatives to share resources in order to take advantage of new technologies and counter the lack of resources that have led to the accumulation of incompatible tools, processes and the degradation of statistical operations due to lack of capacity or resources to keep these tools functioning. These constraints can be countered in the future to some degree by adopting common tools and sharing solutions among countries. Where this has been achieved in developing country regions it has already involved:

- facilitating 'south-south' collaboration in regional initiatives that share statistical solutions and expertise;
- providing forums for collective reflection and encouraging innovation such as PSSC and the Technical Working Groups;
- collaborating regionally on complex issues such as big data and changes to common elsewhere. Access to a personal measurement;
- enriching the knowledge of PICs and development partners of the potential scope and nature of shared solutions and the pathways to their adoption in the region, which is more difficult; and
- finding incentives for countries to adopt common tools and share their experiences in applying them, country by country.

Advances in technology that are occurring in all stages of the 'statistics value chain' (the process of collecting, storing, analyzing and disseminating official statistics) have a huge impact on the value users get from official statistics. This is increasingly being reflected in the management of statistics in the Pacific region. In many fields of endeavor, both public and private, innovation in ICT is capitalizing on the rapidly advancing technological infrastructures of the Pacific region, such as mobile telecommunication systems (3G and 4G extending to provincial as well as national capitals, and 2G mobile systems extending into more remote provinces), new solar technologies, more reliable infrastructure and better backup systems; all have enabled a number of innovations and enhancements at all points of the statistics value chain in PICs. The UN-sponsored Sustainable Development Solutions Network highlighted this in a recent report focused on ways in which data production, analysis and communication can be modernized and take into account emergent technologies, observing: "the unprecedented rate of innovation in data collection techniques and technologies and the capacity to distribute data widely and freely has expanded the horizon of possibility" (Espey 2015, p. 6). These technologies and the widespread adoption of NSDS, the report notes, will make it easier and cheaper for countries to report on the newly adopted Sustainable Development Goals. Technologies noted in the report as having significant impact include satellite and geospatial technologies, biometric data and mobile data collection. Such innovations also include better web-based access to data for planners and decision-makers and to innovative tools for data display to help interpret data.

Also having considerable impact in terms of data capture using server based systems, are some sophisticated tools provided by the United Nations and its agencies to support the collection and management of statistical data. The UN developed programs CensusInfo and ASYCUDA, for example, are freely available to member countries. They are not without costs, since they require a well-supported infrastructure (hardware with up-to-date operating systems, good connectivity, servers of sufficient capacity and good backup) and trained personnel. But, they have a singular advantage over older systems in that they have a user-friendly interface and manage the process of handling statistical outputs, requiring less highly trained staff and fewer staff with an extensive background in statistics.

CensusInfo, the UN Statistics Division's most recent software package for handling and disseminating census data is designed specifically for UN data requirements and made freely available. It is often used in conjunction with the SPC Statistics for Development Division's own census and survey data processing system, CSPro. In addition, in line with TYPSS' focus on accessibility and user-relevance of official statistics, the SDD has implemented an interactive web-based statistical information system PRISM for data presentation and analysis (including the NMDIs) and employs POPGIS2 for data mapping and graphic display. Some countries (e.g. Cook Islands, Vanuatu, Solomon Islands) have also been experimenting with character recognition software, which speeds up data entry although, according to the Cook Islands Government Statistician, it does not reduce the number of errors (Cullen and Hassall 2013).

External trade statistics are mostly captured through PC/Trade, an SQL/Visual Basic database system which was created some 20 years ago by Statistics New Zealand for capturing trade data in PICs, and which continues to be used by the majority of countries.[10] Trade statistics are also captured by ASYCUDA, a web-based customs and border control system developed by UNCTAD (United Nations Conference on Trade and Development) provided at no cost to developing countries, used by some larger PICs which can afford the costs of installation and

[10]The development of PC/Trade was funded by the New Zealand Ministry of Foreign Affairs and Trade, and its capabilities have in recent years been extended to include customs and revenue management, as explained in Chap. 5.

maintenance.[11] As well as managing imports in line with the World Customs Organization's data model, ASYCUDA produces reliable statistics, including bio-security data. In most PICs, proprietary Financial Management Information Systems, HR and Revenue Management Systems with sophisticated reporting and analytic components[12] provide administrative and financial data that can form the basis of the national accounts and contribute to the SPC's NMDIs (See Box 8.2).

However, as the Pacific Island Region Plan for Implementation of Initiatives for Strengthening Statistical Services through Regional Approaches, 2010–2020 noted, while the region has a large 'endowment' of such statistical sources, these need to be better managed so that users obtain more value. The report notes that providing effective access to countries' existing statistical sources will add value to users that will be significantly larger than the costs of doing so, and comments "we see the opportunity costs of not doing this as a continual series of high cost donor-run surveys, not all of which are even accessible from within the targeted countries" (SPC 2010, p. 16). Administrative sources contain much relevant information, but in general, the report notes, administrative records are stored rather than managed, despite their potential. This is exemplified by the following report on Solomon Islands.

Box 8.2: The potential of Solomon Islands financial and HR systems for statistical reporting

Solomon Islands' Ministry of Finance uses a general ERP system, Microsoft's AX Dynamics, operating from a large data center in the Ministry's ICT Support Unit. The Treasury Division uses the Commonwealth Secretariat's CS-DRMS (Debt Recording and Management System), a state-of the art system which integrates well with the World Bank Debtor Reporting System, widely used across the Commonwealth and beyond for managing public debt. Treasury's Payroll, Revenue and Imprest Branch uses a full HRMIS, Aurion's HR and Payroll system (which is linked to the central accounting system).

These recently installed powerful systems have the capacity to extract data that will support and enable the Solomon Islands National Statistical Office to fulfill the main goals of its three-year plan "To advance the quality of national and economic statistics in the country… further enforced though the proposed National Statistical Development Strategy (NSDS)." However, the NSO'S Plan also acknowledges that higher level technical skills are needed for advanced economic compilation and methodological analysis, sampling and advanced demographic analysis in order to provide timely statistics for policy and planning. Currently the NSO lacks capacity to undertake all the key national statistical projects, such as the Census, Household Income and

[11]This is covered in detail in Chap. 5.

[12]These are also covered in detail in Chap. 5.

Expenditure Survey (HIES), Demographic and Health Survey (DHS). Support and technical assistance is still needed from key development partners to provide quality statistics for these core needs, as well as essential economic and social indicators for policy-making and planning. The NSO's Plan includes a case for a National Statistics Development Strategy, the Health Survey planning process within the NSO to be strengthened, as well as systems and training to enable data for the DHS and HIES to be extracted from the financial and payroll systems that have been installed.

Source: Cullen and Hassall (2016); Solomon Islands Government (2013).

8.6 Emerging Technologies

Among the new technologies which are already adding to the statistical value chain in PICs, two in particular stand out—mobile technologies and geospatial technologies. These are increasing the quality of statistics in the region at all points of the value chain.

8.6.1 Mobile Technologies for Data Capture

The introduction of mobile technology, using solar power and satellite or WiMax (wireless microwave)[13] to Small Island Developing States, is widely seen as more transformational than in more developed countries, simply because it brings communications technologies to regions which have neither electricity nor telecommunications systems (Cave 2012). Such technologies enable more rapid communication between families, businesses and government (discussed in more detail in Chaps. 8, 9 and 10); they also provide solutions to the challenges of achieving accurate and efficient data collection in remote rural areas.

Data collection for the broad range of statistics required by the SPC National Minimum Development Indicators is difficult, especially for the smaller countries (i.e. with populations of 20,000 or less, such as Cook Islands, Niue, Nauru, Tokelau and Tuvalu) and for those whose populations are spread over many remote and often inaccessible islands (e.g. Vanuatu and the Federated States of Micronesia). For these countries, the innovative use of tablets and mobile phones, often linked through satellite-based telecommunications systems can be very helpful in collecting NMDI data; mobile technologies enable more timely and accurate data

[13]WiMax is the acronym for Worldwide Interoperability for Microwave Access.

collection and in greater amounts, since the process is more efficient (however, the cost of telecommunications in PICS should not be ignored).

Smart phones and tablets enable operators to collect biometric and geospatial data simultaneously and link text and images with GPS systems, satellite images and geospatial databases; systems using these combined technologies are now widely used in the Pacific for mapping weather events, data on the impacts of climate change as well as data on land ownership, land use, soils, cropping, building structures, water storage tanks and diseases (especially vector-borne diseases). When this data is layered using interactive geospatial visualization software, new insights emerge, adding immediate value for planners and adding to the statistical value chain.

Using such technologies, it is not uncommon for national statistics offices, for example, to collaborate with agencies collecting geospatial data for other purposes (such as determining land ownership in customary land systems, or for disaster preparedness) to create address registers for their entire population and compare this with other datasets (see the example below from the Cook Islands). Address registers of this kind, which link every individual to a known abode, used in conjunction with other population data can be a valuable aid in assessing development progress. They provide richer data than is readily available in many larger developed countries. At the same time, systems of this kind raise major issues related to privacy and the ownership of personal data thus obtained; these concerns highlight the lack of up-to-date and appropriate legislative frameworks for the collection and dissemination of population and health data in the region.

Box 8.3: The Cook Islands' GIS portal—GeoNode

The Cook Islands GeoNode system and portal (http://geonode.emci.gov.ck), funded by UNESCO'S regional development program ESCAP, is a GIS-based system of documents and maps to support emergency response. The system makes use of data from Infrastructure Cook Islands (ICI), which undertakes the mapping of soils, geology, topographical data, land use and crop modeling using GIS technology. ICI also undertakes surveying and mapping of existing building structures to assess their vulnerability to natural disasters and is developing an Infrastructure Asset Database. ICI maintains and services a GPS Station in order to service the data needs of the local and global community and, in particular, the Australian Climate Change project COSPPac (http://cosppac.bom.gov.au) which serves 14 Pacific countries.

GeoNode is an open source geospatial content management system and web-based application platform for developing geospatial information. Registered users can create their own maps overlaying different sets of data from the collection. Emergency Management Cook Islands (EMCI) uses GIS to plot data about geographical environments, roads, rivers, water quality and land use against population data (e.g. house, location, village, owner, house type, whether occupied, the age and state of health of occupants, water tank capacity, Internet connection or cell phone, etc.). For disaster management and emergency response, which is the main aim of the system, this can then be combined with data from the Met Service.

The GIS data collected by ICI and EMCI is used by many agencies, e.g. the National Environment Service maintains the data on land use, water quality and waste management; combining this with climate data can help predict drought and other adverse events. Google Earth, another source of land-use data that is being entered into the system, is also used for presenting data.

The GeoNode network is very reliant on the quality of its metadata and dependent on cooperation between key stakeholders—the disability sector, the National Environment Service, Internal Affairs and the Cook Islands Statistics Office (some of these groups have raised concerns about privacy and security of the data. Access is therefore protected and, even in an emergency, highly restricted).

Source: Cullen and Hassall (2013).

8.6.2 New Technologies for Analyzing and Presenting Data

Communicating the value of statistical data and information and presenting data in a meaningful way to planners and decision-makers is a key part of the statistical value chain. The facility of new technologies to enhance understanding of the meaning of statistics and realization of their value can play a significant part in reducing the capability gap between statistics office staff and their various audiences, especially politicians and policy-makers (often unfamiliar with statistical concepts), but also media, students and development partners. Statistical literacy levels are universally a long way behind what powerful software can now produce. Visualization programs for data presentation that have an immediate impact are revolutionary in bridging this gap and can help persuade budget holders and aid partners of the importance of the work done in the statistics offices in the Pacific and how it can be used in decision-making. Technologies such as GapMinder[14] and the OECD's data visualization tool,[15] both of which now include data on Pacific Island countries, and the Google visualization tools, GoogleVis, GoogleCharts and Google Motion Charts, which can be applied to the user's data, are all powerful ways of communicating the meaning of statistics. This is just the beginning of a new generation of data visualization tools built for the new mobile technology era.

---

[14]http://www.gapminder.org.

[15]http://stats.oecd.org/OECDregionalstatistics/#.

8.7 Strategies and Programs for Capacity Building

Statistical capacity is defined by the World Bank as a nation's ability to collect, analyze and disseminate high-quality data about its population and economy (World Bank 2016). This requires an adequate level of technology and the skills to maintain it, as well as statistical knowledge. As the *Pacific Island Region Plan for the Implementation of Initiatives for Strengthening Statistical Services through Regional Approaches, 2010–2020* noted:

> the management of the existing statistical database[s] reflects the quality and relevance of the ICT systems available, and the support for their ongoing use, including training and expert assistance, as well as the ease of integration of new systems into established processes (SPC 2010, p. 26).

The report also noted that the management and accessibility of statistical information could be enhanced to add significant value to users, were it not that "innovation in processes is severely stifled by existing systems, as is the capacity to integrate statistical outputs with the systems users work with…" (p. 26).

In the intervening years, as a result of a more coordinated regional approach, NSOs in many more PICs are using standard tools and the technology base has been significantly enhanced. But because of that the nature of the skills needed to exploit this new technology is changing. ICT-related capacity needs identified in the original Implementation Plan included not only the capacity to process statistical data and enter and codify survey responses, but to manage metadata and interrogate and manipulate databases with mapping/charting systems (SPC 2010, p. 75). While these skills are still in demand, training needs have changed and skills in data interpretation are needed to continue adding value to a rapidly increasing amount of statistical information and enhance its relevance to evidence-based policy making. Improving data accessibility and utilization has therefore been a major component of the TYPSS and has been included in all three phases of the strategy.

This capacity can be built in various ways and international organizations and development partners make a significant contribution. The United Nations Statistical Institute for Asia and the Pacific (SIAP), based in Chiba, Japan,[16] has taken a lead in organizing face-to-face training programs and online learning programs for new graduates and mid-level statisticians in less developed countries in the region. Programs range from basic training to the use of emerging technologies and are closely linked to the main drivers in the domain, the NSDS Guidelines produced by PARIS21, the MDG reporting requirements and the SDGs. SIAP is also committed to regional partnerships for training and includes the SPC Statistics for Development Division and the University of the South Pacific among its development partners.

---

[16]SIAP (http://www.unsiap.or.jp) is sponsored by ESCAP with financial support from JICA (the Japan International Cooperation Agency).

The SPC Statistics for Development Division (SDD) is actively engaged in building capacity in its own right. Qualified graduates are not as plentiful in Pacific countries as in Asia or the metropolitan members of the SPC and training is a great deal more basic. The SDD runs workshops at various levels on data collection and analysis, databases, management information systems and GIS systems in all member countries. Part of its focus is to develop capacity in the analysis of data from financial, HR and revenue management systems to reduce the reliance on sample surveys to fill gaps in statistics which can be gathered on an ongoing basis and more economically through exploiting systems already in place. A third major contributor to training in official statistics in the region is the University of the South Pacific (USP). USP, as the regional public research university, has campuses in all twelve countries and offers open learning programs across the Pacific. It offers a major in Official Statistics, the first two years of which can be taken at most campuses. The official statistics program and course material have been recently redesigned and updated to meet changing needs.

Despite these opportunities, capacity remains a huge problem. Lack of capacity can be exacerbated by the proliferation of tools left over from a plethora of programs introduced by various development partners over the years and the increasing demands of users, whether governments, development partners or the UN and its agencies. Collaboration and planning, the PARIS21 paper on Regional Statistics Cooperation argues, can help maximize investments and reduce this burden through a staged process of regional cooperation and integration, collectively funded. This model projects a progression from the publication of up-to-date monthly regional statistics through harmonization of statistical practices, to statistical integration and eventually a regional system based on regional investments (PARIS21 Secretariat 2015, p. 7). It is a high-tech model, heavily reliant on some of the data systems outlined above, as well as the electronic transfer of data and sophisticated computer-based modelling for data analysis. It is also dependent on the support of regional development banks (in this case the Asian Development Bank) to support the necessary ICT systems and training. There is clearly a long road ahead given that most of the Pacific is not yet reaching the first stage of publishing monthly statistics; as we observed earlier, the statistics recorded in SPC's NMDI database are frequently out of date or totally missing.

The level of economic cooperation and leadership advocated in the PARIS21 paper cited above is therefore essential to building capacity in the Pacific region. And while the report notes the key role that SIAP plays in the region and the invaluable work of the SPC and its technical working groups focusing on data collection and use (PARIS21 Secretariat 2015), south-south country-country collaboration is also critical, playing a major part in building capacity. Knowledge sharing, where official statisticians from one country provide technical or operational support to another country in the region, is commonly reported. Methods of geocoding in the Agricultural census of Vanuatu have been shared with Samoa, Vanuatu's expertise in scanning census data has been shared with Solomon Islands and Cook Islands, Samoa has lent assistance to neighboring countries in the development of national accounts and the Fiji Bureau of Statistics has done the same in regard to trade statistics and business registers. Such examples strengthen commitment to the more formal regional structures for cooperation and capacity building.

8.8 Sustainability

Given the immense pressures on SIDS that stem from their lack of resources and the demands made on them by international bodies, sustainability is a key issue. Although there is considerable goodwill between PICs and their development partners to pursue regional solutions and make use of technology to enhance the management of statistics in the region, these initiatives also face considerable challenges. Recognizing and managing these is essential to ensure that statistics fulfill their potential to contribute to development in the Pacific Islands.

8.8.1 Regional Collaboration and Cooperation Challenges

Extending both regional and country capacity requires a great degree of mutual cooperation and support. National statisticians in PICs face significant demands in their joint roles as leaders of national statistical systems and custodians of the regional statistical system; and many also play a part in international statistical organizations, such as the IAEG-SDGs. Significant constraints commonly discussed at regional meetings include[17]:

- the impacts on capacity of such meetings due to the sheer distance between countries, the associated costs and inconvenience and time away from routine tasks;
- the difficulties of freeing up staff for regional secondments (exacerbated by considerable disparity in wages between countries);
- the cost of collaboration (not just for travel but also work pressure) for small isolated agencies trying to engage and contribute across government, regionally and internationally;
- significant differences in culture, languages, histories and administrative practices resulting in different information needs and data collection approaches, which present problems in the creation of solutions for use in a range of countries;

[17]Personal communication with V. Galvin, Statistics New Zealand, 19 June 2016.

- different resource levels between countries and development partners, the frequent turnover of development partner representatives and changing and conflicting agendas.

That regional collaboration and cooperation is increasing despite such difficulties illustrates not only the value of such cooperation, but also the need for committed leadership to maintain it. The initiative of PARIS21 in encouraging developing countries to develop and work within national strategies for the development of statistics, and the impetus given to this initiative by the Ten Year Pacific Statistics Strategy (TYPSS) and the TYPSS associated SPC strategic plans (SPC 2011, 2015) have been extremely helpful and will help counter the impact of 'agenda disparity'.

8.8.2 Adapting to the Constraints of Size

The small size and limited economies of PICs have a direct impact on the management of official statistics, affecting everything from financial resources and capacity to the nature of information that it is appropriate to collect and to use. The larger members of the SPC, in particular Papua New Guinea (population over 7 million) and Fiji (over 800,000), are able to maintain relatively well-staffed statistics offices, allowing staff to specialize in areas such as environmental data and to participate in regional and international organizations and decision-making. They are able to use methods for data collection and statistical tools widely used in industrialized countries and make use of statistical data for planning infrastructure and social services and amenities. But the difference between their situation and that of countries with populations of less than 20,000, which account for over a third of the independent member states of the SPC, means that not only are the methods that smaller countries can afford very different, so too are the purposes for which statistics can be used. Their scope of action is more limited and the decisions their governments can make constrained. The data they require needs to be targeted towards the decisions they are able to make within their resources (and those available through development partners) and the activities that support their lifestyle—whether this is based on traditional subsistence agriculture and fisheries or newer industries such as tourism and pearl culture.

Consideration of appropriate measures for such countries, cost-effective methods of data collection and even a simplified form of national accounts that better capture the essence of a small economy have been mooted. So too has the need for some way of making the collection of statistical data more affordable and sustainable in PICS, including those with populations from 100,000 to 250,000. Currently the large sampling fractions necessary to ensure reliability in small populations means even a small set of household sample surveys can have a very high respondent load for the population. Care must be taken with new measures arising from the SDGs and other global initiatives that they do not overwhelm these small countries already burdened with the requirements of statehood, as noted in Chap. 1.

8.8.3 Harnessing Innovation in a Sustainable Way

Good design decisions that take account of the constraints of PICs and that integrate local capability, the methods to be used and available affordable technology lead to sustainable solutions where the value to the country is evident. Appropriate models are not always readily available and considerable astuteness is often needed, as is shown in this chapter's examples, especially the Brisbane Accord Group's CRVS initiative, to develop suitable solutions. These solutions are more likely to close the design-reality gap that threatens sustainability according to Heeks' ICT4D 2.0 Manifesto (Heeks 2014), a concept that informs this book and which is explored in more detail in Chap. 1 and other chapters.

But new systems are in competition with the need to maintain existing systems. Difficult choices must then be made about whether to retain an existing system that works and can be maintained or to migrate to a new system that imposes additional training needs; decisions on how far to invest in technological innovation and common regional systems while ensuring sustainability are also necessary. A pattern of contracted developers applying their own preferred software for projects has led to a plethora of tools and various impacts; the software provided by donor partners focused on short-term projects is often unaffordable long term, so systems either languish unused or are maintained with difficulty by local staff whose endeavors are taken up maintaining the software. This can often require considerable skill and initiative and lead to some innovative solutions, but in the long term maintaining such systems hinders innovation. Open source software is often seen as the solution, but open source software can also bring high maintenance costs and require a great deal of customization—a challenge for an isolated technician with limited experience and support.

And while regional systems and common solutions are beginning to have considerable impact they can sometimes mean that countries lose autonomy and the ability to make their own decisions about software, so they select a suite of tools that suit their own domestic requirements. Countries and agencies which are dependent on development partners, international technical assistance and regional collaborations have less influence on these choices. Sustainable regional solutions must take local needs into account.

8.8.4 The Importance of Enabling and Mandating Legislation

Working with mandated authority is often overlooked but essential to sustainability. Some of the changes being advocated in the TYPSS and by organizations such as PARIS21 cannot be properly implemented without legislation that authorizes the collection of and use of data. For example, the first Implementation Plan (SPC

2010) notes that making further use for official statistics of data drawn from central (or regional) administrative systems usually requires some amendment to legislation. But the Plan also notes "when information is available, its accessibility may be seriously delayed by legislative approval practices" (p. 18). Some years later the Phase 2 report reiterates that there are recognized foundational processes or 'building blocks' for strengthening national statistical systems including the existence of relevant and up to date legislation.

The use of geospatial data and address registers also requires enabling legislation that will protect individuals while not losing the outstanding benefits these systems can bring. The issue of protecting personal privacy in GIS systems has been discussed in the GIS community since the early 1990s and has remained an issue ever since (Onsrud et al. 1994; Nourbakhsh et al. 2006). But the enabling legislation needed to underpin many e-government initiatives is often a casualty of time, lack of resources and failure to understand the need for it. Defining and protecting citizens' rights is not a top priority during an emergency and even afterwards cannot be dealt with until conditions return to normal (which can be some years). This is all the more reason for vulnerable states to ensure enabling legislation is in place well before access to data that will save lives and property is made accessible to a wide range of emergency agencies. As the TYPSS Phase 2 report notes, most Pacific countries have Statistics Acts and/or Census Acts (albeit slightly dated),

> but the mechanism for different data producing agencies to enter into agreements to jointly collect and share data varies considerably. Of the countries visited by the design team, Samoa and Fiji are the standout examples where good progress has been made towards developing these protocols. In Vanuatu some formal mechanisms exist and several informal mechanisms serve to maintain good working relationships between statistical producers. (Kelly et al. 2014, p. 57)

Since that report, Samoa, the Kingdom of Tonga and the Cook Islands have also put in the necessary legislation in place with the assistance of PARIS21.

8.8.5 *Leadership*

Leadership is by far the most critical factor to ensure a continuing development path for Pacific statistics. Leadership was identified as one of the key areas for the Pacific Statistics Steering Committee to work on in the initial Implementation Plan for the TYPSS, and it continues to be emphasized as a key work area in the Phase 2 Implementation Plan which will seek to "strengthen regional leadership and ensure greater coordinated technical assistance" (Kelly et al. 2014, p. 2). This will be done by strengthening the PSSC, which "will be supported to strengthen its capacity to be a regional and international voice for Pacific statistics; and overseeing the implementation and progress of TYPSS" (p. 3). As the original Implementation Plan notes:

> At its most effective, regional leadership of the Pacific Island statistical system will determine the long term share of investment put into infrastructure, frame management, methods and tools. It will play a part in building up the place of statistics in government, and challenge long term direction and mix of statistics in the region. This will then become easier to do at a country level. In all its decisions, respondent trust will be protected as the uses and forms of statistics are extended. This will be done in part by building local commitment to good statistical collections, through producing community statistics in a timely way, building up statistical portraits of island countries, and a commitment to non-market activity measurement and its valuation.
>
> Enhance the NSO capacity to report by building the common elements of a regular national statistical report in all Pacific Island countries, addressing "Who are we?", "How do we make a living?", "How do we participate in the global economy?", "Where do our people, goods, land go to, and come from?", "How is health, education, safety maintained?", "How do we pay for government, and what do we get?" (SPC 2010, p. 73)

The leadership emerging from the PSSC is therefore a critical element in balancing the need for regional leadership with "country owned and driven prioritization" (p. 3). At the same time, governments must also take some of the responsibility. "Political commitment is the major catalyst for bringing about change in the purposeful use of statistics" (SPC 2010, p. 5). Acknowledging the value of statistics for development, recognizing the role of new technologies in the effective management of statistics, balancing the benefits and risks, achieving economies of scale through regional initiatives while minimizing the burden of communications: all these take committed leadership for all stakeholders.

8.9 Conclusion

As this chapter has shown, the use of communications and information technologies has been a major factor in progressing the collection and use of official statistics in Pacific Island countries in recent years. Technology, innovation, collaboration and regional leadership are helping PICs overcome the challenges of their small size, fragile economies, lack of financial and human resources and the increasing demands for official statistics. But achieving the right balance between regional cooperation and leadership, individual country needs and the goals of development partners is the key to sustainability. While the Pacific as a whole is unlikely for the foreseeable future to be able to be adequately resourced to undertake the full range of essential statistical functions and requirements, "with good quality technical support and effective regional governance, Pacific countries will develop to the position of being able to control and manage their statistical collections" (Kelly et al. 2014, p. 42).

The increased demand for statistics, partly as a result of the adoption of the SDGs and the SDG Indicators and partly due to emerging issues that require new statistical data such climate change, disability, gender and culture, mean that more targets are likely to be added in coming years. At the same time, regional collaboration is empowering PICs to focus on statistics that reflect the needs of individual countries and the value of non-market activity, as we noted above. Information and communication technologies therefore must continue to provide gains in the collection and management of statistics in the region in order to deal with these demands; they are also needed to provide advances in data dissemination and analytical capability, helping to ensure that the benefits of the resources and endeavor that are put into official statistics are resulting in significant gains for planners and decision-makers.

The leadership of the PSSC, the vision shown in the Ten Year Pacific Statistics Strategy and the commitment shown to it by Pacific Island countries (and hopefully their development partners) have the potential to bring about some of the transformational change that e-government initiatives are credited with (United Nations 2014, p. 37; World Bank 2015). It will also enhance the contribution of official statistics to planning and decision-making that the TYPSS anticipates, and make national statistics more availability to citizens to promote democracy and accountability. But the impact of regional statistical collaboration and its reliance on innovative and practical applications of technology, is more than that. The leadership of the PSSC has the potential to allow national statistical offices and their staff to shape the development of official statistics in the region, setting their own regional priorities rather than adapting to external demands as they have had to do in the past. While acknowledging the challenges that still lie ahead, this is a more sustainable solution that builds on the commitment and pragmatism of local staff and priorities identified by the Pacific Island states themselves.

References

Cave D (2012) Digital islands: how the Pacific's ICT revolution is transforming the region. Lowy Institute for International Policy, Sydney

Chen S, Fonteneau F, Jutting J, Klasen S (2013) Towards a post-2015 framework that counts: developing national statistical capacity. PARIS21 discussion paper, No. 1, Nov 2013. PARIS21, Paris

Cook L, Paunga M (2010) A Pacific Islands region plan for the implementation of initiatives for strengthening statistical services through regional approaches, 2010–2020. SPC/HOP/working paper 2. SPC, Noumea, New Caledonia. Available under 'Strategy' as CONF_HOPS_2010_WP-2_Implementation_Plan.pdf at: http://sdd.spc.int/en/resources/document-library

Cullen R, Hassall G (2013) e-Government in the Pacific Island states: ICT policy and implementation in small island developing states: Cook Islands country report. Victoria University of Wellington, Wellington. Available at: http://pippr.victoria.ac.nz/handle/123456789/33. Accessed 17 Mar 2016

Cullen R, Hassall G (2016) e-Government in the Pacific Island states: ICT policy and implementation in small island developing states: Solomon Islands country report. Victoria University of Wellington, Wellington. Available at: http://pippr.victoria.ac.nz/handle/123456789/35. Accessed 17 Mar 2016

Espey J (2015) Data for development: a needs assessment for SDG monitoring and statistical capacity development. Sustainable Development Solutions Network, Paris/New York

Haberkorn G (2016) Measuring progress in SIDS: roundtable on SDG indicators and the implementation of the 2030 agenda and the SAMOA pathway. Paper presented at round-table organized by AOSIS, UNDESA and UNDP, New York, 9 Mar 2016

Heeks R (2014) A development 2.0 research agenda. ICTs for development 25.4.14. Available at: https://ict4dblog.wordpress.com/2014/04/. Accessed 3.2.16

Igoe M (2015) Are we ready for a 'universal' development agenda. Devex, Washington. Available at: https://www.devex.com/news/are-we-ready-for-a-universal-development-agenda-86652. Accessed 17 Mar 2016

Kelly I, Cordeiro M, Cook L, Passunga M (2014) Ten year Pacific statistics strategy. Phase 2 design document. Final draft October 2014. SPC, Suva, Fiji. Available at: http://nsdsguidelines.paris21.org/sites/default/files/TYPSS_phase_2_final_draft_Oct_2014.pdf. Accessed 18 Mar 2016

Kroll C (2015) Sustainable development goals. Are the rich countries ready? Bertelsmann Stiftung, Gütersloh, Germany. Available at: https://www.bertelsmann-stiftung.de/fileadmin/files/BSt/Publikationen/GrauePublikationen/Studie_NW_Sustainable-Development-Goals_Are-the-rich-countries-ready_2015.pdf. Accessed 17 Mar 2016

Nourbakhsh I, Sargent R, Wright A, Cramer K, McClendon B, Jones M (2006) Mapping disaster zones. Nature 439(7078):787–788

Onsrud HJ, Johnson JP, Lopez X (1994) Protecting personal privacy in using geographic information systems. Photogram Eng Remote Sens 60(9):1083–1095

Pacific CRVS (2014) The Brisbane Accord Group and the Pacific vital statistics action plan (2011–2014) outline document. SPC, Noumea, New Caledonia. Available at: http://www.pacific-crvs.org/documents/regional-strategy-documents/25-pacific-vital-statistics-action-plan-outline/file. Accessed 14 June 2016

Pacific Islands Forum Secretariat (2005) Pacific Islands plan. Pacific Islands Forum Secretariat, Suva. Available at: http://www.forumsec.org/resources/uploads/attachments/documents/Pacific_Plan_Nov_2007_version1.pdf. Accessed 17 Mar 2016

Pacific Islands Forum Secretariat (2014) The framework for Pacific regionalism. Pacific Islands Forum Secretariat, Suva. Available at: http://www.forumsec.org/resources/uploads/embeds/file/Framework%20for%20Pacific%20Regionalism_booklet.pdf. Accessed 17 Mar 2016

PARIS21 Secretariat (2015) Regional statistics cooperation: discussion paper for PARIS21 cross-regional forum, Paris, 5–6 Oct 2015. PARIS21 Secretariat, Paris

PARIS21 Secretariat (2016) PARIS21 promotes the better use and production of statistics throughout the developing world. Available at: http://www.paris21.org/about. Accessed 6.10.16

Samoa Ministry of Finance (2012) Strategy for development of Samoa, 2012–2016. Ministry of Finance, Apia, Samoa. Available at: http://www.mof.gov.ws/Portals/195/Services/Economy/SDS%202012%20-%202016%20ENGLISH%20VERSION.pdf. Accessed 16 Mar 2016

Secretariat of the Pacific Community (2011) Pacific statistics strategy action plan, 2011–2014. SPC, Noumea

Secretariat of the Pacific Community (2015) Pacific community statistics for development division, strategic plan, 2015–2017. SPC, Noumea. Available at: http://sdd.spc.int/en/about/strategic-plan. Accessed 15 Mar 2015

Solomon Islands Government Ministry of Finance and Treasury (2013) Corporate plan 2013–2015. SIG, Honiara, Solomon Islands

UNICEF (2014) Counting every child: partnerships lifting birth registration numbers in Vanuatu. Child protection case study. UNICEF Pacific, Suva. Available at: http://www.unicef.org/pacificislands/04_CASE_STUDY_Vanuatu_.pdf Accessed 16 Mar 2016

United Nations (1989) Convention on the rights of the child. Adopted and opened for signature, ratification and accession by General Assembly resolution 44/25 of 20 November 1989. Entry into force 2 September 1990, in accordance with article 49. United Nations, New York

United Nations (2015) World statistics day, 20 October. United Nations, New York. Available at: http://www.un.org/en/events/statisticsday/. Accessed 15 Mar 2015

United Nations ECOSOC (2013) Principles governing international statistical activities. United Nations, New York. Available at: http://unstats.un.org/unsd/methods/statorg/Principles_stat_activities/principles_stat_activities.htm. Accessed 16 Nov 2015

United Nations ECOSOC (2014) UN E-government survey 2014. E-government for the future we want. Section 1.4.1. United Nations, New York. Available at: https://publicadministration.un.org/egovkb/Portals/egovkb/Documents/un/2014-Survey/E-Gov_Complete_Survey-2014.pdf. Accessed 18 Mar 2016

United Nations High Commission for Refugees (2013) Submission by the United Nations High Commissioner for Refugees for the Office of the High Commissioner for Human Rights' Compilation Report—Universal Periodic Review: VANUATU. UNHCR, Geneva. Available at: http://www.refworld.org/pdfid/51b829460.pdf. Accessed 10 Mar 2016

United Nations Statistics Division (2015) Statistics division at a glance. http://unstats.un.org/unsd/statcom/brochures/UNSD_Brochure.pdf. Accessed 18 Nov 2015

World Bank (2015) Definition of e-government. Available at: http://go.worldbank.org/JKO5DVDMQ0). Accessed 5 Nov 2015

World Bank (2016) Data on statistical capacity. World Bank. Available at: http://datatopics.worldbank.org/statisticalcapacity/. Accessed 6.10.16

Chapter 9
The Contribution of E-Government to Primary Industries and Rural Development in Pacific Island States

Transform Aqorau, Rowena Cullen, Anju Mangal and Peter Walton

Abstract This chapter focuses on e-government initiatives in the primary sector. Primary industries (agriculture, forestry, fisheries and mining) make a significant contribution to the economies of Pacific Island countries and information systems in this sector have a major role to play in sustainable development. The chapter highlights the value of regional collaboration, the importance of effective policy and the role of ICT in facilitating collaboration in the primary industries sector. This includes gathering statistics which are critical to planning and some regional initiatives for the management of mineral exploration and resources. Drawing together the work of authors working in the region, it highlights some of the benefits and challenges of applying ICT in this area of e-government and some examples of projects with significant potential for economic development, although some have yet to deliver on this potential and some have made a significant contribution to development in the region. The final example outlines the recent development of a Fisheries Information Management System for the billion-dollar Pacific tuna fishery. The chapter concludes by identifying regional cooperation as a critical factor in the success and sustainability of e-government initiatives in the primary sector in the Pacific Islands region.

T. Aqorau
Technical and Legal Support,
Parties to the Nauru Agreement (PNA), Majuro, Marshall Islands

R. Cullen (✉)
School of Government, Victoria University of Wellington,
Wellington, New Zealand
e-mail: Rowena.cullen@vuw.ac.nz

A. Mangal
Land Resources Division, SPC, Suva, Fiji

P. Walton
Peter Walton & Associates, Darwin, Australia

R. Cullen and G. Hassall (eds.), *Achieving Sustainable E-Government in Pacific Island States*, Public Administration and Information Technology 27,
DOI 10.1007/978-3-319-50972-3_9

9.1 Introduction

E-government has a key role to play in the primary industries in Pacific Island countries (PICs), principally because of the importance of the sector for the majority of the population of PICs. According to the most recent population data (Pacific Community (SPC Statistics for Development Division 2016a)) the Pacific Islands region has approximately 11 million people, nearly 8 million of whom live in rural areas. These rural populations are "nearly universally reliant on subsistence production to support their livelihoods and general well being" (Fink 2014, p. 2). And although the extent and nature of commercial agriculture varies widely across the region, it provides a vital source of employment and income in rural areas. Agriculture, forestry, fisheries and aquaculture are the primary industries that account for a disproportionately larger part of the economy of most PICs than in many other developing countries. This remains so even though the sector has declined considerably as a proportion of GDP due to the exponential growth of tourism, the contribution made by the development of infrastructure in the region (World Bank 2014) and the exploitation of mineral deposits on land and in the sea. Not only is the sector critical to food security in PICs; it also has considerable potential for growth in terms of international trade within the Pacific region and globally. Its management for sustainable development, to maintain livelihoods and reach new markets, is highly dependent on both accurate and timely information and strategic communication between key stakeholders. In this chapter, therefore, we look at the contribution that ICT and e-government initiatives make to the management of primary industries in Pacific Island countries, their role in sustainable development and the government and inter-governmental systems that help maintain and develop them.

Agriculture and the other primary industries noted above have had a significant role to play in terms of meeting many of the targets related to the United Nations' Millennium Development Goals (MDGs) that Pacific Island nations adopted, for example, in food security, income, employment, conservation (especially forests) and biodiversity. Among the Sustainable Development Goals (SDGs) that follow on from the MDGs (see Chaps. 1 and 2) those that are directly impacted by agriculture and rural development include: SDG 1, No poverty; SDG 2, Zero hunger; SDG 3, Good health and wellbeing; SDG 15, Life on land; and SDG 13, Climate action, which acknowledges the role of agriculture and farmers in climate change mitigation (United Nations 2015). The impact of more severe climate change events such as cyclones, droughts and floods on agriculture in the Pacific, (including the management of forests and fisheries) makes this SDG also a focus for the primary sector in PICs.

9.1.1 Sustainability a Critical Issue

The critical role of agriculture and other primary industries in sustainable development means that the e-government projects and programs being established must also be sustainable, meeting some of the criteria for sustainability developed in Chap. 1 of this volume. These include:

- being adequately resourced with continuing budget allocation;
- being supported by capacity building and ongoing training;
- developing a critical mass of users to reach the 'tipping point' of sustainability;
- having strong local leadership and ownership; and
- having evident value.

A review of the literature on the use of ICT in support of rural poverty elimination and food security identifies a number of themes that reinforce these criteria for sustainability of ICT projects, highlighting in particular the need to build on existing systems and structures that relate to local needs, ensure systems are demand driven and can secure ongoing funding, provide equitable access to avoid increasing existing inequalities, promote local content, build capacity and use realistic technologies (Chapman et al. 2004). These authors are influenced by Heeks' focus on the 'design–reality gap' that often underpins the failure of ICT projects in the development context (Heeks 2002) and reinforce his emphasis on the need to focus e-government for development projects on information which is integral to the environment, integrated with development, intermediated and indigenized, rather than focus on technology (Heeks, cited in Chapman et al. p. 3). This is a critical issue in this chapter.

9.1.2 The Value of Primary Industries in PICs

In most PICs 70–80% of the population rely on a mix of subsistence farming, supplemented by cash income from a tradeable surplus sold in local markets. In addition, in many areas an increasing number of smallholders grow cash crops (such as copra, coconuts and cocoa) as well as root crops and vegetables (such as taro, sweet potatoes, yams and watermelon) for export to Australia and New Zealand (Malua 2003). However, these potentially valuable export markets are affected by fluctuating prices, unreliable supplies due to the impact of extreme weather events, poor road conditions and the cost of transport, problems related to sea transport (exacerbated by the large number of tiny islands that make up many PICs) and the difficulties of getting small quantities of crops to international ports. Smallholders also struggle to compete and to meet the many requirements of international markets.

Taking both the local and export sectors into account, agriculture (defined by the World Bank as including forestry and fisheries) can account for up to 30% of GDP (compared with five percent in a developed country that exports agricultural products and one percent in most high-income countries). The economic value of Pacific Ocean fisheries to Pacific Island countries and territories, was estimated in 2010 at US$1.04 billion,[1] which represents 3.3% of regional GDP and 1.5% of regional employment (Seidel and Lal 2010). Although data on other ocean-based industries (such as pearl cultivation) is less reliable, tuna fisheries account for more than 50% of this total, followed by subsistence fishing which represents only about 16% of the fish caught.

The number of Pacific Islanders engaged in agriculture, therefore, and its significance in terms of poverty reduction and food security, let alone the considerably larger potential of the entire primary sector to the small and vulnerable economies of PICs, has led to a certain amount of collaboration and investment in information systems to assist in the management of primary industries, although not all are long lived. Some are detailed in this chapter, in particular some regional initiatives, along with projects which have not had the impact that might have been expected. A number of factors that lead to these two outcomes are explored.

9.1.3 Scope of the Chapter

The chapter first discusses some of these regional initiatives and networks and then each subsector, highlighting examples of e-government projects in the domains of farming (cropping and livestock), forestry and fisheries, focusing on the use of information systems to manage and disseminate information in ways that are integrated into and enhance work practices, while opening opportunities for economic development. Some of these technologies are simple mobile applications that can provide local farmers with current crop prices, or let them follow shipping schedules so that crops can be harvested at the optimal time for shipping. Others are networked databases operating at regional level that require considerable political leadership to ensure sustainability, but which can transform the management of an entire regional industry. Some examples (such as forestry) show that the demand for information extends beyond the sector and that a number of information sources/systems (such as land data) must interact to ensure effective management of the resource; others show that the best solutions may well rely on regional cooperation and investment, as the Parties to the Nauru Agreement (PNA) Fisheries Information Management System shows. This system, which makes near real time information on the billion dollar Pacific tuna fisheries available to each partner country, is a high investment and high tech solution to a problem critical to PICs that contains many lessons for successful e-government in the region.

---

[1]This figure includes employment from post-harvest activities.

9.2 Regional Collaboration

The Pacific Islands Forum's[2] Framework for Pacific Regionalism adopted in 2014 sets out a process for "developing and prioritizing regional public policy" with the aim of promoting 'deeper regionalism'. This, the Framework indicates, rests on the recognition of

> the advantages of shared purpose and of close cooperation and coordination. Our nations commit to working together to address our common challenges, harness shared strengths, and ensure that our individual and collective advancement brings practical benefits to all Pacific people. Deeper regionalism will help increase market opportunities, improve service delivery, and ensure good governance for Pacific people. (Pacific Islands Forum 2015)

In the primary industries, the benefits of this cooperation are shown in particular by the role of the Pacific Community (SPC),[3] and its many programs to support agriculture and fisheries, particularly the tuna fisheries information management system, discussed later in the chapter. The SPC's partner organization, the Secretariat of the Pacific Regional Environmental Programme (SPREP), also plays a leadership role in the region, with an overview of the Pacific as an entire ecosystem. Its programs and freely available online resources cover climate change, biodiversity and ecosystem management, environmental monitoring, waste management and pollution control. Its online networks and resources are explored in more detail in Chap. 10; Examples of regional programs that make use of mobile technology are also given below in Sect. 9.3.2.

9.2.1 The Role of the Pacific Community (SPC)

The SPC, the principal and technical scientific organization in the Pacific which is "owned and governed" by its 26 Pacific country and territory members, focuses on sustainable development through the application of science and knowledge in all sectors in the Pacific Islands. This covers all aspects of agriculture and forestry (through its Land Resources Division) and fisheries (through its Fisheries Division), the work plan being driven by the annual meetings of the heads of agriculture and forestry services, the ministers of agriculture and fisheries and the separate meetings of the heads and ministers of fisheries. In these sectors the SPC's work focuses especially on:

- reducing the impact of climate change and disasters;
- ensuring biosecurity risks are well managed to enhance food security and trade;
- the application of innovation in areas such as fisheries sciences and plant genetics; and
- sharing expertise across the region.

---

[2]The Pacific Islands Forum is discussed in more detail in Chap. 3.

[3]More detail is provided in Chap. 3, and on the SPC website at: http://www.spc.int/index.php.

In doing so it collaborates with many similar international organizations (such as the FAO, the World Bank and UNCTAD) and development partners such as Australia, New Zealand, Germany and the United States of America. It also participates in the EU-funded Pacific Agriculture Policy Project, which is implemented by the Pacific Community to improve the livelihoods of smallholder farmers across the 15 PACP countries: the All Africa, Caribbean and Pacific Islands Agricultural Commodities Program which funds a number of SPC-led programs.

The SPC's Land Resources Division (LRD) whose strategic goals include "to assist the Pacific Community to improve food, nutritional and income security and sustainable management and development of land, agriculture and forestry resources" (SPC Land Resources Division 2014a) runs regional and in-country workshops on a wide range of topics from animal diseases and genetics to plant health, forest and agriculture diversification, climate change, biosecurity and trade. It surveys member countries to gather information on plant and animal diseases, offers technical advice and publishes brochures and articles on these topics which are freely available from its website and supports a helpdesk where individuals can ask questions related to any of the Land Resources Division's areas of technical and/or policy expertise. It also supports the regional Pestlist database (see below) and encourages PICs to develop their own databases and utilize data to protect against pests. Individual member countries collect their own data on animal diseases, including epidemiological data and report notifiable diseases immediately to the World Animal Health Information System (better known as WAHIS) and the FAO's Transboundary Animal Disease Information System. The SPC signed a memorandum of understanding in 1999 with the World Organization for Animal Health (OIE) to promote the adoption of OIE standards in the small island developing nations in the Pacific and runs training in the use of these systems.[4] The LRD is also very active in forest management running regional and country-based workshops that are promoted and published online, including forest inventory training in conjunction with the UN-REDD group (see Box 9.3) and supporting a Forest Restoration Network. The Division also covers climate change adaptation and mitigation.

The work of the SPC in all these areas makes significant use of online networks, notices, information sheets and two-way communication with ministries and individuals throughout its members. Its expertise and contribution is beyond what any member country could develop individually, making the primary sector an area (along with the Statistics for Development Division and the Geoscience Division) where the SPC plays an essential regional role that contributes to e-government in the region.

The effect of resource constraints and ongoing dependence on development partners often threatens the sustainability of some these valuable initiatives,

[4]Unfortunately, at the time of writing, only five of the 22 SPC Pacific Island member countries are members of the OIE: Fiji, Federated States of Micronesia, New Caledonia, Papua New Guinea, and Vanuatu.

especially when a program is not renewed due to a shift in political focus in the development partner's country. So, it is perhaps inevitable that information on the SPC website is sometimes out of date or has been moved. The Pacific Agricultural Value Chains Portal, for example, is no longer functional, but a more recent and very useful working tool, *Agricultural Value Chain Guide for the Pacific Islands* (McGregor and Stice 2014) is available on the LRD's Pacific Agriculture and Forestry Policy Network website (www.spc.int/pafpnet), to help farmers, traders and, in particular, policymakers conduct a value chain analysis.[5] PAFPNet itself, along with the Pacific Agriculture Policy Project (PAPP) (see Sect. 9.2.3), makes a considerable contribution to knowledge sharing in the region, facilitating communication and promoting good evidence-based agriculture and forestry policies in Pacific Island countries.[6]

The Division of Fisheries, Marine Aquaculture and Ecosystems' website is a more structured repository of resources (which even includes a catalogue of its information sheets) organized around coastal fisheries and oceanic fisheries. It publishes a highly informative and up-to-date online newsletter with links to in-depth information, current bulletins on a number of species, as well as safety at sea, meeting papers, presentations, manuals, handbooks and posters on the same range of topics and boat plans. It also publishes an address book of key agencies in all member states, international agencies and the roles and contact addresses for relevant SPC staff. The Division also promotes its workshops on its website. It works closely with the Locally Managed Marine Area Network, a broad coalition of government and non-government agents and volunteers involved in various community-based marine conservation projects around the globe, primarily in the Indo-Pacific region,[7] which also provides resources online. Such regional e-government initiatives provide leadership that encourages better use of ICT at the national and local level.

9.2.2 *Agricultural Data and Statistics for Decision Making and Policy Planning*

Good quality data and summary statistics are essential for both good decision making and for reporting on the SDGs. However, having limited resources, many PICs find keeping all required statistics up-to-date a considerable challenge (as we noted in Chap. 8) and several reviews over the past decade have consistently noted a considerable gap in the availability of agricultural statistics in the Pacific region (Walton 2002; Fink 2014). The SPC's Statistics for Development Division

[5]Mention should also be made of the Pacific Island Farmers Organisation network (http://www.pacificfarmers.com) which also offers information, support, and a two-way individual.

[6]More information at http://www.spc.int/pafpnet/.

[7]http://lmmanetwork.org.

Table 9.1 Dates of the agricultural censuses for some Pacific Islands

Country	Most recent agricultural census	Previous agricultural census	Previous agricultural census
Cook Islands	2011	2000	1988
Fiji	2009	1991	1978
Federated States of Micronesia	1960s	n/a	n/a
Kiribati	n/a	n/a	n/a
Papua New Guinea	1963	n/a	n/a
Samoa	2009	1999	1989
Solomon Islands	1990	n/a	n/a
Tonga	2014 (ongoing)	2001	1985
Tuvalu	n/a	n/a	n/a
Vanuatu	2007	1993	1983

n/a = not available/does not exist; *Source* Fink (2014). Reproduced with permission

(SDD) and the Land Resource Division have been active in addressing this gap, particularly through the Ten Year Pacific Statistics Strategy, 2011–2020 (TYPSS), adopted by the heads of planning and heads of statistics group (see also Chap. 8). The importance of agricultural data was given increased urgency by the United Nations' adoption of the SDGs and the need for PICs to report progress against them. As the Sustainable Development Solutions Network (2015) has recently advised, regular surveys of the agricultural sector will be needed, covering data such as number of farms and people operating them, land use, ownership, production practices, crop yields and productivity and income and expenditures (p. 12). Data on changes in forest and land use will also be required (p. 80). However, Fink's review (2014), as did Walton's earlier study, showed that most Pacific countries do not routinely collect data on crop production; only one had collected data on the use of fertilizers and two on irrigation and water use. Those countries that do collect data "typically gained it through agricultural censuses and surveys and some estimation" (p. 4), such censuses being conducted every ten years (see Table 9.1); there are some exceptions, e.g. Samoa and Fiji undertake more frequent agricultural surveys. In many PICs, agricultural censuses being considerably out of date, current estimations are based on seriously outdated data.

The SDD's National Minimum Development Indicators (NMDI) Database of over 200 development indicators,[8] including all of the population-based demographic, economic and social indicators required for the Millennium Development Goals (these are all are available on the SDD's website) contains a large number of agriculture and forestry indicators. These include key measures such as the proportion of household income and the proportion of the labor force primarily engaged in the sector; agriculture and forestry export value as a proportion of total

[8]The SDD's NMDI website is at: http://www.spc.int/nmdi/.

export and import value; and the proportion of national GDP from agriculture and forestry. Data on agricultural and forested land as a proportion of total land area and on the annual rate of deforestation is also reported. In the fisheries sector, the NMDI indicators include: the proportion of the labor force engaged in subsistence fishing and in small and larger scale commercial fisheries and aquaculture; the annual commercial and subsistence fisheries catch in both numbers and dollar value for coastal waters; foreign off-shore catch, fisheries as a proportion of GDP and fisheries exports totals (SPC Statistics for Development Division 2016b). However much of the data is missing. This lack of data on a key industry for the region makes the impact of the tuna fisheries management information system reported below, all the more significant.

The critical lack of data in the primary sector accentuates the problem that PICs face in collecting and using data. As a former Director of the SDD noted, the UN Statistics Commission's announcement of March 2016 that the final set of 230 indicators and 169 targets that countries will use to measure progress against the 17 SDGs has increased from the 60 indicators and 18 targets associated with the MDGs, and PICs will need to focus on identifying data gaps and likely challenges in collecting this data. He also noted the need to focus on those indicators which are more relevant to the economic base of most PICs, in light of the particular challenges they face in collecting this amount of data (Haberkorn 2016). (This issue is also discussed in more detail in Chap. 8.)

However, the SPC is collaborating with the UN Food and Agriculture Organization (FAO) through the FAO's Global Strategy to Improve Agricultural and Rural Statistics (GSARS), and several PICs recently participated in a Pacific capacity building program and Strategic Plan for Agricultural and Rural Statistics that will help improve agricultural, rural and fishery statistics systems (SPC Statistics for Development 2016c). The TYPSS also includes strategies to strengthen agricultural statistics through the inclusion of new indicators in the NMDI databases; these measures are to be included in countries' own National Statistics Development Strategies and in new modules in the household surveys conducted by member countries. Ideally, this would also see agricultural questions included in population censuses to be conducted in PICs in 2020 (Fink 2014, p. 11).

9.2.3 Policy for Development

The Pacific Community (SPC), through the Pacific Agriculture Policy Project (PAPP) has also been collaborating with national ministries of agriculture to establish an online inventory and repository of national agricultural policy documents. The European Union funded Pacific Agriculture Policy Project, SPC and the Technical Centre for Agricultural and Rural Cooperation (CTA) are jointly implementing a partnership to support Agricultural policy processes, value chains and extension support through stronger knowledge management (KM) concepts, approaches and tools. One of the project's goals is

> to address the need for greater food security for the region by encouraging subsistence farmers to move into commercial production through sustained access to competitive markets. Improvement of domestic market infrastructure, local value-added processing and development of regional markets for a diversity of Pacific crops would build on and support the desirable aspects of traditional farming systems (PAPP 2016).

The project also has a focus on capacity building, through regional workshops, which are intended to promote learning from shared experience and generation of knowledge products.

The PAPP project is also intended to assist the staff of the various Pacific ministries of agriculture and forestry to develop effective national evidence-based agriculture sector policies which would focus on interventions to ensure the competitiveness of smallholder farmers' participation in agriculture based development. Farmers' organizations are also targeted with the aim of helping them adopt new farming methods to improve productivity through training and research and to help them counter the effects of climate change. The project also aims to enhance capacity through workshops in policy making, the use of data for evidence-based policy and knowledge management.

The PAPP Policy Bank is an online repository which contains all available agricultural policy, legislation, local resources and contacts from each of the 15 PACP members.[9] It goes a long way towards addressing the difficulty of accessing current policy in the region, makes it easier for researchers and policy-makers to examine policy in countries facing similar challenges and thus encourages knowledge sharing and cooperation as well as highlighting the importance of good policy. The long-term intention is that each country's policy bank will include summaries of policy for farmers and that these will be disseminated through a range of media—print and radio, a website accessible by mobile technology and social media. As more of this material is made available in print format or for download in local languages the more chance there is that demonstrable best practice and more sustainable practices will be applied. The SPC PAPP is now working with Pacific countries on a number of agreed actions to further build the momentum of these ICT and knowledge management initiatives such as the policy bank. These included national outreach plans and initiatives to promote cooperative regional initiatives.[10]

9.2.4 Collaborative Management Contributing to Development of Mineral Resources

The SPC website also hosts online tools to support mineral industries in the Pacific. Currently underground mining takes place in Fiji, with surface and open-pit mining

[9]See http://pafpnet.spc.int/our-areas-of-focus/info-comms-knowledge/573-online-agriculture-policy-banks.

[10]The Pacific Islands Rural Advisory Service (PIRAS) is another such initiative funded by PAPP.

in Solomon Islands and Papua New Guinea. Marine mining has been undertaken in the Pacific Islands region since the 1960s, and opportunities for deep-sea mining are currently being explored in the Extended Economic Zones of many Pacific Island countries. These activities are monitored and supported by the SPC's Geoscience Division (until 2010 known as SOPAC), which has a staff of over 50 scientists, technicians, project managers and assistants and advisors helping PICs apply scientific knowledge "to govern and develop their natural resources, increase their resilience to natural hazards, [and] access data-based approaches to adaptation" through the Geoscience for Development Programme G4D (SPC Geoscience Division 2015). The Geology, Minerals and Hydrocarbons section and the SPC/EU Deep Sea Minerals Project are both concerned to strengthen the ability of PICs to ensure "informed and careful governance" of deep sea mining in accordance with international law, secure equitable financial returns on the exploitation of their resources and protect the environment (SPC Geoscience Division 2016). The Division hopes to achieve this through a range of strategies: by disseminating information online, promoting a regional legislative framework, helping individual PICs develop national legislative frameworks and by developing an online community of practice.

As a scientific institute, the Geoscience Division, especially in its Geoscience for Development section, makes extensive use of advanced ICT such as Geospatial Information Systems (GIS). It maintains an interactive mapping and spatial data repository of digital maps and GIS-related data, accessible through GEONode, which is set up to map a range of environmental and agricultural and oceanic data, and a parallel tool, PacGeo, which stores similar data, supports interactive mapping and makes accessible existing reports and maps. Its online Environmental Vulnerability Index enables the integration of datasets held by the SPC and the Geoscience Division to create analyses of impacts of one sector on another and of risks to sustainable development caused by a variety of factors including land use, fertilizers and introduced species. Some of these tools are significant sources of evidence for planning in the primary sector and the Division has received international recognition for this work.

9.3 Agriculture

ICT applications make a noteworthy contribution in agriculture and the delivery of agriculture products and services to key stakeholders, especially those in rural areas. Internet connectivity and mobile penetration to and within the Pacific have increased dramatically in the past decade (see Chaps. 4 and 5), which has enabled the introduction of efficient and cost-effective ICT tools to support the sector across the Pacific. Examples of the power of the new scientific tools combined with web technology are shown in Boxes 9.1 and 9.2. At the same time, these applications can suffer from the constraints which affect ICT applications across the region, especially in remote and rural areas, namely lack of access to technology (computers, peripherals, software, etc.), lack of technical support, weak or unstable Internet access (which results in slow speeds and relatively high costs) and a lack of basic ICT knowledge, skills and attitude in government and in the community.

Traditionally, government agricultural services in the Pacific have interacted with the public and both farmers as well as traders/exporters, through direct contact or via older communication channels such as radio, newspapers (articles and announcements), leaflets and workshops and training activities. There have been some instances of video programs broadcast on TV or available as DVDs or on the Internet, but mainly these have been low-key and low-cost interventions. More recently, a significant shift to the use of the Internet to disseminate information to agricultural communities has occurred; if possible, the local language is used or at the very least one of the forms of Melanesian pidgin, the most commonly spoken language in Papua New Guinea, Vanuatu and Solomon Islands, in which there are reasonable levels of literacy (PACE 1996). (Other countries either use their own official language, e.g. Samoan, Cook Islands Maori or have higher levels of English or French.) The introduction of digital media has allowed greater dissemination of audio-visual materials (as well as text). However, in rural areas with no mobile coverage broadcast radio is still an effective tool for disseminating information to farmers; although it does not offer the 2-way communication by mobile technology (which enhances the learning process) it is often still the most effective mode of communication for farmers out in the field.

Even so, while ICT can offer opportunities to disseminate information about new scientific findings, potential markets and strategies for dealing with climate, concerns have been expressed that this information is not reaching as wide an audience as it should and that the information disseminated is only a fraction of that produced by agricultural research institutes. There is also the risk that local traditional knowledge may be lost; this 'tacit' knowledge needs to be captured and reconciled with new knowledge to encourage farmers to modify their practice in what is still an environment bound by customary practice. This is one reason why face-to-face communication, through workshops and extension programs, remains the best way to introduce both new ideas and new communication channels in rural communities in the region.

Box 9.1: ICTs help reconcile quarantine and trade: The Biosecurity Authority of Fiji

In many respects, quarantine and trade are strange bedfellows. Quarantine, or to use the more recent terminology, biosecurity, used to have as its prime objective the mission to keep out the incursion of harmful pests and diseases. Since the various trade liberalization agreements, the principal purpose now is to facilitate trade; unless there is a defined biosecurity risk to the importing country, importation is allowed, although in most Pacific Island countries there are still some restrictions. What underlies the change is the introduction of more sophisticated pest risk assessments and information sharing—largely through ICTs. This has enabled PICS to not only improve their systems for biosecurity control over export of fresh produce, but also to set in place effective systems for mitigating risks due to imports. Reducing biosecurity risks helps improve access for fresh produce in export markets and ensures better environmental protection within each country—ultimately supporting farmers in PICs to make a living through exports.

The Biosecurity Authority of Fiji has a dedicated website which provides extensive information on biosecurity threats, regulations and procedures related to imports and exports, targeted both at travelers arriving independently by boat, or with commercial carriers as well as importers and exporters. Current threats to the Fijian agriculture sector and hence both exports and food security, such as taro beetle, fruit fly and American iguanas ('pests not pets!') are strikingly brought to the attention of anyone visiting the website, which provides essential information for anyone arriving or leaving country, or any citizens who may have concerns. Brochures, posters, regulations (for both export and imported products), notifications, feature articles, contact numbers and an increasing number of forms are all available online. Behind all this activity sits a program of support by the SPC's Land Resources Division, the New Zealand Ministry for Primary Industries and the Australian Centre for International Agriculture Research, which includes the creation of databases, information systems, online networks and training in the use of online biosecurity tools (including remote microscopy) for quarantine staff. Free exchange of information is made possible by online access to publications and by such initiatives as the development and dissemination of ACIAR's Cleaner Pathways (e.g. Fiji's Taro Export Pathway, which ensures cleaner products get to market by training farmers in the techniques of soil sampling and the application of standards). This is a work in progress, but with many achievements that contribute to enhanced biosecurity in the region.

Sources: SPC website; ACIAR website; MFAT (2013).

9.3.1 Mobile Services for Agriculture

Many claims have been made about the potential benefits of mobile services in developing countries, including up-to-date access to scientific and market information, shipping schedules, microfinance and online banking services. Although there is some anecdotal evidence of such innovations, hard evidence of these working out in practice is scarce (Rashid and Elder 2009). While several PICs have endorsed proposals to take advantage of the growing uptake of mobile services in rural areas to implement new services in agriculture, (see for example Va'a et al.

2012) there does not seem to be a high level of government involvement in such activities. In countries such as Fiji, where mobile penetration is close to 100%,[11] there is considerable potential for farmers to use a mobile phone or tablet to access information on markets, prices or alerts about specific crop diseases and management. However, while technical bulletins and crop planting guides, both of which contain information of immediate value to farmers, are available on the Fijian Ministry of Agriculture's website, these are not heavily promoted on the website, nor on the Ministry's Facebook page, and there is no evidence that they are routinely 'pushed' to farmers through an SMS messaging service or the like. (The 'market watch' page is over a year out of date at time of writing and 'real-time' information which is of the highest value to farmers to ensure they obtain a fair price for their produce seems to be unavailable.) The independent (EU-funded) Fiji Crop and Livestock Council, however, recently signed up more than 10,000 farmers to its new SMS messaging system with the help of the SPC, to share information such as market prices, expert agricultural information, real-time data support and weather forecasts (Valemei 2016).

In Papua New Guinea the mobile phone provider Digicel and the Fresh Produce Development Agency (FPDA[12]) launched the FPDA Market Information Service in 2009, an automated SMS-based market price service for vegetable crops in eight domestic markets in the region. At a relatively inexpensive cost, it was anticipated that farmers would be able to "maximise their profit and at the same time deliver the best quality goods [to] the right market" (Digicel 2009). Despite Digicel's market dominance in PNG, a subsequent study of rural smallholders and mobile telephony in PNG's Western Highlands and East New Britain found that whereas 76% of rural household heads owned a mobile phone at the time of the study just one had heard about the Service.[13] Confidence is further eroded when searching the Digicel website for information about the Service; there is none, apart from the news release on its inception in 2009. The FPDA website has a banner mentioning the Market Information Service, but currently provides no link to it or information about it. For whatever reasons, the service appears to be no longer operating.

Mobile banking seems to meet a more immediate need. For example, Solomon Islands Telekom, in partnership with Bank South Pacific has introduced a mobile/rural banking service in southern and northern regions of Malaita; this is one of several new products designed to assist farmers and small rural businesses in remote and rural areas to access banking services. Initiatives such as this can be transformational in allowing farmers to leverage the value of their crops in a cash economy and help lift them out of subsistence living (Toby 2013). However, this innovation, like the previous example, is led by the private sector; the government's only role is to facilitate such developments by ensuring that an appropriate

[11]Estimated at 97% according the CIA (2016).

[12]A government-funded organization for highlands-based vegetable growers.

[13]Personal communication between Elizabeth Dumu and Peter Walton, 14 September 2015.

legislative framework (e.g. deregulation of the telecommunications industry or legislation to support online banking) is in place.[14]

Social media is another aspect of technology which, because of its popularity with youth, is likely to have an impact on the development of agriculture as this generation starts to take on leadership roles in the rural community. Tools such as Facebook, blogs and Twitter offer opportunities for information sharing with and among farmers, especially with youth as they begin to engage with the agriculture sector. Young agricultural reporters are changing the way they interact and are using social media tools to communicate with a new audience. In some countries (Vanuatu is a case in point) young reporters are recruited as part of the social media team to assist in the dissemination of agricultural information to a new generation of stakeholders. The SPC's Land Resources Division has encouraged the use of social media and collaborative Web 2.0 technologies (such as blogs, wikis, Google docs, Google+ and video sharing sites) and conducts workshops across the region to increase understanding and the use of these powerful knowledge sharing tools (SPC Land Resources Division 2014b).

Box 9.2: Vanuatu's CertLink

Agriculture accounts for about 20% of Vanuatu's GDP and is critical to its economic growth. To maximize the economic value of its organic agriculture-based exports the Vanuatu Sustainable Agri-Business Initiative (VASABI) developed CertLink in partnership with the Vanuatu Agricultural College and the Vanuatu Organic Cocoa Growers Association (VOCGA)'s Coop Growers community as a tool for smallholder organizations to easily and cheaply access and manage multiple certification systems (e.g. Organic, Fair Trade, Rainforest Alliance, and the UTZ certification for sustainable/ethical coffee production) for increased market opportunities, primarily in the cocoa and coconut markets. CertLink, managed by Vanuatu agricultural consultancy LionFish, has since evolved into a powerful tool for agri-systems management, with applicability to a range of products, including copra, coffee, kava and livestock, which provides members with services such as data management, analytics, and price sharing. Its aim is to mentor and accredit 10,000 professional Vanuatu growers and connect them with international markets and 10,000 customers around the world. The system is based on a customizable cloud-based platform developed by IBM Business partner Wujitech (www.wujitech.com) tailored by CertLink to meet the specific needs of Pacific agribusiness for both capacity-building and track-and-trace technology. Accessible to certified users through a mobile app, CertLink integrates all levels of the supply chain, from smallholder growers through extension and training services, transport and logistics, processing, export and marketing through to point of retail sale.

[14]Deregulation and other legislation is discussed in Chap. 4.

CertLink is effectively a virtual public-private partnership focused on enhancing the livelihoods of Vanuatu smallholder farmers. It is supported by the Vanuatu Commodities Marketing Board which acts as the regulator of specific commodities—copra, cocoa and kava—with responsibility for their quality and quantity, their production, processing, exports and prices. The VCMB uses CertLink as an Information, Marketing and Management System to create a National Agricultural Database, which profiles growers, processors, agents and exporters in order to support product track-and-trace from grower to point of export. Partners also include the Pacific Growers Export Partnership (PGEP), a transparent and sustainable network of companies working towards the common goal of delivering positive development outcomes to Pacific Island producers using fair and transparent international supply chains, the Vanuatu National Provident Fund.

Sources: CertLink website (http://www.certlink.org); Rihai et al. (2013).

9.3.2 Regional Initiatives

Several other web-based tools have been developed to facilitate knowledge sharing in agriculture across the region, most of which are collaborative efforts involving governments, intragovernmental organizations (e.g. the FAO and the World Bank), universities, research institutes and the private sector. One such has been developed by PestNet, an email network (initially focused exclusively on the Pacific but now open to a global membership) that shares information on agricultural pests and diseases. Members can post queries seeking advice on a pest or disease, usually accompanied by a photograph; responses to such queries are moderated and posted in a structured, freely accessible database. Many responses have been turned into factsheets and simplified mini-factsheets for those with limited English. The content of these factsheets has been transferred from MSWord to a set of dynamic pages in the Fact Sheet Fusion database, which can then be used to 'compile' HTML pages and a mobile application provides access to this information from a smart phone (using iOS or Android). The result is the freely available mobile app Pacific Pests and Pathogens that was launched in late 2014, the first agriculture sector example in the region of true web publishing, in that the content is delivered from a database that publishes the information 'on the fly' on mobile devices (smartphones and tablets). This initiative, which compiles well-managed, constantly updated information, from explicit and tacit knowledge, is an exemplar of an effective, sustainable e-government application that meets immediate needs and is undoubtedly the precursor to further mobile publishing initiatives in the region (FBC 2015).

The Secretariat of the Pacific Regional Environment Programme (SPREP) is also introducing online information sharing through a wide range of technologies via the Pacific Environment Information Network (PEIN); this includes making its virtual library of downloadable resources accessible to mobile phones and tablets (with a tag cloud to enhance access to these), a YouTube network of audio and audio-visual resources and Facebook and Twitter. SPREP has recently transferred their large bibliographic database (and associated full text documents) to Inmagic's Presto—a social knowledge networking application that serves as a portal to databases and library collections (of documents, videos, photographs and other materials). Those who access Presto from a mobile device will be automatically switched to responsive screen designs, making the database and its contents more easily accessible in the form most likely to have the greatest number of users on mobiles and tablets. The SPREP website also provides access to several international databases including OARE, AGORA HINARI and ARDI, which have been established to provide free or reduced cost access to researchers and policy makers in developing countries to scientific material that would otherwise be unaffordable.[15]

Box 9.3: The Melanesian Agricultural Information System

The Melanesian Agricultural Information System (MAIS) exemplifies what can be achieved by a group of people that share a common vision and are committed to collaboration. MAIS began in 2001 as a 'library catalogue database system' for the newly-established Papua New Guinea National Agricultural Research Institute (NARI). The brief was for a centralised library catalogue that would serve all NARI's stations and laboratories around the country; although there was no computer network to connect all the sites, a software application was selected that enabled a central database, with records shared and distributed as simple email attachments so that each installation was identical to all the others. The new system, called the NARI/CRI Agricultural Information System (NAIS), was launched in 2001, with some 17,000 bibliographic records derived from various sources.

From the outset, the system was intended to not only enable NARI but other organizations in PNG to share library records. The PNG Coffee Research Institute was an early partner and by 2005, five other agricultural organizations in PNG had joined what was now a *National* Agricultural Information System with around 25,000 bibliographic records. From the outset, the system was scalable and its management diffuse in order enable it to survive the 'boom and bust' cycle of development in the region, where organizations flush with funds one year are paupers the next. With multiple organizations and many pairs of hands carrying the load the system could be maintained in good times and bad.

At the same time the Ministry of Agriculture and Livestock (MAL) in Solomon Islands adopted a similar system in 2009 with the name Solomons National Agricultural Information System (SoNAIS). And in 2011, NAIS and SoNAIS were merged to form MAIS which currently comprises over

[15]These schemes are discussed in more detail in Chap. 11.

37,000 bibliographic records. In addition, MAIS includes a journal index (Pacific Index to Agricultural Journals, 1920–), a publications database, with over 1200 full text documents linked to records in the database, and a web portal, with over 1400 records linked to web resources. A working prototype for a non-bibliographic component, Activities, to manage information on agricultural research and development activities in the region, is underway. Initially, MAIS was only accessible in the libraries of partner organizations. In 2013, a runtime version was piloted in Solomon Islands that enabled the runtime application software, databases and electronic publications to be distributed on a USB stick and installed on any computer. MAIS first appeared on the web in 2011 (with a better quality subset of records launched as SoNAIS in 2015).

Two issues have always dogged MAIS, limited funding and management buy-in. However, in late 2016 MAIS is to become PAIS (Pacific Agricultural Information System) sponsored by the SPC: an interactive web portal and knowledge-sharing platform for agricultural research and development information in the Pacific region, retaining the participatory model for management of the system, as well as both online and offline versions.

Sources: Walton (2009) and author P Walton

9.4 Forestry

The tropical forests of the Pacific Island countries are especially vulnerable to adverse events such as cyclones, tsunamis, climate change and unsustainable harvesting. Although the Melanesian countries are heavily forested, apart from Papua New Guinea their land area overall is small and their forests represent a limited resource. In Papua New Guinea many forest areas are in mountainous territory with very limited access. In the Polynesian countries, some of which are comprised of atolls as well as islands, remaining forest areas are needed for domestic and traditional purposes (such as building, firewood, medicinal purposes and traditional crafts) rather than for harvesting for export. Because in the past PICs have lacked the resources and knowledge to develop an export trade based on forestry, logging rights have often been sold to overseas interests, which have often attracted criticism for not paying a fair price for the timber exported and for exporting whole logs rather than investing in processing in-country. This has particularly affected Papua New Guinea, Solomon Islands and Vanuatu (Brown 1997; Crocombe 2001, p. 328). In addition, forested areas in the region are almost without exception on what are referred to as 'custom' or 'customary' land which is held in communal ownership by a particular clan or tribe. This has often opened up a source of conflict as ownership and rights to land utilization may be determined by familial inheritance lacking any written documentation and being unrecorded.

Because of the importance of forests in the lives of many Pacific Island peoples, for both traditional uses and employment and foreign trade, their sustainable management is critical to the ongoing welfare of PICs. High quality information and its use in resource management, policy and planning is essential. National and regional resources rely on information to improve the management of forests, preserving them, maintaining their good health (forests and plantations are also subject to pests and diseases), harvesting them and replanting. Several sections of the SPC's Land Resource Division deal with forestry, including the Forest and Trees section (which promotes forest management and agroforestry, i.e. enhancing the value of forests for harvesting and environmental protection), the Forest and Agriculture Diversification section (a key issue for sustainability) and the Climate Change section. The Land Resources Division also maintains a Pacific Regional Forest Inventory Facility, part of the sustainable resource management program of the LRD and a critical element of the REDD program. Individual countries participate and contribute (PRFIF 2015).

The Solomon Islands Ministry of Forests and Research has begun developing a national monitoring, reporting and verification system, supported financially by the UN-REDD program (see Box 9.3) to collect the data needed for planning and to determine the extent of deforestation and degradation. Data from its regional repositories is enhanced by remote sensing data supplied by the SPC's Geoscience Division. In contrast, Fiji's forest inventory data goes back to the 1960s (Collins et al. 1991); data, systematically collected several times, has received considerable impetus from the REDD program, which demands forest carbon stock assessment of all species in both indigenous forests and plantations, using site sampling (above and below ground) as well as aerial photographs and satellite images (Payton and Weaver 2011). This data then becomes an invaluable resource for forest management. Fiji is unusual among PICs in having historical data for comparison. Papua New Guinea's Forest Authority which has also undertaken a multi-purpose national forest inventory, supported by the REDD program, (and development partners FAO, JICA and the German international aid agency GIZ) lacks data to determine changes in land use over time, although comparisons of satellite images over the past 20 years partially fill this gap (Turia 2013).

9.4.1 Land Data

Data and information are clearly critical to the management of forests; and because they are such a valuable resource, open to exploitation, land data that provides proof of ownership and entitlement is also critically important. For this reason, initiatives across most PICs to establish databases that verify and record land ownership and tenure contribute significantly to the effective and equitable management of forestry resources. This is especially urgent in PNG because of its large land mass, most of which remains forested. A National Land Development Program (including necessary legislation) has been underway for some years, integrating the work of several agencies and the Department of Lands and Physical Planning, with the intention of freeing up land for development purposes—a process known as Voluntary Customary Land Registration. First a Land Court was set up to establish ownership of customary land, (since many historical documents that might be used for this purpose have been mislaid or destroyed this often involves capturing information from older generations orally, verifying and storing it. Once title to an area of Customary Land has been established, a tribal group can apply to become Incorporated Land Group. The land may then be developed or leased.[16] At the same time the Department of Lands and Physical Planning has been updating its Land Information Services and has an active program for creating a new lands database and information system that will record Customary and Alienated land and integrate this with the GIS data provided by its National Mapping Bureau, which has a considerable collection of digital and GPS data on roads, bridges, villages and health facilities.

Several other PICs have followed a similar path, depending on the status of their land data. Vanuatu's Ministry of Lands and Natural Resources has a Land Lease Registry that fulfils a comparable function. Once ownership of customary land is determined it is registered with the Customary Land Management Office in the Ministry of Justice. It is then able to be leased (for periods of 3–75 years) and the customary owners pass control to the leaseholder. The Lease Lands Register's Titles Information System holds electronic copies of all relevant documents and links to satellite-based GIS data held by the Lands Survey Department (Cullen and Hassall 2016a). Samoa, which has had a Land and Titles Court since the beginning of the 20th century, has a digitization project to preserve court records of titles assigned and make them more widely accessible. It uses the FAO's Solutions for Open Land Administration (SOLA) program that combines land titles with GIS cadastral data to manage the twenty percent of land not held in customary ownership (there is no land title system for customary land at the present time) (Cullen and Hassall 2016b). As these examples show, the introduction of land title information systems has made a significant contribution to the management of both custom and leased land in most PICs to land tenure and to more equitable and sustainable management of forests, plantations and other agricultural endeavors.

Box 9.4: UN-REDD and REDD+ programs offer incentives for good forestry information management in PICs

Five members of the SPC (Fiji, Papua New Guinea, Samoa, Solomon Islands and Vanuatu) are members of the United Nations REDD program (which stands for Reducing Emissions from Deforestation and forest Degradation in Developing countries), which is a collaborative program of the FAO, UNDP

[16]The first ILG to be incorporated was the Kum Kopi Kumaipana ILG, whose certificate was awarded in July 2015 (NRI 2015).

and the UNEP. Degradation and deforestation of the world's tropical forests are estimated to be responsible for about 11% of net global carbon (greenhouse gas) emissions and appear to be continuing unabated. Parties to the United Nations Framework Convention on Climate Change determined that preventing the destruction of forests is an essential part of the concerted global effort to combat climate change. The REDD+ program incentivizes developing countries by rewarding them financially for any emissions reductions achieved that is associated with a decrease in the conversion of forests to alternate land uses.

A three-stage process is to be followed. Having identified current and/or projected rates of deforestation and forest degradation, countries can take remedial action to effectively reduce those rates by developing national strategies, policies and actions plans with relevant stakeholders, building their capacity and designing demonstration activities. In this initial 'readiness' phase, a lot of financial and technical support is available. Critical to this phase is the development of high quality baseline data and what is termed a 'safeguards information system' (using a standardized REDD set of indicators) to manage the data; SPC works with the FAO to provide training in this area, developing skills in measurement, reporting and verification. In the second phase the policies and action plans are implemented, tested and the proposed REDD 'results-based' formulas validated. In the third phase results-based actions are implemented at the national level, fully measured, reported and verified. At this point countries can access results-based payments. As well as encouraging conservation of forests as carbon stocks the scheme brings several benefits to PICs: a model of collaborative action on an issue of major importance to PICs, especially through the Pacific Regional Forest Inventory; capacity building in data collection and information management in the industry is enhanced; and the scheme helps finance sustainable forest management. As the REDD Desk website notes, economically valuing the role forest ecosystems play in carbon capture and storage "allows intact forests to compete with historically more lucrative, alternate land uses resulting in their destruction." It adds, "clarifying tenure rights is central to an effective and equitable REDD+ mechanism as it can help identify who the key REDD+ stakeholders are, who should participate in decision-making processes as well as who should obtain benefits. In turn, developing REDD+ can provide an opportunity to recognize customary rights and to empower local rights-holders."

Sources: UN-REDD Programme website (http://www.un-redd.org); The REDD Desk (http://theredddesk.org)

9.5 Fisheries: The Contribution of Regional Organizations to the Management of Fisheries in the Pacific Islands

When the United Nations Convention on the Law of the Sea was adopted in 1982 and ratified in 1994 (Crocombe 2001) the Pacific Island countries were able to claim Extended Economic Zones (EEZs) based on their acknowledged territories. Because most PICs comprise many scattered tiny islands or atolls their EEZs cover large areas of the Pacific Ocean, areas that give them the right to license access to the valuable tuna that migrate through those waters. Such a wealth of resources not only requires sophisticated management; it also attracts considerable international interest. Few PICs are able to manage this resource and defend their territorial rights. Through several regional organizations, however, the Western and Central Pacific Fisheries Commission, the Pacific Islands Forum Secretariat's Forum Fisheries Agency (FFA), the Pacific Ocean Commissioner and the SPC's Division of Fisheries and Marine Ecosystems (FAME), which deals with both oceanic and coastal fisheries, they are attempting to do so.

In this section, our main focus is on a detailed report on the PNA Tuna Fisheries Information Management System and what can be learned from the development of a robust and comprehensive management scheme to support the billion-dollar global tuna industry by a small group of Pacific Island countries working together. However, coastal fisheries for subsistence fishing and local markets ('artisanal fisheries' which use traditional methods and small boats) are an important part of the economy of PICs. These fisheries are also prone to over-exploitation and vulnerable to climate change; accurate current data is essential to the management of this resource, as well as the more extensive oceanic fisheries. Various ICT initiatives by the SPC and by fisheries agencies within PICs promote better management of this industry. One such example is the development of a mobile app that allows mobile phones and tablets to collect data on the artisanal tuna catch in coastal fisheries across the region (SPC Division of Fisheries, Aquaculture and Marine Ecosystems 2016a). Another is the use of satellite imagery and GIS positioning to assess populations of sea cucumbers to enable the Fisheries Department of Vanuatu to determine appropriate limits on catch size (Leopold et al. 2015). Many such innovations being piloted in individual PICs, show the powerful use of ICTs for development; a library of online scientific reports is shared on the SPC FAME website, to enable others to follow suit. The subsection below indicates some of the impacts of small country-based e-government projects in fisheries management and larger oceanic projects.

9.5.1 Developing an Information Management System for Tuna Fisheries in the Pacific

The final example of the transformative value of applications in the primary industries in the Pacific Islands region is a management information system for commercial tuna fishing in the Pacific. The system, known as the PNA Fisheries Information Management System (FIMS) was put in place to support the Vessel Day Scheme (VDS) implemented by the Parties to the Nauru Agreement (PNA),[17] a coalition of those members of the Forum Fisheries Agency most involved in tuna fisheries. The PNA Vessel Day Scheme was implemented in 2007 as part of the Western and Central Pacific Fisheries Commission's Vessel Monitoring Scheme specifically to constrain and manage the catches of target tuna species within the EEZ's of participating PICs and to increase the rate of return from fishing by what are known as Distant Water Fishing Nations (DWSNs) through the payment of 'access fees' (FFA 2016). It sets overall limits on the number of days purse seine[18] and long line fishing vessels can be licensed to fish in PNA waters; days are allocated to each country and can be traded between countries (Aranson 2014). The VDS scheme is one of the largest, most complex and most lucrative tuna fisheries management arrangements in the world and one of largest underpinned by an FIMS. (For a full analysis of the system, see Banks 2016.)

The Pacific tuna fishery industry is multidimensional with vessels fishing in multiple exclusive economic zones (EEZs) across the Pacific (see map, pxviii), transshipping their catch to carrier vessels in multiple ports. Crews and vessels are internationally diverse coming from many countries, with many geopolitical interests at stake. The world's major political and economic powers, namely the United States, China, Japan, Korea, the European Union, Taiwan and Philippines all have flagged vessels fishing in the region. As a result, the Pacific Islands' tuna fishery is not only of political interest because of the difference in the political and economic status of the key actors, namely rich and poor, powerful and weak, but it is also becoming a theatre of political competition between China and the United States, and China and Taiwan.

An FIMS can refer either to paper-based records sent by fax from some of the older boats used by China, Korea, the Philippines and Taiwan, or to the most up-to-date forms of communication technology which support real time fisheries information delivered to a server via on-board handheld devices, or stationary reporting devices that send data either from vessels directly or from the company headquarters of the vessels. This investment in technology has been influenced by three key drivers. The first has its origins in the establishment of Pacific Island countries' EEZs or Fisheries Zones (FZ) in the mid-1970s, culminating in the adoption of the UN Convention on the Law of the Sea 1982. The second driver is

[17]The countries that are Parties to the Nauru Agreement concerning Co-operation in the Management of Fisheries of Common Interest 1982 are: Federated States of Micronesia, Kiribati, Marshall Islands, Nauru, Palau, Papua New Guinea, Solomon Islands and Tuvalu. Tokelau is an Observer with the PNA. It is however a participant in the PNA VDS Scheme.

[18]A purse seine consists of a long wall of netting framed with float line and lead line having purse rings hanging from the lower edge, through which runs a purse line made of steel, wire or rope which allow the 'pursing' or drawing together of the net. Purse seine fishing is commonly used for tuna fishing in the Pacific.

the development of fisheries management and verification systems which are aimed at improving the collection of biological data, improving economic returns and providing an enabling environment for enforcement. The third driver is market and consumer imperatives, driven in part by efforts to reduce illegal, unreported and unregulated (IUU) fishing and compliance with food safety standards of the major markets. Each of these, the way they have affected the evolution of the fisheries information systems and the value of these systems as exemplars of e-government in the region, is discussed below.

When the Pacific Islands declared EEZs or FZs in the mid-1970s they had no way of knowing what was being caught in their region; there was no baseline economic or biological data about the fishery. They needed basic information about the vessels operating within each EEZ, their identification features, the names of owners and where they were registered, to distinguish vessels from each other. They needed to know about the patterns of fishing, when vessels entered their waters, when they left, how much fish these vessels had in their hold when they first entered their waters and how much fish they had when they exited from a particular EEZ. This need for data led to the first attempts to establish a fisheries information management system in order to determine the value of the catch (fees due to each country were calculated as a percentage of the value of the landed catch in the vessel's home port), to verify information on where the fish was caught and how much was landed. A Regional Register of Foreign Fishing Vessels was established by the Pacific Islands Forum Fisheries Agency (FFA) and all foreign fishing vessels seeking a license were required to be registered on it, providing information about the vessel name, its international radio call sign (IRCS), flag state registration number, the number of fish aggregating devices (FADs), name of the captain and charterer, details about the vessel's Inmarsat Automatic Locations Communicator (ALC) and the vessel's Unique Vessel Identifier (UVI) issued by the International Maritime Organization (IMO). A recent A4 sized photograph of the vessel is also required.

The Parties to the Nauru Agreement then instituted measures to collect information about the actual activities of vessels operating in the EEZs to determine how much revenue was due from the licenses issued and, in particular, information related to zone entry and exit and a weekly updated report on catch in zone. Initially this information was paper based, sent indirectly by telex, then fax, from the vessels via their companies; it suffered from the same problems of verification, in particular the possibility that catch information might be under-reported in order to reduce the fees paid. As technology advanced, it became possible to send near real time information directly from the vessels.

The second driver in the development of fisheries information management systems was the application of limits in the fishery. The initial period after the establishment of EEZs was characterized by open access fishery in which no restrictions were placed on the quantity of tuna that could be taken by vessels. In the mid-1990s, new arrangements were made to impose limits on the catch. The first of these, the *Palau Arrangement for the Management of the Western Pacific Purse*

Seine Fishery[19] restricted the total number of purse seine vessels that could be licensed in the region. This was followed by limits imposed by the Western and Central Pacific Fisheries Commission (WCPFC) that required the development of fisheries information management systems to support those measures (Western and Central Pacific Fisheries Commission 2012). The development of the WCPFC's satellite based Vessel Monitoring System (VMS) also contributed to the evolving fisheries management framework; having both flag-based and zone-based limits (through the Vessel Day Scheme (VDS)) enabled information management systems that could verify the activities of vessels and ensure that they complied with their limits, and the establishment of e-Reporting, e-Monitoring and e-Observer reports to provide near real time catch and positional data from fishing vessels.

The third driver is market and consumer driven, particularly in Europe which is still the largest tuna market in the world. The evolution of eco-labels, increased environmental consciousness among consumers in developed markets and the use of market measures to address illegal, unreported and unregulated or IUU fishing has given impetus to the development of catch documentation schemes. These have now become an important plank in the management of the fishery by ensuring that fisheries managers within PICs have a more comprehensive overview of the fishery —how many vessels have fished in their zone, where those vessels are from, which areas of their zone they have fished, how many days they have fished and the composition of the catch of each respective vessel.

9.5.2 *The Fisheries Information Management System (FIMS) and iFIMS*

The drivers discussed above have led to the development of databases and systems that are held by different regional organizations and agencies. Databases used to support compliance, monitoring and surveillance are held by the Honiara based Pacific Islands Forum Fisheries Agency (FFA). These are the Regional Register of Foreign Fishing Vessels, a database of all the foreign fishing vessels licensed to fish in the region, the Vessel Monitoring Scheme (VMS) which provides positional data of fishing vessels, a prosecutions and violations database which holds a list of cases from the Pacific Islands and a regional surveillance center which supports compliance. Scientific databases are maintained by the Pacific Community SPC's Fisheries, Acquaculture and Marine Ecosystems Division in Noumea based on catch and observer data that are provided by the fisheries agencies in PICs. This data is used by SPC scientists to undertake stock assessment and to provide scientific information and advice for regional and national fisheries management

[19]Palau Arrangement for the Management of the Western Pacific Purse Seine Fishery, (Majuro: PNA Office, 2012). See Aqorau and Bergin (1997).

authorities on the status of fishery stocks and the impact of regional oceanic fisheries; data on the region's resources of tuna, billfish and other oceanic species; and improved understanding of pelagic ecosystems in the western and central Pacific Ocean. The TUFMAN (Tuna Fisheries Management) database enables individual countries to enter and extrapolate license and catch information which they can use to complete required reports for the WCPFC.

These useful databases serve the purpose for which they were created. However, the PNA Fisheries Information Management System (FIMS) was established specifically to support the implementation of the Vessel Day management system. In this system all purse seine and longline vessels that wish to fish in any of the EEZs or FZs of PNA countries must register their vessels and intention through an online vessel registration (OVR) portal on the PNA website. The VDS Register is also linked to the FFA Regional Register and WCPFC Record of Fishing Vessels. The VDS Register is more than just a register of fishing vessels. A precondition of registration is that the vessel must agree to provide eReports of catch by zone and by composition (the e-catch log). In addition, vessels must provide a list of the number of radio buoys attached to their Fisheries Aggregating Devices (FAD)s, which are used for GPS tracking. The components of the system include the OVR portal and the VDS Register; an electronic licensing system; VDS monitoring system; an e-catch log or documentation system; observer support and transmission of data required for SPC modules through Observer Tabulates and Delorme InReach two-way satellite communicators; provision for the Monitoring and Control System (MCS) database including boarding and inspection tools; FAD tracking and monitoring. These are all part of an integrated system as shown in Fig. 9.1; data is transmitted via satellite to each country's Fisheries agency and reported through the PNA data center to the FFA and the SPC.

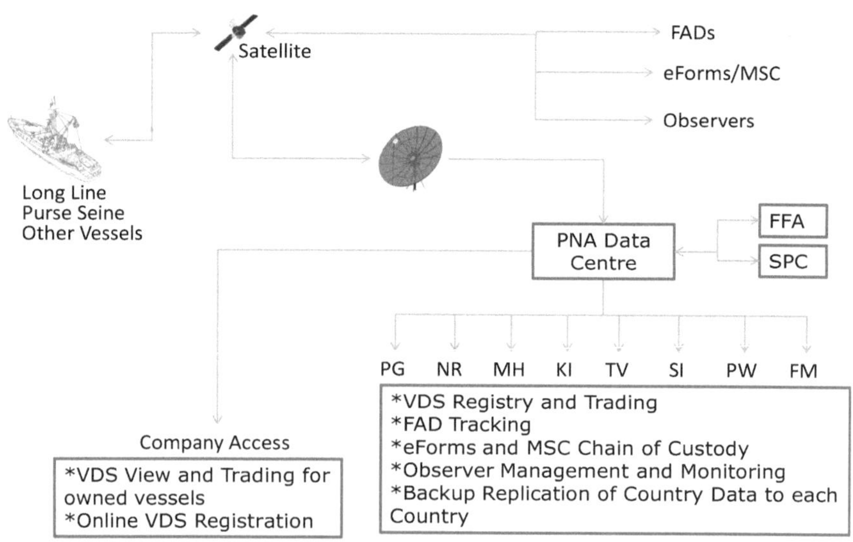

Fig. 9.1 Schema of the PNA vessel day system

An analysis of the benefits of the scheme has shown improved compliance and reporting and enhanced fisheries sustainability, including in non-target species (Banks 2016). Electronic tracking, monitoring and reporting real time data and intelligence has strengthened awareness of fishing activities for both fisheries managers and industry and has enabled in-country fisheries agencies and the FFA to monitor more fishing events while keeping the cost of increased monitoring relatively manageable. It has also shown itself to be more cost effective than costly aerial monitoring. The system has produced savings in manpower due to electronic data storage and transmission, reduced health and safety risks, promoted voluntary compliance in reporting catch data and provided evidence to support prosecutions. It has also improved the quality of stock assessment because of the timeliness and accuracy of data reported to the SPC's FAME Division and created better traceability and catch quality, all of which contribute to sustainable economic development. Although the PNA Tuna Fisheries FIMS has required a high level of development partner investment and considerable commitment and leadership from Pacific partners, the return on investment has been substantial. As the Poseidon report on the system notes, a well-designed fisheries information management scheme provides many benefits, including;

> improved functionality within the workplace and better staff morale from new employment opportunities arising out of introduction of [electronic reporting and monitoring]. External spin offs include improvements in perceptions of the efficacy of fisheries management by, inter alia, international stakeholders. This may result in positive spinoffs such as maintaining access for seafood exports to international markets.
>
> Based on experience in other situations and fisheries, the financial scale of benefits that improve compliance and the sustainability of fisheries, or which provide greater access to export markets, are generally orders of magnitude larger than the costs involved in deploying such solutions. (Banks 2016, p. 78)

The PNA purse seine Vessel Day Scheme and FIMS, the largest and most complex tuna fisheries management system in the world at this time, is a powerful example of the value of ICT in fisheries management and an excellent example of sustainable e-government. As well as the drivers underpinning the scheme which are outlined above, it meets many of the criteria outlined in Chap. 1 for sustainable e-government: building on existing systems and structures, meeting local needs, is demand driven, generates income that can sustain it long term, and ensures equitable use of a very valuable resource.

9.6 Conclusion

In the primary industries sector it seems clear that because of the technical demands regional collaboration is not only essential but has provided many benefits for PICs. This chapter includes some equally effective e-government innovations at the country level, which despite the reservations noted above, certainly make a contribution to development. But they are not mainly driven by central government, although governments in PICs are clearly supportive. They also struggle to survive, as the examples of the Fiji Ministry of Agriculture market watch system and the Vanuatu CertLink systems show. Such failures are most commonly due to lack of resources, capacity and the lack of a critical mass of people engaged in the initiative which will see it through lean times. It is the regional initiatives undertaken by the SPC and SPREP, and examples such as the PestNet service, that stand out: they show significant benefits resulting from investment in ICT and the value of current information to the primary sector. They carry the hallmarks of sustainable e-government, meeting the same criteria as the PNA FMIS. Regional collaboration and cooperation is dependent on ongoing leadership; these projects are also well supported by the regional meetings of heads of department, whether the heads of policy and planning the heads of agriculture and forestry, or ministers responsible for fisheries.

Unlike some of the other examples noted in preceding chapters, in the public finance sector (Chap. 6), for example, or the justice sector (Chap. 7), it is clear that in the primary sector an additional set of drivers have led to regional, rather than national, solutions. The primary sector, more than any other, is dependent on technical expertise for economic development; thus we see that regional organizations that can build teams of experts, committed to knowledge sharing and the dissemination of information are leading sustainable e-government in those sectors. The coalescence of these critical primary sector industries—agriculture, forestry and fisheries—with regional intra-government organizations and ICTs is shown here to be a highly effective model of e-government and there are many lessons to be learned from these projects. The technical demands made by development in the sector and the fact that these industries are part of a regional and global ecosystem that face challenges beyond their control, have been drivers for some impressive and sustainable e-government innovations.

References

Aqorau T, Bergin A (1997) Ocean governance in the Western Pacific purse seine fishery: the Palau arrangement. Mar Policy 21(2)

Aranson R (2014) Independent review of the PNA vessel day scheme (VDS). PNA Office, Majuro

Banks R (2016) Analysis of the costs and benefits of electronic fisheries information systems applied in FFA countries and identification of the legislative, regulatory and policy supporting requirements. Poseidon Aquatic Resource Management, Port Douglas, QLD. http://consult-poseidon.com/fishery-reports/1155%20WWF%20EFIS_11032016_Final_7%200.pdf

Brown C (1997) Regional study—the South Pacific. Asia-Pacific forestry sector outlook study working paper series, working paper no. Apfsos/Wp/01. FAO Forestry Planning and Statistics Branch, Rome

Central Intelligence Agency (CIA) (2016) The world factbook. https://www.cia.gov/library/publications/the-world-factbook/geos/fj.html. Accessed 26 Feb 2016

Chapman R, Slaymaker T, Young J (2004) Livelihoods approaches to information and communication in support of rural poverty elimination and food security: the literature update.

Overseas Development Institute, London. https://www.odi.org/resources/docs/208.pdf. Accessed 1 Sept 2016
Collins NM, Sayer JA, Whitmnore TC (eds) (1991) Western Pacific Islands, in the conservation atlas of tropical forests: Asia and the Pacific. McMillan, London, pp 240–249
Crocombe R (2001) The South Pacific. University of the South Pacific, Suva
Cullen R, Hassall G (2016a) E-government in the Pacific Island states: ICT policy and implementation in small island developing states. Vanuatu country report. http://pippr.victoria.ac.nz/handle/123456789/30
Cullen R, Hassall G (2016b) E-government in the Pacific Island states: ICT policy and implementation in small island developing states. Samoa country report. http://pippr.victoria.ac.nz/handle/123456789/31
Digicel Papua New Guinea (2009) Digicel provides access to market information for local farmers. http://www.digicelpng.com/en/about/news/digicel-provides-access-to-market-information-for-local-farmers. Accessed 14 Sept 2015
Fiji Broadcasting Corporation (FBC) (2015) Pacific pests and pathogens mobile application. FBC, Suva. http://www.fbc.com.fj/fiji/30455/pacific-pests-and-pathogens-mobile-application
Fink A (2014) Gap in the figures? Gap analysis of agricultural statistics in the Pacific region. http://pafpnet.spc.int/attachments/category/75/Gap-in-the-figures-analysis-of-agricultural-statistics-in-the-pacific-region.pdf. Accessed 10 Oct 16
Haberkorn G (2016) Pacific regional sustainable development goals indicator availability assessment. SDD Newsletter. SDD, Noumea
Heeks R (2002) Information systems and developing countries: failure, success and local improvisations. The Information Society 18(2):101–112
Leopold M, Ham J, Kaku R et al (2015) Spatial sea cucumber management in Vanuatu and New Caledonia. SPC Beche-de-mer Inf Bull#35, March. http://www.spc.int/coastfish/en/component/content/article/62/440-beche-de-mer-information-bulletin-35.html
Malua M (2003) Case study. The Pacific Islands. UNCTAD, Geneva. http://unctad.org/en/docs/ditctncd20031p5_en.pdf
McGregor A, Stice K (2014) Agricultural value chain guide for the Pacific Islands. KokoSiga Pacific, Suva. http://pafpnet.spc.int/attachments/article/504/Agricultural%20Value%20Chain%20Guide%20for%20the%20Pacific%20Islands.pdf
Ministry of Foreign Affairs and Trade (MFAT) (2013) Enhancement of biosecurity and quarantine services in the Pacific. MFAT, Wellington. https://www.mfat.govt.nz/assets/_securedfiles/Aid-Prog-docs/CASE-STUDY-PF-MPI-Case-Study.pdf
National Research Institute (NRI), Papua New Guinea (2015) Kum Kopi Kumaipana incorporated land group awarded ILG certificate. NRI, Port Moresby. http://www.nri.org.pg/index.php/design-idea-competition/212-kum-kopi-kumaipana-clan-awarded-ilg-certificate
PACE (1996) Special report: PACE in Melanesia. PACE 7. Pidgins and Creoles in Education (PACE). http://www.hawaii.edu/satocenter/pace/pace_news/7-special.htm
Pacific Agriculture Policy Project (PAPP) (2016). What we do. PAPP, Suva. http://pafpnet.spc.int/about-papp/what-we-do
Pacific Islands Forum (2015) Framework for Pacific regionalism. http://www.forumsec.org/resources/uploads/attachments/documents/Framework%20for%20Pacific%20Regionalism_booklet1.pdf. Accessed 10 Oct 16
Pacific Islands Forum Fisheries Agency (FFA) (2016) Pacific Islands Forum Fisheries Agency (FFA). Pacific Islands Forum Fisheries Agency, Honiara. http://www.ffa.int/vds
Pacific Regional Forest Inventory Facility (PRFIF) (2015) Booklets 1–4. http://lrd.spc.int/pubs
Payton I, Weaver S (2011) Fiji national forest carbon stock assessment. Version 1. Compiled by Carbon Partnership Ltd for SPC/GIZ Regional Programme Coping with Climate Change in the Pacific Island Region and the Fiji Forestry Department. Global Canopy Programme, Oxford. http://theredddesk.org/resources/fiji-national-forest-carbon-stock-assessment-version1
Rashid AT, Elder L (2009) Mobile phones and development: an analysis of IDRC-supported projects. Electron J Info Syst Dev Countries, EJISDC 36(2). http://www.ejisdc.org

Rihai N, Thomas C, Lombadozzi A (2013) CertLink—working with smallholder farmers to bring Vanuatu to the world. Paper presented at the IFOAM Asia Pacific symposium: entrepreneurship and innovation in organic farming, Bangkok. http://certlink.org/a/presentation-certlink-working-with-smalholder-farmers-to-bring-vanuatu-to-the-world

Seidel H, Lal PN (2010) Economic value of the Pacific Ocean to the Pacific Islands countries and territories. International Union for the Conservation of Nature, UCN, Gland. https://cmsdata.iucn.org/downloads/economic_value_of_the_pacific_ocean_to_the_pacific_island_countries_and_territories_p.pdf

SPC. Land Resources Division (2014a) LRD strategic plan. SPC LRD, Suva. http://lrd.spc.int/about-lrd/lrd-strategic-plan. Accessed 10 Oct 2016

SPC. Land Resources Division (2014b) SPC facilitates training in the use of web online applications to promote information sharing. SPC LRD, Suva. http://lrd.spc.int/our-work/information-communication-and-extension/training-and-workshops-/spc-facilitates-training-in-the-use-of-web-online-applications-to-promote-information-sharing. Accessed 10 Oct 2016

SPC. Geoscience Division (2015) Geoscience for development program. SPC, Noumea. http://gsd.spc.int/gsd-programmes/geoscience-for-development-programme. Accessed on 16 Aug 2016

SPC. Geoscience Division (2016) Geology, minerals and hydrocarbons. SPC, Noumea. http://gsd.spc.int/geologymineralshydrocarbons. Accessed on 16 Aug 2016

SPC. Division of Fisheries, Aquaculture and Marine Ecosystems (2016a) Mobile phone data collection app for artisanal fisheries makes its debut in Funafuti, Tuvalu. PC Fish Newslett #149. Jan–Apr 2016. http://www.spc.int/Coastfish/en/component/content/article/463-spc-fisheries-newsletter-149.html

SPC. Statistics for Development Division (2016a) Population Projections by Pacific Island Countries and Territories. SPC, Noumea. http://prism.spc.int/regional-data-and-tools/population-statistics. Accessed 13 January 2017

SPC. Statistics for Development Division (2016b) National Minimum Development Indicator Database. SPC, Noumea. http://www.spc.int/nmdi/en

SPC. Statistics for Development Division (2016c) Better Pacific fisheries and agriculture statistics to help improve livelihoods of millions. Press release, Apr 2016. SPC Statistics for Development Division, Suva, Noumea. http://sdd.spc.int/en/news/press-releases/108-better-pacific-fisheries-and-agriculture-statistics-to-help-improve-livelihoods-of-millions. Accessed 10 Oct 2016

Sustainable Development Solutions Network (2015) Data for development. A needs assessment for SDG monitoring and statistical capacity development. United Nations, Geneva

Toby S (2013) Mobile banking boost to rural Solomon Islands businesses. Pacific Trade and Invest. https://pacifictradeinvest.com/mobile-banking-boost-to-rural-solomon-islands-businesses/

Turia R (2013) Proceedings of the national preparatory workshop on the national forest inventory for Papua New Guinea. Papua New Guinea Forest Authority, Port Moresby. http://www.forestry.gov.pg/site/files/Proceedings%20of%20the%20National%20Preparatory%20Workshop%20on%20the%20National%20Forest%20Inventory%20for%20Papua%20New%20Guinea.pdf

United Nations (2015) Sustainable development: knowledge platform. Sustainable Development Goals. https://sustainabledevelopment.un.org/sdgs. Accessed 30 Mar 2016

Va'a F, Va'a T, Fuatai F, Chan Mow I, Amosa D (2012) Aspects of economic development in Samoa's journey, 1962–2012. Aspects of history, Victoria University Press, Wellington, pp 99–140

Valemei R (2016) 10,000 join council. Fiji Times Online, 10 May 2016. http://fijitimes.com/story.aspx?ref=archive&id=353160. Accessed 10 Oct 2016

Walton P (2002) Collection, access and use of agricultural statistics in the Pacific Islands: report of a study. ACIAR impact assessment program. Working paper series, no. 45. Australian Centre for International Agricultural Research, Canberra. http://aciar.gov.au/files/node/2125/wp45.pdf . Accessed 10 Oct 2016

Walton P (2009) Networking for sustainable development: The Papua New Guinea National Agricultural Information System. Agric Inf Worldwide 2(1):19–24. http://journals.sfu.ca/iaald/index.php/aginfo/article/view/83

Western and Central Pacific Fisheries Commission (2012) Statement describing purpose and principles of the WCPFC. Pohnpei, WCPFC. https://www.wcpfc.int/doc/tcc-03/statement-describing-purpose-and-principles-wcpfc-vms

World Bank (2014) World Bank development indicators. http://wdi.worldbank.org/table/4.2#

Chapter 10
ICTs in Pacific Islands' Climate Change and Disaster Risk Reduction Policy and Programs

Netatua Pelesikoti and Joseph Kim Suwamaru

Abstract Pacific Island countries are among the countries most vulnerable to climate change and to disaster risks and variability. This chapter discusses lessons learned from selected PICs on the role of e-government in implementing climate adaptation and disaster risk management, with a focus on two separate regional policy drivers for climate change: the *2005–2015 Pacific Islands Framework for Climate Change Adaptation* and the *2005–2015 Pacific Disaster Risk Management Framework for Action*. The chapter also highlights the transition to a new regional *'Framework for Resilient Development in the Pacific: An Integrated Approach to Address Climate Change and Disaster Risk Management* 2017–2030. These regional policies are relevant to PIC commitments to the United Nations Framework Convention on Climate Change, the Paris Agreement, the Sendai Framework and the Sustainable Development Goals and PICs use information and communication technologies in early warning systems, adaptation, mitigation and disaster response and recovery initiatives. The challenge for individual PICs has been to establish national policies and programs that made beneficial use of the ICT capabilities available to them to drive the achievements of these policy/framework targets as relevant to their national priorities.

N. Pelesikoti (✉)
Secretariat of the Pacific Environment Programme (SPREP), Apia, Samoa
e-mail: netatuap@sprep.org

J. K. Suwamaru
PNG, DATACO, Port Moresby, Papua New Guinea

J. K. Suwamaru
Independent researcher, formerly at Divine Word University,
Madang, Papua New Guinea

R. Cullen and G. Hassall (eds.), *Achieving Sustainable E-Government in Pacific Island States*, Public Administration and Information Technology 27,
DOI 10.1007/978-3-319-50972-3_10

10.1 Introduction

Pacific Island Countries (PICs) are among the most vulnerable to climate change[1] and to disaster risks and variability. This high level of vulnerability, as has been noted in Chap. 1, results from a combination of physical and geological factors, including: being located in a disaster-prone region of the world (in addition to being near the "ring of fire" and a traditional cyclone belt); being surrounded by the ocean; being scattered and remote and in some instances comprising small, low-lying land; having limited economic and technological capacity; being highly dependent on primary resources; and facing governance and management challenges. Added to these characteristics of vulnerability are the physical impacts of climate change, which include adverse weather events of increasing intensity, frequency and coverage that result in accelerated coastal erosion, flooding, disease, water and food shortages, damage to infrastructure and properties, biodiversity degradation and, potentially, species extinction.

These conditions are confirmed in the findings of the Fifth Assessment Report of the Intergovernmental Panel on Climate Change (Intergovernmental Panel on Climate Change 2014) concerning current and future climate-related drivers of risk for small islands. Water and food insecurity result from storm surges, droughts and floods. Sea level rise is increasing coastal inundation, erosion and salt water intrusion. The effect of ocean acidification on coral reefs and shellfish is yet to be fully understood. In such contexts PIC governments and communities are implementing adaptation, mitigation and disaster risk reduction and disaster management strategies within the context of achieving their respective programs of national sustainable development.

Since 2005/6, two programs have operated regionally to assist policy making, planning, implementation and institutional strengthening in PICs to address climate adaptation and disaster risk management: the 2006–2015 Pacific Islands Framework for Action on for Climate Change Adaptation (PIFACC) (Secretariat of the Pacific Regional Environment Programme 2006) and the 2005–2015 Pacific Disaster Risk Management Framework for Action (Secretariat of the Pacific Community 2005). As a consequence, PICs have made considerable progress in implementing integrated approaches to managing disaster and climate risks.[2] From 2017 they will transition to an integrated framework for climate change and disaster risk management known as the Strategy for Resilient Development in the Pacific: An Integrated Approach to Climate Change and Disaster Risk Management (SRDP).

[1]The use of 'climate change' in this chapter refers to anthropogenic climate change caused by the excessive emissions of carbon dioxide and other greenhouse gases into the atmosphere. 'Disaster' refers to natural disasters where consequences and impacts are exacerbated by climate change. More detailed definition of concepts used in this chapter is provided in its list of acronyms.

[2]These risks are related to all natural hazards, including weather extremes and climate variability and change.

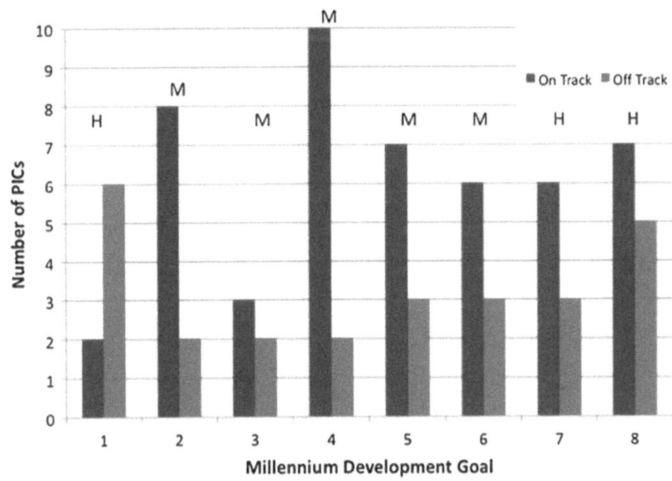

Fig. 10.1 Progress in the Pacific to achieve the millennium development goals, and their sensitivity to climate change and disasters. Note: *H* high; *M* medium. A grade of "low" was available, but not given

10.1.1 The Context of Vulnerability

Climate change and disaster preparation and response have received much attention from international forums and indeed have become key issues facing PICs. Now that they are well aware of the extent to which climate change increases water and food insecurity, for instance, governments and communities are implementing adaptation and mitigation strategies to build and strengthen their resilience against climate change and disaster risks within the context of achieving sustainable development. They are doing this, however, with all the vulnerabilities of small island developing states. The Pacific Islands region (after the sub-Saharan African region) was the most "off-track" for achieving all of the Millennium Development Goals by 2015. Significantly, as shown in Fig. 10.1, progress was poorest for Goal 1 (eradication of extreme poverty and hunger)—the Goal judged to be the most adversely affected by weather, climate extremes and variations and natural hazards.

In 2012, SPREP articulated the risks in the Pacific from natural hazards such as cyclones and drought (Secretariat of the Pacific Regional Environment Programme 2012). For example, the subsistence lifestyles common in the Pacific region are highly vulnerable to both rapid (e.g. cyclones) and slow (e.g. drought) onset events, as well as to inter-annual and longer-term variations and climate change. But even if climate change was not occurring, the vulnerability of PICs is increasing due to economic and social changes such as population growth, internal and external migration, poorly planned coastal development, unplanned urban growth and land use, and environmental and ecosystem degradation including contamination of sub-surface and coastal waters.

Table 10.1 Consequences of the disasters in the Pacific Islands region 1950–2013

Hazard type	Number						Total damage
	Disasters	Killed	Injured	Affected	Homeless	Overall	USD thousand
Climatological	11	60	0	956,635	0	956,635	61,650
Geophysical	58	5964	1499	267,502	65,400	334,401	401,575
Hydrological	50	683	56	728,983	118,085	847,124	322,375
Meteorologica	165	1592	2629	2,003,396	238,891	2,244,916	2,728,360
Total	284	8299	4184	3,956,516	422,376	4,383,076	3,513,960

Source EM-DAT—the OFDA/CRED International Disaster Database www.emdat.be Universite catholique de Louvain Brussels Belgium

The majority of the 284 disasters that occurred in the Pacific Islands region between 1950 and 2013 were caused by severe storms, including tropical cyclones (Table 10.1). Both individually and collectively, these disasters had enormous social and economic consequences. In many Pacific countries, the accumulated impacts of small and medium-sized hazard events are equivalent to, or exceed, those of single large disasters. Low-intensity events are typically more widespread, affecting a comparatively larger number of people. They are also likely to involve damage to housing, land and local infrastructure, rather than major mortality or destruction of nationally-critical economic assets. As the poor and other vulnerable, marginalized groups tend to live in more hazard-prone areas, increases in the frequency of these lower-intensity hazards have large effects on poverty (World Bank 2012). Importantly, the consequences of these small- and medium-sized disasters are far more amenable to being reduced through investments in disaster risk management (DRM) and climate change adaptation (CCA), including prevention preparedness and building back better after a disaster (FRDP 2017).

Vanuatu is at the top of the World Risk Index,[3] which assesses exposure to natural hazards: susceptibility as a function of public infrastructure, housing conditions, nutrition and general economic framework; coping capacities; and adaptive capacities to future natural events and climate change. Four other Pacific Island countries (Tonga, Solomon Islands, Papua New Guinea and Fiji), are also in the top 15 of 173 countries for which the Index is calculated. Eight of the 20 countries with the highest average annual disaster losses scaled by gross domestic product are in the Pacific (Asian Development Bank 2011).

Climate change is exacerbating the current vulnerabilities of PICs, with climate change adaptation and disaster risk reduction and disaster management fast-becoming the key risk management response options. Climate change is affecting food, water, health, economic, environmental and political security. Responses must extend beyond concerns related to the anticipated increases in

[3]http://weltrisikobericht.de/english/.

weather- and climate-related hazards, to considering the underlying causes of vulnerability to hazards such as hardship and isolation, poorly informed planning, or no planning at all.

The 1998 Aitape Tsunami on the northern coastline of PNG, presents one example. It was caused by a local earthquake and comprised three successive waves 10–15 m high, which destroyed three villages and caused 2200 deaths, left a further 1000 people with serious injuries and displaced 10,000 people from their homes (Davies et al. 2001). Subsequent investigations found lack of effective warning systems for tsunamis originating within 40 km of the coast along the inner wall of the New Guinea trench, which was identified as the source region. It also found that the villages along frontal sand barriers have poor escape routes. While rescue efforts began after disaster struck, with survivors helping each other, the first external help arrived 16 h after the event. The major rescue effort only began 40 h after disaster struck and only 72 h later were all of the injured transferred to Vanimo hospital.

10.1.2 Climate Projections

Climate change projections (Australian Bureau of Meteorology and CSIRO 2011) predict large increases in the occurrence of extremely hot days and warm nights and an increase in the number of heavy rain days (20–50 mm), such that extreme rainfall events which now occur on average once every 20 years will occur (under a high emissions scenario) four times per year, on average, by 2055 and seven times per year, on average, by 2090. In the South Pacific sub-basins, the frequency of tropical cyclones will decrease, but the proportion of more intense storms will rise, while in the North Pacific sub-basin a decrease in both the frequency of tropical cyclones and the proportion of more intense storms is expected. While not uniform across the region, sea-level rise will generally be similar to the global average, between 0.18 and 0.59 m by 2080–2099, relative to 1980–1999, with potential contributions from ice sheets resulting in an additional 0.1–0.2 m. Ocean acidification is expected to increase, with greatest rates occurring in the eastern equatorial Pacific and affecting the easternmost islands of Kiribati, while droughts are projected to occur less often.

Climate projections also anticipate that by the mid- to late-twenty-first century there will be an intensification of both El-Niño-driven drying in the western Pacific Ocean and rainfall increases in the central and eastern equatorial Pacific (Kirtman et al. 2013). Significantly, the number of people affected by each disaster and the economic losses per disaster is increasing. These trends are consistent with population growth, development patterns and, in many cases, movement of people to more hazard-prone, peri-urban areas. Therefore, the Pacific continues to experience an unfortunate combination of circumstances and poorly planned and implemented socioeconomic development initiatives are increasing the already significant exposure to natural hazards, including extreme weather and climate events (World Bank 2012). While this situation is not unique to the Pacific, statistics from many sources demonstrate that the Pacific Islands region is an extreme case (Foundation of the Peoples of the South Pacific International 2013). Numerous studies for the Pacific, as for other regions (Hay and Pratt 2013 and Global Facility for Disaster Reduction and Recovery 2013), have highlighted the changes that need to be made in order to increase the resilience of development outcomes.

10.1.3 Estimated Costs of Natural Disasters

PCRAFI (Pacific Catastrophe Risk Assessment and Financing Initiative 2013) compared current and future tropical cyclone risk for 14 Pacific Island countries using global climate change models to calculate projected financial losses from cyclone damage to buildings, infrastructure and crops in each country, for both mid and end of this century. No adjustment was made for future changes in any of these assets. The findings indicate increasing losses for the region, largely as a result of the projected increase in category 5 cyclones. While, for the region as a whole, the increase in average annual losses is relatively small, the end-of-century increases for many individual countries are large—for example, 25.4% for Samoa, 14.8% for Niue and 7.6% for Vanuatu. Increases are anticipated to be greater for buildings than for infrastructure and crops, with wind being the main contributor to building loss. Flooding is likely to be the main contributor to increasing infrastructure losses. The total value of the infrastructure, buildings and cash crops in the Pacific Islands region at risk from tropical cyclones, earthquakes and tsunami[4] is estimated to be over USD 112 billion. This translates into annual average economic losses being as high as 7% of gross domestic product (GDP), with natural disaster losses in any single year sometimes exceeding the GDP (World Bank 2012 and cited from the FRDP 2017).

10.2 Policy Drivers for Climate Change and Disaster Risk Management in the Pacific

The vulnerability of the PICs to climate change and natural disasters is clearly established in the introduction above. How they manage and reduce this vulnerability and how they incorporate ICTs into disaster risk reduction and disaster management is considered in the following sections.

Rising sea levels are already causing significant damage along Pacific Island coastlines and chewing away at the fabric of peoples' existence through land erosion and destruction of food crops through salt water intrusion (Secretariat of the Pacific Regional Environmental Programme 2006). For the majority of the low-lying islands, this is a helpless situation that requires urgent attention through a concerted

[4]Tsunami is induced by geological events such as earth quake and is not a climate related event.

effort of governments, donors and other development partners. The importance of a concerted effort aligns with Heeks' ICT4D 2.0 Manifesto recommending guidance of stakeholders through corroborated and contingent design strategies to ensure sustainable projects for successful development outcomes. In this context, PICs governments are encouraged to lead in showing their resilient leadership roles towards the vision *Pacific island people, their livelihoods and the environment resilient to the risks and impacts of climate change* (Secretariat of the Pacific Regional Environmental Programme 2006). An important and integral aspect in the overall strategy on climate change in PICs is the creation of a strong enabling environment where government should demonstrate leadership in governance, partnerships and cooperation. A practical starting point is the formulation of policy and institutional frameworks which may guide climate change specific interventions.

Disaster risk management entails all activities that aim to avoid, lessen or transfer adverse effects of hazards, including reducing disaster risks, preparing for disasters, providing emergency relief and undertaking reconstruction (Hay 2013). This forms an important connection with the overarching idea of 'Pacific island people, their livelihoods and the environment resilient to the risks and impacts of climate change' (Secretariat of the Pacific Regional Environmental Programme 2006). More generally, DRM comprises disaster risk reduction (DRR) and disaster management (DM), the former involving climate change adaptation and mitigation and the latter involving disaster preparedness and response (Hay 2013). The demarcation between DRR and DM can be conceptually challenging because recent discussions in PICs often combine DRM and CCA, as attested by several joint national action plans. The following example, which aligns well with Heeks' (2009) idea of participatory approach for beneficial outcomes, demonstrates practical outcomes achieved by collaborating with regional, national and community players in Fiji.

> The regional disaster reduction program promoting national DRM capacity across PICs has demonstrated greater involvement in supporting community based DRM with resources in community-based DRM initiatives. In a partnership with UNDP Fiji national disaster management office, Fiji Red Cross and Live and Learn, a program has been successfully rolled out in communities on the Navua flood plain in Fiji, to reduce the risk to flooding. The project has brought together local government and the community to improve flood response which included the installation of a flood warning system. A flood response plan was developed together with community-based first aid and disaster preparedness workshops. (Hay 2013)

Such international, regional and national collaboration also necessitate better alignment of international financial resources. Recent trends have seen a shift from funding mitigation to increased assistance for adaptation programs due to the increasingly visible efforts to respond to the effects of climate change. Clear policy interventions harmonizing the programs of various players are necessary to ensure that tensions between agencies at regional, national and community levels are minimized.

Several Pacific regional policies and plans reference and highlight the importance of access to data and information. Table 10.2 lists key regional plans and

Table 10.2 Key Pacific regional plans and polices related to climate change/DRM information management

Plan/policy	Summary
SPREP strategic plan 2011–2015	Aims to gather, store, provide access to, and analyze regional environmental data and information
SPC climate change engagement strategy 2011–2015	Aims to assist PICTs through identification of risks and provision of relevant climate change knowledge, technical assistance and resources to enable informed policy and operational decisions
Pacific Islands framework for action on climate change (PIFACC) 2006–15	Recommended the establishment of a regional clearing house on climate change information and the promotion of improvements in telecommunications capacity across the region
Hyogo framework for action (HFA) 2005–2015	Priority action 3: Use knowledge, innovation and education to build a culture of safety and resilience Recommends the collection, compilation and dissemination of relevant knowledge and information on hazards, vulnerabilities and capacities
Draft strategy for disaster and climate resilient development in the Pacific (SRDP)	Integrates climate change adaptation and disaster risk management into one strategic approach Highlights the importance of collection and dissemination of information and knowledge for decision-making at all levels
Framework for Pacific regionalism (endorsed 2014)	PIFS master strategy for strengthening cooperation and integration between the states and territories of the Pacific region Recommends regional coordination to achieve open consultation and access to information; and coordinated application of shared, best-practice norms and standards
Framework for action on ICT for development in the Pacific 2010 (FAIDP)	Outlines seven themes for action aimed at effectively utilizing ICTs for sustainable development, governance and improving the livelihood of Pacific communities Recommends the pooling of resources and expertise and a 'whole-of-sector' approach, based on the concept of 'many partners, one team'
Draft Pacific regional ICT strategic action plan (PRISAP) 2015–2020	Describes the critical activities that promote open consultation and access to information and the coordinated application of shared, best practice norms and standards

policies which relate broadly or specifically to climate change/DRM information management. With the exception of the Pacific Regional ICT Strategic Action Plan (PRISAP) 2015–2020, none of the policies summarized in Table 10.2 actually provide a framework or guidance on the role of ICT.

The two main regional frameworks guiding climate change responses and disaster risk management for the last decade are the *2005–2015 Pacific Islands Framework for Action on Climate Change* (PIFACC) and the *Regional Framework for Action on Disaster Risk Management* (RFA). These policy frameworks are contextualized regional priorities for action aligning to the UNFCCC and the Hyogo Framework for Action (HFA). The HFA was the global framework on disaster risk management that is being now replaced by the Sendai Framework. All independent PICs are party to the UNFCCC and have endorsed the UNISDR Global Hyogo Framework for Action on disaster risk reduction.

While PIFACC and RFA provided opportunities for a more coordinated governance, PIFACC was coordinated by the Secretariat of the Pacific Regional Environment Programme (SPREP) and RFA by the Pacific Islands Applied Geoscience Commission (SOPAC)[5] (which, as explained in Chap. 3, is now a division of the Pacific Community). This resulted in separate implementation structures and institutions, with SPREP and the Pacific Climate Change Roundtable (PCCR) coordinating implementation of PIFACC with environment or climate change and meteorology ministries and SPC leading coordination of the RFA in collaboration with national disaster agencies or ministries of public works or water. Both policy frameworks will be replaced by a more integrated framework from 2017 (elaborated below).

PIFACC aimed at ensuring that Pacific Island people, their livelihoods and the environment are resilient to the risks and impacts of climate change by promoting coherent and coordinated regional approaches in supporting PICTs, to provide an 'entry point' to prioritized areas where support is needed both at regional and national levels, to mobilize resources and to act as an advocacy framework (Secretariat of the Pacific Regional Environment Programme 2011).

The RFA, on the other hand, aims for safer, more resilient Pacific Island nations and communities in disasters so that Pacific peoples may achieve sustainable livelihoods and lead free and worthwhile lives. Its mission is to build the capacity of Pacific Island communities by accelerating the implementation of DRR and disaster management policies, planning and programs through: development and strengthening of DRR and disaster management including mitigation, preparedness, response and relief/recovery systems; integration of DRR and disaster management into sustainable development planning and decision making processes at all levels; and strengthening partnerships between all stakeholders in DRR and disaster management.

Both frameworks have similar and related themes as shown in Table 10.3. For example, PIFACC theme 2 is similar to RFA theme 1. PIFACC theme 1 needs to be guided by information from PIFACC themes 2 and 3 and RFA themes 3, 5 and 6.

Each theme has several actions that could be carried out either at regional or national level. For example, expected outcomes for PIFACC theme 1 for national actions are enhanced resilience to the adverse effects of climate change through the

[5]SOPAC became the GeoScience Division of the Pacific Community (SPC) in 2011.

Table 10.3 PIFACC and RFA themes

PIFACC themes	RFA themes
1. Implementing tangible on-ground adaptation measures	1. Governance—organizational, institutional, policy and decision making frameworks
2. Governance and decision making	2. Knowledge, information. Public awareness and education
3. Improved understanding of Climate change	3. Analysis and evaluation of hazards, vulnerabilities and elements at risks
4. Education, training and awareness	4. Planning for effective preparedness, response and recovery
5. Contributions to global greenhouse gas reduction	5. Effective integrated people focuses early warning systems
6. Partnership and cooperation	6. Reduction of underlying risk factors

implementation of best practice adaptation and risk reduction measures (SPREP 2011). The RFA does not have a specific monitoring and evaluation framework; but it clearly outlines that the responsibility for implementation and monitoring, led by the SOPAC (now a division of SPC), is to utilize the set of indicators and monitoring platforms from the global *Hyogo Framework for Action 2005–2015: Building the Resilience of Nations and Communities to Disasters* (HFA).

The regional reports for the HFA are submitted to the Global Platform for Disaster Risk Reduction on a voluntary basis. The Global Platform for Disaster Risk Reduction is a biennial forum for information exchange, discussion of latest development and knowledge and partnership building across sectors, which aims to improve implementation of DRR through better communication and coordination among stakeholders. Clearly, the effects of climate change can best be reduced or minimized if outcomes of both the PIFACC and RFA are achieved and both frameworks promote mainstreaming across sectors, across different levels of decision-making, at policy level and at all levels of the development processes.

Within the timeframe of PIFACC and RFA (2005–2015) many national strategies, policies and action plans sought to mainstream actions for climate change resilience and for disaster reduction in national and sectoral policies. Also, PICs developed and improved national governance structures to manage climate change and DRM issues, include the following:

- Cook Islands National Strategic Development Plan 2011–2015 and the Cook Islands The Caving Nui: National Sustainable Development Plan 2011–2015
- Kiribati Development Plan 2012–2015
- Solomon Islands National Development Strategy 2011–2020
- National Climate Change Policy Framework 2011, the Solomon Islands National Climate Change Policy 2012–2017
- Tonga Strategic Development Framework 2011–2014
- Tonga Joint National Action Plan on Climate Change and DRM 2010–2015
- Strategy for the Development of Samoa 2012–2016, Tokelau National Strategic Plan 2010–2015

- Tuvalu Climate Change Policy and Integrated Action Plan (Tuvalu 2012)
- Strategic Tuvalu Infrastructure Strategy and Investment Plan 2012
- Government of the Republic of Vanuatu's Priorities and Action Agenda 2006–2015
- Republic of Fiji National Climate Change Policy 2012
- National Climate Change Policy for Niue 2009
- Te Kaniva: Tuvalu Climate Change Policy—Charting Tuvalu through the Challenges of Climate Change (2012)
- Federated States of Micronesia Nation Wide Integrated Disaster Risk Management and Climate Change Policy (2013)
- Palau National Disaster Risk Management Framework (2010) and its new integrated Climate Change and DRM Policy 2015–2019 (Palau 2015)
- Papua New Guinea National Disaster Risk Management Plan 2012
- The Western Pacific Regional Framework for Action for Disaster Risk Management for Health (DRM-H) produced by WHO for the western Pacific region[6]
- The Pacific Youth in Agriculture Strategy 2011–2015[7]

Good Governance is central to the achievement of national development goals, as it is an enabler for efficient and effective approaches, management of climate change resources and for the timely achievement of resiliency and DRR outcomes. What are the implications? The impacts of climate change and disasters are felt at the immediate level of the resources and livelihoods that communities depend on—such as coral reefs which are the habitat for fisheries, or precipitation which has direct effect on subsistence crops.

PIFACC's monitoring and evaluation framework outlines clear and measurable indicators of national and regional outputs. Reporting takes place via a number of mechanisms, which include the PCCR Working Group Reports, Country Profiles, SPREP Annual Report and Meeting and the Pacific Meteorological Council Reports. The Action Plan (SPREP 2005) for the PIFACC identifies the PCCR as the reporting mechanism to monitor the implementation of the framework. It is clear that the PIFACC and RFA did not explicitly established a role for ICTs. However, ICTs can assist in achieving most of the themes of the two frameworks. The following section therefore explores the role of ICT in achieving the outcomes of selected themes from PIFACC and RFA.

[6]This calls for the implementation of the health component of risk management prevention, preparedness, response and recovery and recognizes the key role of the health sector in managing risks related to all hazards.

[7]This identifies the need for young men and women to be trained on the implications of current patterns of agricultural development and resource use and the potential impacts of climate change.

10.3 The Role of ICTs in the Implementation of PIFACC and RFA

The preceding sections have noted that the United Nations Framework Convention on Climate Change (UNFCCC) has triggered several Pacific regional organizations to cooperate in progressing adaptation and mitigation interventions using appropriate programs, and that the PICs have been implementing CC and DRR activities jointly or separately under two regional policy frameworks (PIFACC and RFA). It has noted, furthermore, that whereas these frameworks did not specify a role for ICTs, such roles flowed as a consequence of their need for managing, sharing and delivering data and information. Table 10.4 provides some examples of climate change and DRR information services being provided in the Pacific.

Table 10.4 highlights the extent to which the landscape of climate change ICT services in the Pacific is highly populated. Most key regional organizations operating in the Pacific region (most notably SPREP and SPC) host either a climate change or DRR related data and information website, repository or portal. Numerous inter-governmental organizations (e.g. UNOCHA, UNISDR, UNEP, UNDP, World Bank, FAO), NGOs and Pacific neighbors (e.g. BOM in Australia, NIWA in New Zealand) also provide web-based knowledge banks and data repositories.

Table 10.4 RFA and PIFACC theme outcomes

RFA Theme 1—expected outcomes by 2015	PIFACC Theme 2—expected outcomes by 2015
(a) Disaster risk reduction and disaster management mainstreamed into national policies, planning processes, plans and decision-making at all levels and across all sectors (b) Partnerships and organizational arrangements with and between government agencies, civil society, development partners, communities and other stakeholders strengthened (c) CROP agency partnerships coordinated, harmonized and strengthened to ensure country and outcome-focused delivery of services[a] (d) Good governance by all stakeholders in disaster risk reduction and disaster management at regional, national and local levels strengthened	2.1 Strengthened national and regional climate change governance mechanisms (i.e. policy and institutional frameworks) 2.2 Enhanced cross sectoral and multi-disciplinary coordination, collaboration and decision making around climate change 2.3 Enhanced integration of climate change risks into development decision making process and assessment cycles, sectoral planning and management at all levels

[a]RFA outcomes (b) and (c) align with the Theme 6 of the PIFACC, and outcome (c) is the same as outcome 6.2 of the PIFACC. The RFA outcomes will be reported together with the Partnerships and Cooperation theme in Sect. 10.2 of this chapter

10.3.1 ICTs and Strengthening Governance and Decision Making: The Case of Vanuatu

Initiatives by the Government of Vanuatu are presented as a case study indicating the use of ICTs to coordinate climate change policy and management. PIFACC 2.3 and RFA (a) focus on mainstreaming climate change and disaster risk reduction into policies, planning processes, plans and decision-making at all levels. Vanuatu's case approaches PIFACC 2.3 from various levels. To ensure that mainstreaming occurred at ministerial level and in each sector the government established the National Advisory Board of Vanuatu (NAB) in 2012 for joint governance of climate change and disaster risk management. NAB convened workshops and policy development with various ministries. It also managed all DRR and climate change programs, projects and initiatives and established the Vanuatu National Climate Change Portal.[8] NAB serves as an effective forum for information sharing among relevant ministries, departments and sectors and provides an example of national action to integrate climate change and DRR, increases collaboration and strategic management and reduces duplication. A summary of Vanuatu's ICT policies in the area of climate change and disaster risk management is given in Tables 10.5 and 10.6.

10.3.2 Institutional Barriers[9]

There are, however, some institutional barriers. Although the role of ICTs is explicitly established as illustrated in Table 10.5, responsibility for overseeing and adding information to the NAB Portal rested with a single Communication and Outreach Officer who had multiple other responsibilities. The incumbent Officer did not have formal training in information management and, with the exception of a brief manual on use of the NAB Portal, had no guidelines or standard operating procedures for information management activities. Technical assistance for the NAB Portal and all other information management activities is provided by the Vanuatu Meteorological and Geological Department IT manager; however, this falls outside their core role and the IT manager had no specific training in the Drupal content management system used to build the Portal.

Given the breadth of the NAB PMU's role in information gathering and dissemination, in the longer term an appropriately trained, ongoing Information/Knowledge Manager position will be needed. There is currently minimal formal information sharing between the NAB and other government and non-government agencies. Some inter-governmental data sharing takes place in other government agencies, a notable example being the National Statistics Office that has formal data

---

[8]http://www.nab.vu/pacific-climate-change-portal-1.

[9]This section draws on Brown et al. 2015, p. 26.

Table 10.5 Selected organizations providing climate change-related information in the Pacific region

Host	Information source	Information type
SPREP (climate change division)	Pacific climate change portal www.pacificclimatechange.net	Climate change related documents and resources, projects database, donor directory, details on the RTSM, events calendar and a glossary of terms. Also provides access to the Pacific Environmental Information Network
SPREP	Pacific Environment Information Network (PEIN) https://www.sprep.org/virtual-library	A collection of materials located in the SPREP Library and Information Resource Centre, plus digital documents drawn from a network of environment libraries throughout the Pacific
SPC (economic development division)	Pacific regional data repository Prdrse4all.spc.int	National and regional energy data and project information, plus related publications
SPC	Pacific DisasterNet www.pacificdisaster.net	Disaster risk management information resource for the Pacific. Provides documents, events, contacts, forum, training and tools
SPC (land resources division)	Maintains a number of information resources and knowledge sharing networks for agriculture, farming and forestry www.spc.int/lrd/	Statistics, publications, knowledge sharing
SPC (statistics for development division)	PRISM—(including PopGIS 2.0 interactive mapping application) www.spc.int/prism	Statistics of the Pacific Island countries and territories
SOPAC/SPC GeoScience division	PacGeo www.pacgeo.org	Geophysical, geodetic and marine spatial data
SOPAC/SPC GeoScience division	GEONETWORK	Interactive maps, GIS datasets, satellite imagery related to oceans and islands
Australian Government	Pacific climate change science www.pacificclimatechangescience.org	Publications, tools and educational resources developed under the Australian Government's PCCSP and PACCSAP programs
Australian Bureau of Meteorology	COSPPac www.bom.gov.au/pacific/	"One stop shop" for Pacific Island weather and climate data; country homepages for 14 PICs

Source Brown et al. (2015)

Table 10.6 Key plans and polices related to information management, ICT and climate change adaptation/disaster risk management in Vanuatu

Plan/policy	Summary
National Climate Change Draft National Climate Change and Disaster Risk Reduction Policy—DRAFT January 2015	Highlights 'knowledge and information' as one of six key priorities Knowledge and information objectives are 'to meet knowledge information needs, improve understanding and communicate to empower action' Specifies the establishment of an accessible and up-to-date database Provides a policy framework for integrated CC/DRR activities, strategies and priorities in Vanuatu
Priorities and Action Agenda 2006–2015	Encourages reform to enable the "free flow of information on government programs and services to Members of Parliament, citizens, civil society and development partners"
Draft National ICT Policy 2011	Sets out a vision of "ICT for all" Includes the aim of educating all people about environment and climate change issues
Right to Information Policy (approved 2013)	Aims to provide a free flow of information on government programs and services to Members of Parliament, citizens, civil society and development partners To be followed by Right to Information legislation which will formalize the policy arrangements to ensure that the Government actively releases information and provides information where it is requested by the public

Source Brown et al. (2015)

sharing agreements in place with other government agencies and makes a variety of statistics and documents available online. The following barriers are common in Vanuatu and may be similar in other Pacific countries:

- Lack of staff with prior experience and/or formal training in information management Sub-standard quality of meta data by sector
- Standardized data sharing agreements are not available to assist in brokering partnerships within government departments and between government and non-government organizations
- Lack of national-level guidelines for government departments regarding required information management procedures such as collection, storage, documentation and legal and ethical sharing of information
- Lack of division-specific policies and procedures to guide storage, filing and back up of information and data and ensure information management is integrated into core day-to-day activities

- Divisional staff with multiple responsibilities and limited time to dedicate to information management and training activities
- Difficulties in securing funding for ongoing information/ knowledge management positions (as opposed to short-term, project based positions)
- Lack of IT support staff with the necessary technical expertise and/or time to support national climate change portals
- Lack of user needs assessments to inform the development of climate change portals and websites and lack of long-term monitoring and evaluation of their effectiveness
- Lack of consideration of national and global copyright laws when making resources available via portals

The foregoing challenges show that while ICTs can play an integral role in climate change, adequate training is important to ensure that relevant officers are adequately skilled. It is through training and upskilling that national objectives can be achieved.

10.3.3 Improving Understanding of Climate Change and the Analysis and Evaluation of Hazards, Vulnerabilities and Elements at Risk

PIFACC's Theme 3, *Improving understanding of climate change,* identifies the development of scientific capacity as foundational to its implementation. For instance, improving the quality, access and analysis of meteorological data is essential to understanding climate processes. RFA Theme 3, *Analysis and evaluation of hazards, vulnerabilities, and elements at risk* also recognizes that promoting a culture of resilience lies in a greater understanding of the cause and effects of hazards and the physical, social, economic and environmental vulnerabilities to disasters.

Improving understanding of climate change as a result of *Analysis and evaluation of hazards, vulnerabilities, and elements at risk* are two separate outcomes. Although the use of ICTs to carry out *Analysis and evaluation of hazards, vulnerabilities, and elements at risk* and to collect and assess hazards data are not explicitly established in either the PIFACC or RFA, which is key to connecting these two outcomes, efforts have been made to establish systems to predict, model and assess climate variability and measure the impact of natural disasters on infrastructure and assets, which includes environmental assets. Similarly, improving the quality, access and analysis of meteorological data is essential to understanding climate processes. Furthermore, RFA Theme 3 also recognizes that promoting a culture of resilience lies in a greater understanding of the cause and effects of hazards and the physical, social, economic and environmental vulnerabilities to disasters (Table 10.7).

Table 10.7 RFA and PIFACC Theme 3 expected outcomes

RFA Theme 3 expected outcomes by 2015	PIFACC Theme 3 expected outcomes by 2015[a]
(a) An integrated framework for disaster risk reduction planning developed and implemented in Pacific island nations and communities (b) Estimates of disaster risk and vulnerability in place that will enable informed decisions regarding the impact of disasters on physical infrastructure, social, economic and environmental conditions in Pacific island nations and communities (c) Data and statistical information on disaster occurrence and impacts available for the region (d) Implementation of a comprehensive scientific and technical regional database enabling spatial analysis of hazard-prone areas, and establishment of magnitude frequency relationships and loss functions	3.1 Strengthened technical and institutional capacity to collect, store and analyze data used to detect climate change signals in the Pacific (current variability) 3.2 Strengthened technical and institutional capacity to apply analytical frameworks, models and tools to project future climate changes in the Pacific (future climate change) 3.3 Strengthened capacity to apply analytical frameworks, models and tools (appropriate to the Pacific) to assess national and regional climate change vulnerability (impacts/vulnerability to current and projected climate changes)

[a]The aims of the PIFACC outcomes 3.1 and 3.2 are inter-related with the aims of the RFA outcomes (b), (c) and (d)

10.3.4 Scientific and Technical Data, Statistical Information and Databases for Spatial Analysis of Natural Hazards and Climate Change

The Pacific Risk Information System (PacRIS) was established under the Pacific Catastrophe Risk Assessment and Financing Initiative (PCRAFI[10]—a joint initiative of SOPAC/SPC, World Bank and a range of financial donors). PacRIS constitutes the most comprehensive regional database of baseline exposure (buildings, infrastructure and crops) and probabilistic risk assessment results for 15 countries.[11] PacRIS also includes a regional historical hazard catalogue of earthquake and tropical cyclone events and a historical loss database for major disasters. It includes country-specific hazard models that simulate earthquakes, tsunamis and tropical cyclones. Risk maps showing the geographic distribution of potential losses for each PIC as well as other visualization products of the risk assessments can be accessed through the open-source web-based platform. PacRIS enables proactive regional integration with the Open Data for Resilience Initiative that advocates the

[10]http://pcrafi.sopac.org/.

[11]Cook Islands, Federated States of Micronesia, Fiji, Kiribati, Republic of the Marshall Islands, Nauru, Niue, Palau, Papua New Guinea, Samoa, Solomon Islands, Timor Leste, Tonga, Tuvalu, Vanuatu.

sharing and use of information for building resilience to natural hazards in a changing climate.

Disaster losses are also systematically collected on the Pacific Damage and Loss (PDaLo) information system, maintained and updated by SPC with support from UNISDR. PDaLo allows its users to visualize and analyze hazards, damage and losses in space and time (reported nominal value for damage from an event), that have occurred since 1967, to support decision making processes related to mitigation actions and risk management. Relevant sectors and decision makers at the national level are trained on the meaning, application and understanding of PDaLo which ultimately leads to better decision making—as advocated by the by PIFACC and RFA.

In the fields of Forestry and Agriculture, SPC and the German Development Agency (GIZ) are working with the relevant PIC Departments to map forestry and agricultural resources so as to build their capacity to manage the challenges and opportunities posed by climate change. Temporal mapping of vegetation and forest cover using remote sensing techniques have begun with field verifications being undertaken by in-country teams.

Collectively, Fiji, Kiribati and Niue have all recognized the importance of mainstreaming DRR and DM into their respective development plans. They continue to collaborate with international development partners to install information sharing mechanisms on disaster risk reduction and disaster management. For example, the Pacific Disaster Net[12] has become an important information sharing portal and repository where PICs may store, retrieve and access DRR and DM related information. This portal also offers invaluable resource for climate change and disaster related research and is an excellent example of e-governance in motion.

10.3.5 Regional Severe Weather Forecasting Demonstration Project (SWFDP)

The World Meteorological Organization (WMO) has supported Pacific Island countries to improve severe weather forecasting by:

- Improving their ability to produce severe weather forecasts and warnings;
- Increasing the lead-time for issuing warnings;
- Improving the quality of interactions between each National Meteorological and Hydrological Service (NHMS) and its media, disaster management and civil protection authorities; and
- Raising the visibility, credibility and value of meteorological series to all users.

The WMO's Regional Severe Weather Forecasting Demonstration Project (SWFDP), commenced in southern Africa in 2006 and in the Pacific region soon

[12]www.pacificdisaster.net.

after. SWFDP emphasizes the importance of liaising closely with those receiving the severe weather forecasts and warnings and implemented in Cook Islands, Fiji, Kiribati, Niue, Samoa, Solomon Islands, Tonga, Tuvalu and Vanuatu. These and other WMO projects utilize a chain of orbital weather satellite systems supplementing terrestrial or ground stations which coexist to monitor, report and forecasts weather patterns.

SPREP has been supporting the training of weather forecasters from the National Meteorological Services through partnership with WMO and the United States' National Ocean and Atmospheric Administration (NOAA) International Pacific Training Desk. The initiative sees the attachment of Pacific weather forecasters to the National Weather Services, Pacific Region Headquarters (NOAA) which resulted at strengthening of strengthening NMSs capacity for analysis of climate change and disaster risks. These programs have strong ICT components often overlooked and unsupported and by decision makers. For example, weather forecasting is carried out electronically not only in assessment but in the communication of warnings, another example is geo-data, such as sea level rise, collected to tide gauges from around the regional; it is done through tide gauges sending signals to satellites and from the satellites to land computers. More examples from are region are as follows:

- Vanuatu and Solomon Islands National Disaster Management Offices (NDMOs) have developed disaster loss databases of disasters and related impacts using the DesInventar methodology. Geo-data management training is also conducted across PICs in support of DRM.
- The Cook Islands Geo Portal Project is an online information center for disaster risk response and management. It aims for collaboration and continual improvement of emergency response working in partnership with the public, government ministries and key stakeholders, to promote best practices before, during and after disasters. Developed by Emergency Management Cook Islands in partnership with Government of New Zealand, ESCAP, UNDP and SPC.

10.4 Lessons Learned from the Experiences of Pacific Island Countries

ICTs play a crucial role in strengthening national capacities to inform hazard, vulnerability and risk assessments, as well as in understanding the impacts of "slow onset" climate change—in particular, data collection, analysis and communication to the public in relation to meteorological, hydrological and seismological hazard and risk information. They also improve avenues for national and sub-national agencies to link information to evidence-based planning and decision making that take into account climate and disaster risks. In this section the experiences of some Pacific countries are examined.

10.4.1 Papua New Guinea

Papua New Guinea has established a vision statement to guide national development in the coming decades. One pillar of *Vision 2050* (Papua New Guinea. National Strategic Plan Task Force 2009) addresses *environmental sustainability and climate change*, and requires an appropriation of 5% of the country's budget to fund relevant programs, as set out in the following table.

In support of the government's vision, some public and private organizations have implemented activities towards environmental sustainability. Divine Word University (DWU) has embarked on a paperless policy in the delivery of its educational programs. DWU's paperless policy is aimed at reducing the use of paper in printing and photocopying and the associated wastage in ink, toner and cartridges. Although but a small contribution in minimizing deforestation, such efforts mirror Heeks' ICT4D 2.0 Manifesto 'ICTs are an integral part of greener technologies and in assisting communities adapting to climate change' (Heeks 2009, p. 11) (Table 10.8).

In a second example, Lihir Gold Limited became the first company in Papua New Guinea to be granted the right to accumulate and trade carbon credits, on the basis of its environmentally friendly geothermal electricity generation facility (Yaru 2009). This is in addition to the use of hydroelectricity in many parts of PNG supplying over 65% of total electricity demand in the country. These two cases exemplify renewable energy approaches that may benefit the overall emissions reduction goal.

Table 10.8 Environmental sustainability and climate change pillar in PNG vision 2050

Environmental sustainability and climate change	
Clause number	Description
1.17.9.1	Reduce greenhouse emission by 90% to 1990 levels
1.17.9.2	Assist citizens to become resilient to natural and human disasters and environment changes
1.17.9.3	Establish a sustainable development policy in all sectors, especially forestry, agriculture, mining, energy and oceans by 2015
1.17.9.4	Develop mitigation, adaptation and resettlement measures in all impacted provinces by 2015
1.17.9.5	Conserve biodiversity at the current Ave to 7% of the world's biodiversity
1.17.9.6	Establish a total of 20 national reserves, wilderness areas and national parks
1.17.9.7	Establish at least one million hectares of marine protected areas
1.17.9.8	Conserve and preserve cultural diversity
1.17.9.9	Provide 100% power generation from renewable energy sources
1.17.9.10	Provide 100% of weather and natural disaster monitoring systems in all provinces
1.17.9.11	Integrate environment sustainability and climate change studies in primary, secondary and national high school curricula; and
1.17.9.12	Establish an institute of environmental sustainability and climate change

Source Government of Papua New Guinea (2009)

10.4.2 Integrated Water Management—Tuvalu

Tuvalu's integrated water resource management plan was devised to manage water problems through a coherent effort involving all sectors of the Island's economy. The ICT sector played an indirect role to support the successful outcome of the integrated water management project, which became part of the national development agenda, aimed at protecting all water resources and supplying quality freshwater for all citizens (Seleganiu et al. 2009). The identification of major challenges such as dependency on rain water, lack of storage capacity, droughts and high consumption of available water for flush toilets, served as indicators to formulate the plan. The first step included recognizing that rain water dependency required legislation and policy interventions. This agrees with Heeks' ICT4D 2.0 Manifesto which encourages actors and governance to be guided by aligned and contingent design strategies to ensure sustainable projects. The situation analysis confirmed that scarce ground water was being contaminated by salt owing due to global warming and sea level rise. The dilemma was aggravated by lack of domestic expertise to assess the suitability of existing water resources for human consumption.

The Tuvalu action plan was given legislative support and driven by an appropriate policy that mandated involvement of stakeholders from all sectors of the economy. Legislative support was complemented by donor support for the policy framework. Efforts to increase water storage capacity through tank distribution and installation of desalination facilities were financed (Seleganiu et al. 2009). Composting toilets were introduced to save scarce water.

The acceptance of composting toilets by the citizens required education and awareness as it was a totally new practice in Tuvalu. Radio programs were used to improve public education and awareness on water related issues and to overcome resistance to change. The advantages of composting toilets include mitigating health risks connected with faulty septic systems and reducing the risks of groundwater contamination by septic systems. Contaminated groundwater systems could also lead to degraded coastal resources. These plans required discussion, education and awareness-raising for citizens to accept compositing toilets.

Continuing efforts in the integrated water management system require education and training in areas such as building domestic capacity for testing and maintaining of water quality. Authorities and citizens also need to know that the integrated water management system is a national development agenda to be continuously researched for better understanding and action. Continuous education and awareness on these aspects are important, with ICTs being the main communication medium in this regard. The Tuvalu integrated water management received legislative support from the outset and was ably driven by an appropriate policy which received community acceptance for a successful outcome.

10.4.3 Food Security—Palau

Palau has a population of up to 20,000 with likely increases (Palau 2008). The urban population has been increasing due to rural-urban migration placing demands on agriculture production. Migration to towns has resulted in farming manpower shortages and diminished the quantity of food production, requiring a food security agenda (Palau 2008). Extreme weather conditions induced by climate change may potentially aggravate the existing pressure on food security caused by diminishing farming manpower. Supported by laptops and computers, a team has used traditional knowledge and scientific practice to adapt to problems caused by saltwater intrusion.

Balance is needed between subsistence farming and the demands of the cash economy that has been developing (Palau 2008). This is particularly true with hotel and restaurant visitors and most of the population preferring to consume local nutritious food. ICTs such as radio and television advertisements have been used in a campaign titled ‘Go local’ to diversify and promote food security in Palau. However, with the population increases and the decrease in farming manpower, food security may be a concern. Foreign imported food is expensive in Palau where wages may be limited. Root crops, which form common diet of the population, are crucial to Palau’s food security.

An array of traditional root crops exists in Palau, which has sustained the livelihood of the population over many years; but there are new challenges. The potential of climate change-induced extreme weather conditions due to El Niño and La Niña may adversely impact these traditional root crops. Climate change is also giving rise to environmental effects such as droughts, pests and disease (Palau 2008).

There is also evidence of genetic erosion of some root crops because of urbanization and negative attitude towards agriculture and farming (Palau 2008). Regarding genetic erosion, accurate survey and inventory can be enabled by policy direction which may stimulate improvements in the availability of skilled staff and technology, given financial support. Notwithstanding this, farmers in Palau plant different varieties of roots crops such as taro, cassava and sweet potato, ensuring the conservation of traditional varieties. This practice has been supported by extension services to farmers through technical advice and provision of planting materials with economic incentives (Palau 2008). As a food security initiative, Palau Farmers’ Association has embarked on promoting and ensuring adequate root and fruit crop production and marketing for self-sufficiency to reduce over reliance on imported food products. Despite lack of on-farm participatory plant breeding programs in Palau, localized seed production is mainly for farmers’ own use and is ably supported by technical advice of extension officers from the agriculture department. Farmers also have access to a range of planting materials provided by the department with assistance of international partners such as Taiwan Technical Mission. With the help of such development partners, adding value through processing and marketing of traditional food crops is being promoted (Palau 2008). These programs are commendable in the face of Palau’s climate change-induced food security challenges that may reduce development if left unrecognized. It is important to improve governance issues to cement ongoing assistance from international development partners engaged in food security programs which are important in the face increasing climate change-induced turbulence.

10.4.4 Coastal Development—Cook Islands

Most of the Cook Islands population lives within 5 km of the coastlines on 15 islands and two atolls which are susceptible to the effects of climate change such as sea level rise (Carruthers, n.d.). As in many PICs, coastal areas are exposed to erosion and land loss, reduced agricultural viability, saltwater intrusion, damage to infrastructure and health hazards due to climate change. This could adversely affect Cook Island's economy which is based on tourism, pearl farming and fisheries in the near-shore marine and coastal areas. ICT strategies and programs for improving information and communication between institutions and citizens play important enabling roles.

In addition to adverse economic impacts of climate change on the economy, the disruption of coastal architecture may diminish the happiness citizens experience from pristine beaches. Given that this culture of enjoyment and happiness is implanted among the people, moving inland is not considered an option due to complexities in the customary land tenure system. Hence, the Cook Islands has initiated coastal development programs specific to individual island environments to better understand potential vulnerabilities and adaptation strategies. These aspects are discussed in the following paragraph.

Each of the Cook Islands has peculiar geographical compositions. There are also contrasting seasonal climatic conditions between the main island groups. Due to the varying geomorphology, the coastal development programs need to be contextualized to meet adaptation requirements. Predictions show that over the next fifty years, temperatures could increase by 1–3 °C with a corresponding sea level rise of 10–40 cm. This is worrying. Recent events such as droughts, cyclones, floods and wind direction changes have the potential to affect coastal processes, influencing livelihoods and economic development for Cook Islanders. There is room for improvement in ensuring that ICT, data acquisition and the capacity to undertake studies to improve understanding on the aspects affecting coastal environment in Cook Islands (Carruthers, n.d.).

Notwithstanding the foregoing, Cook Islands has embraced a traditional form of marine management as the modus operandi of preserving and improving coastal resilience. The practice of selective harvesting of Trochus shell from the reefs lends direct benefits to the islanders, while sand mining for construction purposes continues with no environment impacts assessments. Cook Islands exercise the local ra'ui which declares an area 'out of bounds' from harvesting for a designated time period (Carruthers, n.d.). The cultural institution or practice of ra'ui allows marine ecosystems time to recuperate from over-harvesting and other pressures contributing to the preservation of environment. This example demonstrates well the use of traditional environment management system that may help preserve and improve coastal resilience. Much of the education and awareness required to promote this cause in Cook Islands have been effected through the availability of ICTs. It is important to recognize the use of local institutions, context and ownership which may sustain the preservation of the environment in the longer term (Heeks 2009).

10.4.5 Renewable Energy—Tonga

The Kingdom of Tonga has embarked on a long-term program to harness renewable energy as a safeguard to global oil price shocks and to enjoy quality access to modern environmentally-sustainable energy services (Pandey et al. 2013). The program set out to the harness photovoltaic solar energy to supply electricity in selected locations. Reliable and sustainable energy sources is required to ensure that ICTs function reliably with acceptable quality of service. Having acquired policy support from the government, Tonga plans to produce 50% of the country's electricity through renewable resources. The renewable energy program has both financial and environmental benefits by helping to reduce greenhouse gas (GHG) emissions. Renewable energy also ensures that appropriate ICT infrastructures are installed and operated with ease across Tonga. Issues connected with politics, the lack of country-wide implementation and a business plan, as well as few domestic technical experts, may have delayed the program. With government commitment and more awareness and training these issues may be minimized. Heeks (2009) recommends transparent and accountable multi-stakeholder partnerships for successful project outcomes.

The outcomes from Tonga's photovoltaic renewable energy program showed that there were reductions in the sales of diesel fueled electricity, leading to lower fossil fuels burnt. The reduction in diesel generation per month amounted to 106 MegaWatt hours (MWh) due to solar-based generation of 115 MWh (Pandey et al. 2013). Schools that operated the solar energy program also experienced economic value because they exported surplus energy to the common grid in support of the national plan. We can imagine that if more schools could be engaged in similar projects, the aggregate value to the overall economy and the schools may be large due to achieved economies of scale. Schools also benefit by solar energy to power ICTs such as computers, printers and projectors which are essential in supporting education programs.

The total renewable energy generation by MWh was only 3.2% per month, requiring more thorough planning to realize economies of scale (Pandey et al. 2013). Nonetheless, the energy program in Tonga showed the potential of solar-based technology to contribute positively to the economy in the face of climate change challenges. The solar energy program should provide increased motivation to investigate with development partners other renewable energy sources such as wind power, sea power and geothermal power. Depending on the outcomes of such investigations, a mixture of renewable energy sources feeding a common grid may meet Tonga's long term energy needs in an environmentally friendly manner. Tonga has opportunities to create a niche market for the generation and sale of energy which might contribute toward economic growth. This will lay the foundation for a reliable electricity supply to support general development programs. It would contribute to 'new, low-cost devices for local electricity generation and better ways to store, carry and transmit electricity' (Heeks 2009, p. 6).

10.4.6 Vanuatu

Emerging technologies, in particular ICTs have enjoyed increased usage in strengthening adaptation and disaster risk reduction in some regions of PICs. This section delineates some examples concerning usages aspects of such technologies in climate change adaption and DRR and the various actors. ICTs include diverse portfolios having certain roles to play as collection, storage, processing, retrieval and dissemination mediums of data and information using communication technology (Suwamaru 2015). Radio and television broadcasting, hardware and software computing and Internet services, fixed landline phones, mobile phones cable television and duplex radio communication services are all part of the ICT ecosystem. In Samoa, while portfolios such as radio and television broadcasting were liberalized earlier, landline services remained under monopoly control but in PICs mobile phone services were deregulated from 2006 (Chan Mow 2014). Following the introduction of competition in the mobile phone sector in Solomon Islands, Vanuatu and PNG the mobile phone sector soon expanded as citizens in both rural and urban areas were connected, many for the first time. In Vanuatu, mobile phone diffusion has encouraged other forms of ICT services, owing to their affordability and utility in diverse situations, including for climate change-related emergencies (Vanuatu 2007). What follows is an example of how simple and contextualized SMS messages were used by Vanuatu authorities in a mobile phone-based early warning system.

> The Vanuatu national meteorological service (VNMS) collects appropriate data, processes it and disseminates appropriate information in consumable form to stakeholders and the citizens. In 2015 the formation, direction and speed of cyclone Pam was observed by VNMS and processed into easily understandable form for use by the populace. The limited coverage of radio and television broadcasting services in Vanuatu meant that mobile phones were ideal to broadcast short and clear messages to even the remotest islands. First early warning SMS text messages disseminated and as the cyclone intensified and drew nearer, SMS were sent at three hour intervals. This gave the citizens opportunities to take appropriate preventive measures and as the cyclone gained momentum and drew nearer, VNMS sent the SMS texts at hourly intervals (Bolitho 2015).

Aspects of climate change are manifested through extreme weather and climate conditions, such as tropical cyclones and droughts (SOPAC 2009). The 2015 El Niño period that lasted from July 2015 to March 2016 in PNG, Vanuatu and Solomon Islands, is an example. Such extreme events more often than not may lead to declaration of disasters. Current literature also shows that unless sustainable development strategies are systematically planned, implemented and monitored by individual PICs, adaptation and DRR efforts may stagnate. So it is important to articulate adaptation and DRR interventions into overarching development plans and policies, supported by firm political commitment and complemented by robust and benchmarked public finance management (SOPAC 2009).

"Vanuatu has mainstreamed appropriate interventions such as the 'Disaster risk reduction and disaster management national action plan 2006–2016" and the 'Priorities and action agenda 2006–2015' as national development strategies (Vanuatu 2007). This reflects the recognition and political commitment by the government in drawing future safeguards in support of sustainable development. It confirms the recognition of the DRR and DM interventions as essential components for sustainable development in Vanuatu.

Effective and integrated appropriate early warning systems and information dissemination relating to volcanic activity, cyclones, earthquakes, droughts and floods (among others) are especially important. Collaborative efforts are commendable for practical progress on this aspect as has been shown in the case of Vanuatu. Like other PICs, Vanuatu's DRR and DM planning, implementation and monitoring programs require collaborative efforts, for there are major challenges connected with limitations in financial and technical resources (Pelesikoti 2007).

Against the backdrop of weak information systems concerning early warning facilities, collaboration between international meteorological and donor agencies, local authorities in Vanuatu have been able to issue timely and appropriate alerts for citizens. A network of meteorological agencies operates a collection of remote sensors collecting data and transmitting it via weather satellites to observatory centers (SPREP 2011). The collaboration of international, regional and national observatory facilities operates as a web of information repositories where early warning communication may originate. At the national level, Vanuatu deploys a mixture of newspapers, word of mouth, radio and television stations or even mobile phones to issue alerts to the populace.

For example, the resilience of Vanuatu to natural hazards such as Cyclone Ivy (2004) with only one confirmed death, and Cyclone Pam (2015) with eleven confirmed deaths show the success of collaboration (Pelesikoti 2007). Heeks' Manifesto 2.0 concerning aligned and contingent techniques is well applied as harnessed remote sensing technologies feed collective systems of ICT networks to support the flow of timely information for early warning purposes.

Cyclone Pam caused large-scale destruction to food crops and properties across Vanuatu, but amazingly only eleven deaths were confirmed. The Prime Minister's office released a statement claiming that the timely and contextualized SMS texts helped citizens to take preventive measures. The statement from the Prime Minister's office was confirmed by citizens interviewed on Shefa and other outer islands (Bolitho 2015). Among ICT services, mobile phones are the most pervasive communication devices with up to 90% diffusion in Vanuatu (Vanuatu 2007). In Cyclone Pam's inundation, mobile phones proved to be the easiest portable device where reliable connectivity was established to disseminate cyclone information. This exemplifies the idea of harnessing 'technology that has already reached many communities' to support development programs (Heeks 2009, p. 7).

Parsing the Cyclone Pam experiences in Vanuatu, one can harvest lessons concerning ICTs and how they may be harnessed to strengthen adaptation and disaster risk reduction. A combination of different ICT services could be leveraged to monitor weather conditions and to process and disseminate timely information to the populace. For example, remote sensors used to monitor and collect data on tsunami activity could then be processed in observatory stations and then disseminated when needed in easily consumable form to the public (Suwamaru 2015). For Vanuatu, VNMS was able to monitor Cyclone Pam by collaborating with regional monitoring stations using satellite facilities which support Internet and email, complemented by local weather stations which allowed VNMS to disseminate information in a timely manner. In Vanuatu, the mobile phone network, as the most extensive network, ably supported the transmission and reception of SMS texts which then helped citizens to prepare in advance. The success of this modus operandi was reflected by the low number of confirmed deaths.

10.4.7 Kiribati

Like many PICs, Kiribati has embraced the Hyogo Framework for Action and the Pacific DRR and DM framework for Action 2005–2015, as evidenced by its participation in a number of initiatives. Kiribati's action plan addresses tsunami early warning and response and tsunami capacity assessments (SOPAC 2009). Kiribati has embarked on a project to improve seasonal climate predictions for multi-sector applications consisting of equipment and hardware driven by the appropriate software. The software tailored for use on location with the capability of seasonal climate prediction services are applied to support climate-sensitive decision making in private and public sectors. In Kiribati, representatives from agriculture, water management and disaster mitigation departments have received training in the effective use of climate prediction in a risk-management context. In the face of adverse effects of climate change, this program is aimed at improving food security, improved public health, better managed water resources contributing to an improved Kiribati economy.

10.4.8 Niue

Niue has welcomed training opportunities to improve knowledge, innovation and education to build a culture of safety and resilience across at all levels (SOPAC 2009). An integral aspect of these training opportunities involves ICTs and, importantly, recognizing information technologies in all aspects of development. Disaster management courses have been offered in collaboration with Swinburne University of Technology (Melbourne) and the local campuses affiliated with the University of South Pacific. Collaborative efforts between the PICs and USP have progressed to institutionalize emergency health and disaster management training over five years.

10.4.9 E-Government and Decision Making

Harnessing ICTs to support decision making, implementation and monitoring processes on climate change and related matters may ensure that DRR and DM interventions are situated at the appropriate level of government. As noted earlier, political commitment from leadership is critical to ensure the success of national DRR and DM interventions. These could by shaped by harnessing appropriate ICTs to support accountability and good governance which may garner confidence from stakeholders (Vanuatu 2007). This is consistent with Heeks Manifesto 2.0, emphasizing actors and governance in partnership to maximize synergies.

PICs face additional challenges because of lower standards of transportation, electricity and communication infrastructure that may pose impediments to DRR and DM efforts (Cave 2012). To overcome the infrastructure issues in PICs, relevant policy instruments recognizing climate change from a development perspective is required. This requires a paradigm shift. Against this backdrop ICTs may play an important role in many aspects of climate change and disaster preparation and response, including information dissemination, education and awareness on prevention strategies and coordination efforts, among others (Suwamaru 2015). National policy instruments form important starting indications, as they are at the core in recognizing the importance of climate change and related issues.

E-governance initiatives positioning DRR and DM as vital agendas for sustainable development are emerging among PICs. For example, in Samoa the first steps were taken in 2007 to open the ICT sector to competition in hopes of increased choice in price and value. Similarly in Vanuatu, legislative and policy changes promoted competition within the ICT sector enabled the expansion of mobile phone coverage and access to diverse applications: mobile phones are now indispensable, offering anytime and anywhere communication connectivity, including access to government information (Cave 2012).

10.4.10 *Transition to an Integration Strategy in Climate and Disaster Resilient Development*

An integrated strategy in climate change and disaster-resilient development is necessary to achieve the overall objective of 'Pacific island people, their livelihoods and the environment resilient to the risks and impacts of climate change' (SPREP 2011). This requires integration of DRR and DM programs with national development plans. With transparent policy guidance, ICTs should be engaged in DRR and DM programs along the value chain, including remote sensing and monitoring to enable early warning to disseminate timely usable information. DRR and DM interventions require competent actors and prudent governance for workable partnerships relying on ICTs to attain sustainable projects. The PICs' varying geographic contexts demand a mixture of ICTs to ensure reliable coverage that supports the collection, processing and dissemination of DRR and DM-related information. It is therefore important that regulatory authorities impose ICT benchmarks or indicators for service providers specifying service coverage and quality. ICT service coverage and quality standards should enable availability, accessibility and affordability to the populace, thereby supporting reliable connectivity to be enjoyed by citizens (Fig. 10.2).

Climate and disaster resilient development in PICs requires coherent policy instruments, planning and action based on recommendations of organizations, including IPCCC and SPREP, between actors led by relevant government authorities. This could lead to sustainable societies that are environmentally resilient

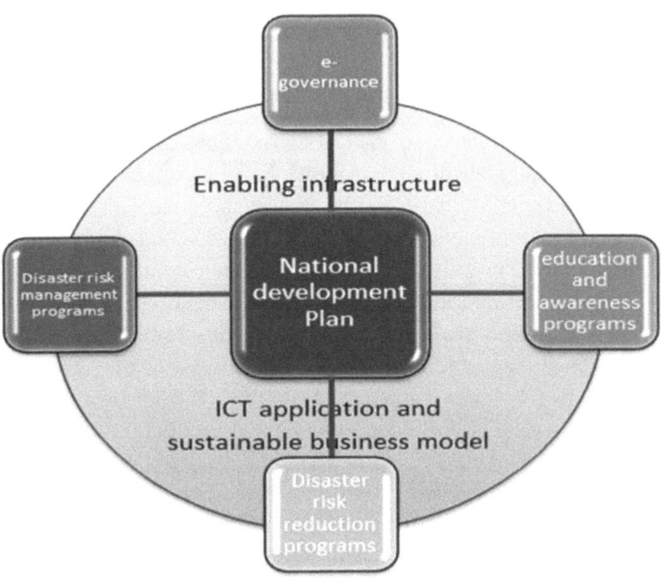

Fig. 10.2 Integrated framework for climate change and disaster risk management

regarding the risks and impacts of climate change. It is essential that ICTs are recognized and positioned as indispensable enablers or direct players in DRR and DM programs. The cross-cutting nature of ICTs has been proven to have a timeless, value-forming web of networks through which invaluable information may be instantaneously disseminated, accessed and used for the intended purposes. This aspect is important from various perspectives such as e-governance and sustainable development encapsulating DRR and DM programs.

A coherent strategy concerning climate change and disaster resilient development is congruent with Heeks' ICT4D 2.0 Manifesto. The Manifesto envisions actors and governance to be guided through aligned and contingent design strategies to achieve sustainable projects for successful ICT4D outcomes. In essence, this demands a strong and healthy ICT sector whereby actors may harness services and applications for all manner of DRR and DM programs. Needless to say, a forward-looking policy and regulatory environment can pave the way for a strong and healthy ICT sector necessary to support e-governance initiatives. An example is timely dissemination of government decisions through websites that may be accessible from both fixed and mobile devices. In other regions of the world such government websites host and support access to information concerning DRR and DM initiatives and other information about climate change. In PICs, such opportunities would increase participation by citizens. These websites would also offer information on practical and simple ways for citizens to engage in the fight against climate change. There could even be application forms for the willing and able to pursue climate change and related studies or even to access website repositories to gather recent DRR and DM information. Moreover, e-governance may increase efficiency and effectiveness of governments in PICs, leading to mainstreaming of DRR and DM programs into appropriate development plans. This can promote improvements in quality of life by building resilience to risks and impacts of climate change.

Observation systems and geo-spatial databases involve long-term activities to ensure data is being collected and managed. As such, databases and information systems should be designed for sustainability long after funding for initial implementation has ceased. In such cases, the phased approach to project design can promote sustainability and incorporate changes in line with technological upgrades. Guidelines development in the collection and determining of baseline data could be beneficial. Guidelines should include disaggregated data (e.g. gender and age) and information sources, particularly when working with communities or when analyzing the impacts of disasters.

It is important for national agencies to share data with each other. This can include sharing geospatial NDMO data with land, agriculture, forestry and fisheries ministries; sharing of statistics, including raw data, by bureaus of statistics; and the feeding of interpreted data to their central planning agencies. Collection of damage and loss data as well as updating and strengthening information systems is critical to support the display and analysis of climate and disaster impacts across different sectors. In terms of scientific and technical data, there are gaps in collection, monitoring and data access, which often result from funding limitations in data management projects; this lack of project sustainability can prevent further analysis by management.

While there have been innovative developments in databases and portals, and progress has been made towards the type of data collected and analyzed, issues remain concerning the localization of data collection. Access to data can be locked in a country or a system; therefore, data sharing needs to be encouraged, even by supporting systems of hardware and protocols. To make the role of ICT explicitly clear as a tool for responding to climate change and disaster vulnerability and impacts, governments should consider policies or strategies to outline a vision for how information should be managed and shared within and between government institutions and with non-government agencies and wider public.

In the case of the Aitape Tsunami, monitoring and timely dissemination of information messages was clearly lacking. In 1998 ICT connectivity in PNG was limited to urban areas, hence the delayed rescue operations. Suffice it to argue that if there had been government foresight in installing appropriate monitoring and early warning systems the death toll would have been minimized. The following anecdote is an excellent example where simple early warning via SMS texting can help minimize the number of deaths in disaster-stricken PICs. It aligns beautifully with Heeks' ICT4D 2.0 Manifesto concerning 'harnessing existing functionality such as the use of SMS' for an array of tasks on hand (Heeks 2009, p. 42).

> The 2015 tropical cyclone Pam caused one of the worst disasters in Vanuatu. It was a category-five cyclone with winds of up to 250 km per hour, causing catastrophic damage to shelter and destruction of 95% of food crops. About 75,000 citizens were left homeless, but remarkably seeing only eleven deaths which experts attributed to a combination of traditional knowledge, mobile phone based SMS warning system and disaster preparedness. (Bolitho 2015)

This example illustrates the use of mobile phones for an SMS based warning system that enabled people to prepare for disaster. Mobile phone diffusion levels in Vanuatu are estimated to be over 90%; they were used to communicate critical information to citizens (Bolitho 2015). Moreover, the few casualties manifests the difference that simpler, cheaper ICT functionality can make. This responds well to the assessment question in Heeks' ICT4D 2.0 Manifesto (p. 8): 'whether the simpler, cheaper technologies already in use can deliver sufficient ICT functionality to make a difference?'

10.5 Conclusion

As an 'ocean of islands', the Pacific Islands region is highly vulnerable because of the high frequency of, and exposure to, a wide variety of natural hazards (cyclones, droughts, earthquakes, electrical storms, extreme winds, floods, landslides, storm surges, tsunami and volcanic eruptions), geographical remoteness and isolation and dispersion across the vast Pacific Ocean. The rapidly emerging concept of 'risk governance' has positioned the management of climate change and disaster risk at the heart of development. However, the challenge of coordinating the implementation of diverse but inter-related frameworks for climate change policy can be overwhelming for PIC governments.

Vanuatu's Climate Portal provides an example of how to address this. However, the cases presented above suggest a continuing lack of effective regional and national information management policies and frameworks to guide priority actions, inform policy development, influence funding decisions and support the implementation of national policies and plans. Agreement and guidelines on data and information sharing are lacking, particularly concerning issues of ownership, copyright and licensing, access to traditional knowledge and commercial information reuse. To earn the respect of international development partners, PICs have to improve their level of governance by recognizing climate change as an integral part of national development plans. This calls for the installation of legislative and policy instruments to guide DRM and DRR programs.

The work of the UNFCCC has encouraged a number of Pacific regional organizations to cooperate in progressing adaptation and mitigation interventions using appropriate programs. In particular, the Pacific Islands Framework for Climate Change Adaptation and the 2005–2015 Pacific Disaster Risk Management Framework for Action serve as guidelines which member countries may follow in implementing DRM and DRR programs. Much effort and progress has been noted in this chapter. Those PICs that have implemented DRM and DRR programs with enabling legislative and policy support have seen much success and progress.

References

Asian Development Bank (2011) Workshop report—the millennium development goals in Pacific Island Countries: taking stock, emerging issues, and the way forward. Retrieved from Manila

Australian Bureau of Meteorology and CSIRO (2011) Climate change in the Pacific: scientific assessment and new research. Volume 1: regional overview. Volume 2: country reports. Retrieved from Australia

Boase R (2009) Final report—outcome evaluation of the information & communication technology for development (ICT4D) project of the UNDP multi-country office (MCO) based in Samoa. UNDP, Apia, Samoa

Bolitho S (2015) Tropical cyclone Pam: why the Vanuatu death toll was so low. www.abc.net.au/news/2015-04-01/. Viewed 03 Aug 2015

Brown R, Mackay S, McGregor K, Mackey B, Gonelevu M, Pelesikoti N et al (2015) Barriers to effective to adaptation and resilience planning in the Pacific: an information management perspective. Griffith University, Brisbane; SPREP, Apia, Samoa. https://www.pacificclimatechange.net/sites/default/files/documents/iCLIM-Barriers-Report-2015.pdf

Cave D (2012) Digital islands: how the Pacific's ICT revolution is transforming the region. Lowy Institute, Sydney. https://www.lowyinstitute.org/publications/digital-islands-how-pacifics-ict-revolution-transforming-region

Carruthers P (n.d.) Cook Islands coastal vulnerability assessments: a small Pacific Island nation's experience. Rarotonga, Cook Islands. http://www.pacificdisaster.net/pdnadmin/data/original/research_materials_cooks_islands_coastal.pdf. Accessed 20 Oct 2016

Chan Mow I (2014) Issues and challenges, strategies and recommendations, in the development of ICT in a small island developing state: the case of Samoa. Electron J Inf Syst Dev Countries 63:1–24

Cook Islands. Office of the Prime Minister (n.d.) Te Kaveinga Nui: national sustainable development plan 2011-2015. www.adb.org/sites/default/files/linked-documents/coo-nsdp-2011-2015.pdf

Davies H, Davies J, Lus W, Perembo R, Joku N, Gedikile H, Nongkas M (2001) Learning from the Aitape Tsunami. Paper presented at the ITS 2001 proceedings, Seattle, Washington

Federated States of Micronesia (2013) Nation wide integrated disaster risk management and climate change policy 2013. http://www.pacificdisaster.net/pdnadmin/data/original/FSM_2013_Nation_DRM_CC_policy.pdf

Fiji (2012) National climate change policy. www.sprep.org/attachments/Climate_Change/Fiji-National-Climate-Change-Policy.pdf

Foundation of the peoples of the South Pacific International (2013) Views from the frontline: 2013—Pacific region summary report. The Foundation, Suva, Fiji. http://www.pacificdisaster.net/pdnadmin/data/original/PPDRM2012_S4_FSPI_FronlineViews.pdf

Global facility for disaster reduction and recovery (2013) Managing disaster risks for a resilient future: a strategy for the global facility for disaster reduction and recovery 2013–2015. GFDRR, Washington D.C. https://www.gfdrr.org/sites/gfdrr/files/publication/GFDRR_Strategy_Endorsed_2012.pdf

Government of Tuvalu. Tuvalu Infrastructure Strategy and Investment Plan 2012. http://www.pacificdisaster.net/pdnadmin/data/original/TUV_2012_Infrastructure_Plan.pdf

Hay J, Pratt C (2013) Strategy for disaster and climate resilient development in the Pacific (SRDP), Background information and guidance on rationale and possible approaches. Suva, Fiji

Hay JE (2013) Roles of Pacific regional organizations in disaster risk management—questions and answers. Brookings-LSE, London. https://www.brookings.edu/wp-content/uploads/2016/06/Brookings_Regional_Orgs_Pacific_July_2013.pdf

Heeks R (2009) The ICT4D 2.0 Manifesto: where next for ICTs and international development? www.oecd.org/ict/4d/43602651.pdf

Intergovernmental Panel on Climate Change (2014) Climate change 2014: synthesis report. Contribution of Working Groups I, II and III to the Fifth Assessment Report of the Intergovernmental Panel on Climate Change [Core Writing Team, Pachauri RK and Meyer LA (eds)]. IPCC, Geneva, Switzerland https://www.ipcc.ch/report/ar5/syr/. Accessed 20 Oct 2016

Kirtman B, Power SB, Adedoyin AJ, Boer GJ, Bojariu R, Camilloni I et al (2013) Near-term climate change: projections and predictability climate change 2013: the physical science basis. IPCC working group I contribution to AR5. Cambridge University Press, Cambridge

Niue (2009) National climate change policy. http://www.preventionweb.net/files/28164_niueccpolicy.pdf. Accessed 24 Oct 2016

Pacific Catastrophe Risk Assessment and Financing Initiative (2013) Current and future tropical cyclone risk in the South Pacific. PCRAFI, Nabua, Fiji. https://www.terranova.org.au/repository/paccsap-collection/current-and-future-tropical-cyclone-risk-in-the-south-pacific-south-pacific-regional-risk-assessment/current-and-future-tropical-cyclone-risk-in-the.pdf/view

Palau (2008) Country report on the state of plant genetic resources for food and agriculture. Koror, Palau. http://www.fao.org/docrep/013/i1500e/Palau.pdf

Palau (2010) National disaster risk management framework 2010. National Emergency Management Office, Koror, Palau. http://www.pacificdisaster.net/pdnadmin/data/original/PLW_2010_NDRM_Framework.pdf. Accessed 4 Oct 2016

Palau (2015) Palau climate change policy. For climate and disaster resilient low emissions development 2015. Government of Palau, Koror, Palau

Pandey S, Bodger P, Lapthorn A (2013) Renewable energy in the Kingdom of Tonga; National plan & PV generation systems. University of Canterbury, Christchurch, New Zealand

Papua New Guinea. National Strategic Plan Task Force (2009) Papua New Guinea national vision 2050. National Planning Committee, Papua New Guinea, Port Moresby. http://www.treasury.gov.pg/html/publications/files/pub_files/2011/2011.png.vision.2050.pdf. Accessed 20 Oct 2016

Pelesikoti N (2007) Government of the Republic of Vanuatu. Disaster risk reduction and disaster management national action plan (2006–2016). http://reliefweb.int/report/vanuatu/disaster-risk-reduction-and-disaster-management-national-action-plan-2006-2016

PIFS, SPC, SPREP (2016) Framework for Resilient Development in the Pacific. An Integrated Approach to Address Climate Change and Disaster Risk Management (FRDP) 2017–2030, Geoscince Division, SPC, Suva, Fiji. http://www.pacificdisaster.net/dox/FRDP_2016_Resilient_Dev_pacific.pdf. Accessed 20 Oct 2016

https://www.pacificclimatechange.net/sites/default/files/documents/PalauCCPolicy_WebVersion-FinanceCorrections_HighQualityUPDATED%2011182015Compressed.pdf. Accessed 4 Oct 2016

Secretariat of the Pacific Community (2005) Disaster risk reduction and disaster management: a framework for action 2005–2015—an investment for sustainable development in the Pacific Island countries. SPC, Noumea. http://www.preventionweb.net/files/34617_mr06131.pdf. Accessed 28 Oct 2016

Secretariat of the Pacific Regional Environment Programme (2005) Performance monitoring and evaluation report: work programme & budget, SPREP

Secretariat of the Pacific Regional Environmental Programme (2006) Pacific Islands Framework for Action on Climate Change 2006–2015. Apia, Samoa. https://www.sprep.org/publications/pacific-islands-framework-for-action-on-climate-change-2006-2015-action-plan-pifacc

Secretariat of the Pacific Regional Environment Programme (2011) Pacific Islands framework for action on climate change 2006–2015 2nd ed. http://www.sprep.org/attachments/Publications/PIFACC-ref.pdf

Secretariat of the Pacific Regional Environment Programme (2012) Pacific environment and climate change outlook. https://www.sprep.org/publications/pacific-environment-and-climate-change-outlook-report

Secretariat of the Pacific Regional Environment Programme (2014) Decision support tools for climate-resilient coastal development: a case study from the Cook Islands Pacific adaptation to climate change (PACC) project. www.sprep.org/attachments/Publications/CC/PACCTechRep12_lr.pdf

Seleganiu P, Moulogo C, Duncan D (2009) Dry sanitation technology: the solution. Cited in: Kinrade P, Arold N, Pickering P et al. Pacific adaptation (costs and benefits) scenarios: water security in Tuvalu. Technical report (Final). AusAid, Canberra. https://terranova.org.au/repository/paccsap-collection/pacific-adaptation-scenarios-costs-and-benefits-water-security-in-tuvalu-technical-report/cba-tuvalu-water-security-technical-report.pdf

Solomon Islands (2011) National development strategy 2011-2020. http://www.nationalplanningcycles.org/sites/default/files/planning_cycle_repository/solomon_islands/cobp-sol-2015-2017-sd.pdf

Solomon Islands. Department for Environment (2011) National climate change policy framework: 2012–2017. http://www.gcca.eu/sites/default/files/catherine.paul/si_climate_change_policy.pdf

SOPAC (2009) Implementation of the Hyogo framework for action and the Pacific Disaster risk reduction and disaster management framework for action 2005-2015. Regional synthesis progress report. www.preventionweb.net/publications/view/10641

SPC. The Pacific Youth in Agriculture Strategy 2011–2015. file:///C:/Users/Vic%20Lipski/Downloads/Pacific%20Youth%20&%20Agri%20Strat%20report_web.pdf. Accessed 24 Oct 2016

Samoa (2012) Strategy for the Development of Samoa 2012–2016. http://www.pacificdisaster.net/pdnadmin/data/original/WSM_2008_Strategy_dev.pdf

Suwamaru JK (2015) Cost effective ICT solutions for the geohazards division—department of mineral policy and geohazards management, consultancy report to geoscience. The author, Port Moresby

Tokelau (2010). Tokelau National Strategic Plan 2010–2015 http://www.tokelau.org.nz/Strategic+Plan.html

Tonga (2010) Tonga Joint National Action Plan on Climate Change and DRM 2010–2015. https://sustainabledevelopment.un.org/content/documents/1335tongaDisaster%20Management%20Strategy.pdf. Accessed 24 Oct 2016. Kiribati Development Plan 2012–2015. https://dfat.gov.au/about-us/publications/Documents/kiribati-development-plan-2012-2015.pdf. Accessed 24 Oct 2016

Tonga (2011) Tonga Strategic Development Framework 2011–2014. http://prdrse4all.spc.int/node/4/content/tonga-strategic-development-framework-tsdf-2011-2014. Accessed 24 Oct 2016

Tuvalu. Te Kaniva: Tuvalu climate change policy—charting Tuvalu through the challenges of climate change 2012. http://faolex.fao.org/docs/pdf/tuv143764.pdf

Tuvalu. Ministry of Foreign Affairs, Trade, Tourism, Environment and Labour (2012) Tuvalu National Climate Change Policy 2012–2021, Te Kaniva

Tuvalu (2012) Tuvalu National strategic action plan for climate change and disaster management 2012-2016. Government of Tuvalu, Funafuti. http://reliefweb.int/sites/reliefweb.int/files/resources/TUV_2012_NSAplan_CC_DRM_2012_16.pdf

Vanuatu (2006) Government of the Republic of Vanuatu's Priorities and Action Agenda 2006–2015. www.adb.org/sites/default/files/linked-documents/cobp-van-2015-2017-sd.pdf

Vanuatu (2007) Disaster risk reduction and disaster management national action plan (2006–2016). Port Vila, Vanuatu

World Bank (2012) Acting today, for tomorrow: a policy and practice note for climate and disaster resilient development in the Pacific Islands region. World Bank, Washington. http://documents.worldbank.org/curated/en/231221468288339561/pdf/694550REVISED00for0tomorrow0WEB0010.pdf

Yaru BT (2009) Engineering opportunities to benefit from a global emissions trading environment. Paper presented at the 2009 Conference on the impact of climate change on national development, Port Moresby, Papua New Guinea

Chapter 11
The Use of ICT in the Health Sector in Pacific Island Countries

Rowena Cullen

Abstract The health of their people and the delivery of better health services are core development goals for all Pacific Island countries (PICs). Information and Communication Technology has a critical role to play in this sector, and e-health is a recognized part of e-government. But in PICs this role is largely limited to the collection and management of health data; where ICT plays a major role in improving health outcomes in industrialized countries, for example, clinical decision support for clinicians, and telemedicine, the use of ICTs in PICs is more sparsely adopted. Examples of applications tend be local, ad hoc projects, although valuable in the limited contribution they make. This chapter outlines the burden of disease and other challenges, such as lack of funding, that PICs face in delivering health services. It explains how ICTs are used in the collection and analysis of health data, the contribution made by regional organizations and the high level of regional collaboration this entails. In the second half of the chapter some ICT projects to support clinicians and other health workers (e-learning, e-resources and telemedicine) and the use ICTs in the delivery of health services, are outlined. In the final section, the use of mobile technology in both health care and public health initiatives is discussed. The chapter concludes that the health sector in Pacific SIDs is so resource-sensitive that any technological 'intervention' must clearly add considerable value to justify the allocation of resources that would otherwise be spent on staff, medicines, equipment and facilities.

R. Cullen (✉)
School of Government, Victoria University of Wellington, Wellington, New Zealand
e-mail: Rowena.cullen@vuw.ac.nz

R. Cullen and G. Hassall (eds.), *Achieving Sustainable E-Government in Pacific Island States*, Public Administration and Information Technology 27,
DOI 10.1007/978-3-319-50972-3_11

11.1 Introduction

The health of individuals and populations and the provision of adequate health services[1] are recognized as critical to development and both the Millennium Development Goals and the Sustainable Development Goals include health goals as major outcomes. The Health and Development Division of the World Health Organization (WHO) notes:

> Better health is central to human happiness and well-being. It also makes an important contribution to economic progress, as healthy populations live longer, are more productive, and save more (WHO 2016).

Information and Communication Technology (ICT) has a crucial role to play in helping deliver these outcomes. The International Telecommunications Union (the UN's specialized agency for ICT) argues that ICTs have a key role to play in achieving the UN's Sustainable Development Goals and in relation to SDG Goal 3, Health and Well-being have "the potential to deliver incredible benefits across the whole of the global healthcare ecosystem" (ITU 2016). These benefits, it suggests, will come through the use of ICTs:

- to connect community health workers with information services and with doctors, nurses and other health services;
- to enable communication with patients, the collection of health data, connect community health workers with expert advisors and receive payment for services (through mobile technology);
- to keep populations informed about disease outbreaks and help them access information about health services through social media; and
- to help identify trends in mortality and morbidity, help plan the allocation of resources in health services, and manage disease outbreaks using health information systems.

These and many other applications of health information technology (health IT), including the management of clinics and hospitals, the transfer of patient information between health services, the ongoing training of health professionals and the management of many common non-communicable diseases (NCDs) like diabetes certainly warrant the claims often made for 'incredible benefits'. But as with all technological innovations, the realization of these benefits brings many challenges, especially for Pacific Island countries (PICs) with their limited resources, tropical location, dispersed and diverse rural populations and limited human capacity.

While the implementation of ICT in the health sector is a very important aspect of e-government in many countries (especially those with state-funded health systems), in most PICs it is one of the least developed aspects of e-government. The

[1]The WHO defines health services as all services dealing with the diagnosis and treatment of disease, or the promotion, maintenance and restoration of health. This includes personal (health care) and non-personal health services (http://www.who.int/topics/health_services/en/).

health sector is also one of the least supported areas for international development partner ICT investment, as noted in Chap. 1, despite the enormous sums spent globally on official development aid (ODA) in the health sector each year (Schafferhoff et al. 2015). Given the potential for ICTs to contribute to the health and well-being of Pacific people, the question must be asked: Is the global burden of disease so great, especially in developing countries, that addressing it directly through funding medicines and health workers takes precedence? Or is it that the benefits are too hard to realize in the problematic environment that PICs present to health IT initiatives? This issue underpins the chapter; the health services available in PICs and the health status of the populations of Pacific SIDs (Small Island Developing States) are so inadequate and so resource-sensitive that any technological 'intervention' must clearly add considerable value to justify the allocation of resources that would otherwise be spent on staff, medicines, equipment and facilities (including clean running water).

Furthermore, although the health IT literature is full of claims for significant efficiencies and better managerial and clinical decision-making from health IT systems, over the years there has also been considerable evidence of waste, frustration and systems failure (Berg 2001; Kaplan et al. 2009). No Pacific Island country can afford this level of wastage, let alone in the health sector. E-government for development champion Richard Heeks claims this failure is widespread and that it is due to a misfit between systems design and contextual reality, between the priorities of users and practitioners and those of system builders, between public and private sector realities and, of most importance in this context, between countries. The gap between industrialized and developing countries on virtually all the dimensions in Heeks' diagnostic tool, the ITPOSMO model—information, technology, processes, objectives and values, staffing and skills, management systems and structures and resources (Heeks 2006)—is especially marked in PICs with the unique conditions that affect health services the region.

The chapter starts by outlining the burden of disease and other challenges that PICs face in delivering health services and the overall lack of funding for health care in the region. It describes how technology is used in the collection and analysis of health data and the contribution made by regional organizations, in particular the WHO's Western Pacific Regional Office and the networks supported by the Pacific Community (SPC)'s Public Health Division, and the high level of regional collaboration across the Pacific that this entails. It outlines some of the benefits of using ICT to support clinicians and other health workers (e-learning, e-resources and telemedicine) and the more limited contribution of ICT in the delivery of health services. The final topic covered is the use of mobile technology in both health care and public health initiatives. Throughout the chapter, questions are raised about wise ICT investment in health in light of scarce resources for health in PICs and the sustainability of health ICT innovations. (Health technology, tools and devices used in *clinical care*, as opposed to in *management of care*, lie outside the scope of this chapter.)

11.2 Context and Background

The Pacific SIDs share many common problems that impact on the overall health status of the population and the delivery of health services. In addition to the universally scarce financial resources and small workforces, these include the fact that a significant proportion of the population of some PICs, comprising rural populations, live in isolated traditional villages with a mainly subsistence economy, speak many different languages as well as the local pidjin and have limited English language literacy. Even with an increasing level of urbanization, which is sometimes not taken into account when development issues are being considered (Lal and Fortune 2000, p. 88), provincial towns and regional and national capitals are often surrounded by agglomerations of villages with poor housing conditions. Apart from a wealthier elite, urban dwellers, who maintain strong links with their rural village communities, share many problems with them: limited income, poor sanitation and endemic tropical diseases, especially vector-borne diseases such as malaria, dengue and more recently, the chikungunya and Zika viruses. The presence of these endemic diseases is exacerbated by poor nutrition (in rural villages through crop failure due to tropical cyclones, rising sea levels and drought; in towns and cities through cheap imported food) and the use by both public and private providers (including street vendors) of pharmaceuticals, especially anti-malarials and antibiotics, which are past their use-by date or are contaminated, due to lack of enforcement of inadequate regulations (Hetzel et al. 2014).

In addition, despite significant improvements in the past two-to-three decades, maternity and perinatal health data and services fall below global standards (WHO 2014); maternal, infant and child mortality rates remain unacceptably high across the region (UNFPA 2014). Estimating infant and child mortality (and indeed mortality rates generally) is made more difficult by the fact that many people in remote villages are not registered at birth and not all deaths are notified. (The Pacific Civil Registration and Vital Statistics program being introduced by the Brisbane Action Group, outlined below, is helping remedy this.) At the same time PICs face a rise in non-communicable diseases (NCDs), especially cardiovascular diseases due to changing diet, substance abuse and increasing longevity (Hayes 2009). Obesity, accompanied by a dramatic rise in the incidence of Type 2 diabetes (Curtis 2004; Parry 2010), is considered to affect over 50% of the population across the region. One of the unfortunate consequences of this is that not only are PICs dealing with mortality and morbidity burdens and the loss of potential contributors to the workforce that this epidemic brings, but also the very high demand that treatment for NCDs makes on already inadequate budgets for medicines, a circumstance which Pacific Island Forum Leaders and the World Bank have defined as an 'NCD crisis' (World Bank 2013).

11.2.1 The Health Workforce in PICs

The lack of financial resources and of skilled personnel means that the standard of health care provision is inevitably far below that in industrialized countries. Government-funded primary health services are usually provided through community clinics, which offer varying levels of care in a network of regional services of which a regional or provincial hospital is usually a part. Provincial and regional hospitals[2] offer limited secondary services such as maternity services, basic surgical and medical treatment, and other services such as gynecology, ophthalmology, ear nose and throat and cancer services. Most PICs have a national referral hospital to which patients requiring more specialist care are referred. National and regional referral hospitals depend on visiting specialists or are staffed by foreign doctors with limited English (and no knowledge of pidgin or the local language); they have limited medical equipment, few pathology or radiological services and some do not even have adequate supplies of clean running water or electricity (Capelle 2016; Carrick 2014). Staffing issues compound the lack of other resources: staff resist working in rural and remote island communities that are hard to reach and where leave and promotion opportunities are non-existent. Qualified health workers are at risk of being enticed by work in private clinics in urban centers, or fail to return from training overseas. Thus, there is a heavy reliance on paramedical workers, community nurses and even staff without qualifications, especially in more remote areas; and both health workers and patients can have long difficult journeys to make to reach a rural clinic (Buchan et al. 2011).

Recent data collected by the WHO's Western Pacific Regional Office (WPRO 2016) shows that the numbers of trained doctors and nurses in the most recently surveyed PICs fall far below ratios in their industrialized neighbors, Australia and New Zealand.[3] Lack of financial resources and of IT skills means that it is a struggle to keep systems functioning when they are installed and to maintain equipment in the moist tropical environment. Internet connections are unreliable and expensive and electricity supply is erratic outside urban centers. Funding for ICT must compete with scarce resources needed for basic patient care (medications, bandages, clean water and basic equipment such as thermometers). In addition, several countries in the region (e.g. Marshall Islands, Solomon Islands and Papua New Guinea) are affected by high rates of corruption, which results in resources, in particular the US$2 billion in development aid spent in the region in the past decade, not reaching their targets (Johnson 2015).

---

[2]In PICs where there are many small remote islands that comprise a number of provinces, several provincial hospitals may be linked to one regional 'referral' hospital, which in turn refers on to a national referral hospital.

[3]OECD figures for 2013 are: Kiribati, 0.2 doctors and 2.79 qualified nurses per 1000 population (2014); Marshall Islands, 0.45 doctors and 2.18 qualified nurses per 1000 (2014); Vanuatu, 0.19 doctors and 1.33 qualified nurses per 1000 (2013); in Australia and New Zealand doctors are 3.4 and 2.78 per 1000 population, and nurses 11.52 and 10.4 per 1000 population respectively (http://www.oecd.org/els/health-systems/oecd-health-statistics-2014-frequently-requested-data.htm).

11.2.2 Economics of Health in the Region

In fact, resources are the major problem. While most PICs expend only between 3 and 5% of GDP on health (compared with an average of 9–11.5% spent by their industrialized neighbors Australia and New Zealand and former colonial powers Germany, and the UK); their average GDP per capita in 2014 was Int'l $5427.00[4] (World Bank 2016).[5] (New Zealand on the same measure had a GDP per capita of Int'l $36,679.00 and Australia $45,925.00). The annual expenditure on health per capita in most PICs ranges from Int'l $100–400,[6] and even though this varies from 10 to 20% of all government expenditure (including ODA) in most PICs, there is little opportunity to increase health budgets and little room to increase the level of 'out of pocket' expenses incurred by users of health services given the levels of poverty in the region.[7] And in a vicious circle, the inability of most PICs to provide tertiary level hospital care means that a significant proportion of the budget must be used to fund tertiary care off-shore (Anderson et al. 2014, p. 35). In this context, the use of ICT must be able to deliver immediate value and contribute directly to health outcomes and the economic benefits of improved health status.

11.3 ICT and National Health Strategies

11.3.1 National Health Strategies

As a consequence of the resource constraints outlined above, ICT does not feature strongly in the national health strategies of PICs. Surveying those whose national health strategies are available on the website of the WPRO (not all are available and not all are current), shows that if they feature at all, goals related to the use of ICT are mainly focused on existing health information systems (HIS) which collect health data for planning purposes and which is required by the WHO (more detail is provided in Sect. 4.1 below.) Palau, one of the smaller PICs with 17,455 people at

---

[4]The international dollar (Int'l $) is a hypothetical currency that, in the cited country and at the same time, would buy a comparable amount of goods and services a U.S. dollar would buy in the United States.

[5]This figure which is affected by relatively high GDP per capita figures for Fiji and Samoa (Int'l $8792 and Int'l 5789 respectively) disguises the fact that the majority of PICs have GDP per capita figures between Int'l $1000 and 3000.

[6]GDP figures do not take into account the fact that in most PICs there is a large non-monetary subsistence rural economy which is not part of GDP; at the same time, it cannot contribute to the cost of modern health services (although traditional health practices may be a part of it).

[7]Out of pocket expenses are charges paid directly by patients and any other non-reimbursable costs of health care. In PICs, especially for those living in remote villages a long way from health services and reliant on a subsistence economy, such charges can be prohibitive. Even for poorly paid urban dwellers they can be a substantial burden.

the last census, speaks of creating such a tool that will standardize data collection, streamline data management to avoid duplication of effort and create a system that will enable data to be used to 'make priority decisions' (Palau 2008). Fiji's Strategic Plan for 2011–15 *Shaping Fiji's Health* (Fiji 2010) refers to the use of data in its Health Information Unit database to assess progress against the Millennium Development Goals (MDGs) and of the lack of data to adequately assess the prevalence of disease in certain key areas, as well as the need for more detailed data:

> An assessment of Fiji's progress towards achieving its health outcomes depends on a well-functioning health information system with access to age, sex and geographical, time series disaggregated data, some of which were not available. Efforts are being made to address the data gaps to enable planning for prevention and response to emerging health issues (Fiji 2010, p. 9).

Tuvalu's Health Strategy (for only 10,564 people at the last census) also comments on the value of its health management information system, noting the importance of accurate, timely and "decision-friendly" information to identify priority needs and "make optimal use of limited resources." Importantly, it notes that "donors are accountable to their stakeholders, particularly taxpayers in their own countries and the justification for funds can more easily be made with up-to-date data and reporting on expenditure of current funding" (Tuvalu 2009). (Papua New Guinea. Department of Health 2010) Papua New Guinea's strategy openly acknowledges that the country has an extremely underdeveloped health ICT structure (apart from the use of radio for communicating essential information),[8] largely due to its size and terrain. But it does note that it has a well-established national finance and payroll system, a national health information system (mainly covering primary health care services) and a national discharge information system for the hospital sector. Its focus is on better communications between health facilities and better use of technology in its regional hospitals. The Solomon Islands Strategy makes good use of data from its health information system to set priorities, but acknowledges that the data is inconsistent, incomplete and out-of-date, which makes the HIS a priority for expenditure; the Strategy assigns other forms of 'medical and public technology' a far lower priority score (partly because, as the Permanent Secretary notes in his overview, the Ministry is 'top-heavy' and locked into a 'medical model' of health services delivery and, despite increased funds in recent years, has delivered sub-optimal health outcomes.) Financial planning, expenditure procedures, accountability and better HR management are needed to produce more efficiencies before further use of ICTs beyond the HIS become a priority. This level of financial planning and accountability involves connecting all

[8]Radio (both broadcasting and HF/HVF radio) is used by a number of PICs as a means of disseminating public health messages and for communications with isolated communities. While it must be acknowledged as one of the ICTs used in the health sector, it is not discussed in detail in this chapter (see Chap. 1 for a more detailed discussion of this issue).

parts of the health system to the centralized finance and HR management information systems, which, to the credit of Solomon Islands Government, has since been achieved (see Chap. 5).

11.3.2 Planning and Budgeting

As the Solomon Islands Strategy makes clear, there have been significant gains through the use of IT systems in many PICs in the financial, HR and payroll management areas. Though not considered in detail in this chapter, the introduction of financial management and HR management systems (detailed in Chap. 5) has brought significant gains to the health sectors of PICs, as they have to the education sector, by introducing robust public finance accountability (including payroll and procurement), benefits to employees such as prompt payment and access to their remuneration through mobile banking and the removal of non-existent or absentee employees from the payroll. Such accountability measures encourage development partners to continue contributing to development projects.

However, while such systems facilitate the integration of policy, planning and prioritization with budgeting and resource allocation, the lack of such integration remains an area of concern for the many organizations that have financial monitoring roles in the region such as the World Bank (Anderson et al. 2014, p. 43) and the Pacific Technical Advisory Commission (PFTAC) of the International Monetary Fund (IMF). In particular, ministries in PICs tend to overlook the need for medium-term plans, funding for multi-year projects and innovations and the need to absorb ongoing staffing and equipment costs for donor projects that are reaching the end of their external funding (Anderson et al. 2014, p. 43). This especially applies in the health sector where, if ICT innovations are to be sustainable and bring long-term benefits, attention must be paid throughout the policy, planning and implementation cycle to the need for ongoing budgeting for ICT.

11.4 ICT Use in the Health Sector in the Pacific Islands Region

In this and subsequent sections of this chapter, we examine the particular domains in which ICTs are utilized in the health sector in the Pacific Islands region and the impact they are having. As the PRIF report on Economic and Social Impact of ICTs in the Pacific notes, although the use of ICTs in health is 'particularly relevant' in the Pacific region because of the lack of doctors and nurses, the primary use of ICTs is in the area of Health Information Systems (HIS); the report lists several countries (Fiji, Samoa, Solomon Islands, Tonga and Vanuatu) which have installed networked HIS systems, despite the financial, geographical, infrastructural and capacity-related challenges faced by most (Minges and Stork 2015). The report notes some additional ICT systems: email and video-conferencing to support communication between physicians, some use of ICTs to manage inventory of pharmaceuticals,[9] other supplies and mobile-based public health initiatives. However, most of the applications of IT deployed in industrialized countries are either unaffordable, depend on infrastructure that is not available, involve the collection and processing of information that is not generally collected in the region, or require resources that are beyond the capacity of PICs. That is, systems deployed in the region must, of necessity, meet criteria that Heeks (2006) outlines in the ITPOSMO model and 'fit' the local context in terms of information, technology, processes, objectives and values, staffing and skills, management systems and structures and resources. Those that are used must be able to justify their value and address immediate needs in the sector.

11.4.1 Health Information Systems and the Role of Health Data in the Management of Health Care and Health Services

The purpose of an HIS is to deliver "sound and reliable information" that can be used as "the foundation of decision-making" across all parts of a country's health system (WHO 2008). This information includes individual patient data, population data at regional and national level, health surveillance data[10] and administrative data on facilities, supplies and workforce. The quality of the systems depends on the quality of the data and their ability to capture it from a range of sources, harmonize, integrate, analyze and synthesize the data and assist the health sector to make use of it. The PHIN comments:

> Sound decision-making at all levels of a health system requires reliable health statistics that are disaggregated by sex, age and socioeconomic characteristics. At a policy level, decisions informed by evidence contribute to more efficient resource allocation and, at the delivery level, information about the quality and effectiveness of services can contribute to better outcomes (PHIN 2011).

However, collecting, managing and applying timely and accurate health data is a significant challenge for most PICs. Although most of this data is collected in some way, few countries have integrated systems that can bring it all together. In addition, the timeliness and accuracy of data may not be of a high standard, meaning that even when it is brought together, it cannot be fully utilized. As the WPRO notes:

---

[9]A critical issue in the Pacific in light of research by Hetzel et al. (2014), and others.

[10]Health surveillance is the systematic and continuous collection of health-related data primarily to provide warning of public health emergencies such as outbreaks of infectious diseases.

> countries and areas in the Western Pacific Region are diverse in terms of their level of health information systems development and the challenges they face. In countries with weak health information systems there are fundamental problems in data collection, data quality, timely generation of quality information, data analysis and utilization of information. … data in the region are often unreliable at best, and sometimes completely lacking (WPRO 2016b).

This is problematic. High quality health data is needed to monitor the incidence and burden of disease, to determine the ability of the country to meet health needs, to estimate population growth and the life expectancy of the population, to plan for the provision of health facilities, equipment and workforce to meet these needs, to identify epidemics as early as possible and to monitor progress of public health measures aimed at reducing the impact of infections and non-communicable diseases. High quality health data is also needed to assess a country's progress against the Millennium Development Goals and the Sustainable Development Goals and assess where development assistance would be most useful.

But despite the value of health information for evidence-based decision making in the areas of planning, management and policy, health information systems in the Pacific have not been able, in the past, to produce the data needed for effective decision-making, and questions have been raised about the ability of PICs to adequately resource the collection and analysis of health data (Landry et al. 2012; Hodge 2012). The PRIF report, for example, states that "authorities in the countries studied have stated that health information systems have helped to improve health outcomes, but lack the concrete evidence and models to prove this" (Minges and Stork 2015, p. 42). The report notes, for example, that the Fiji Ministry of Health tracks almost 90 indicators in its HIS, but cannot yet show how these are used or how ICT affects health outcomes. Linking this data with resource allocation is essential, yet there is no evidence of it.

The WHO plays a key role in monitoring the health status of the Pacific Island countries and territories[11] through the collection and dissemination of health data, as it works towards its goal of the highest attainable level of health for all people. It tries to achieve this by ensuring that each member country has a "solid national health plan" and by supporting member countries in gathering statistics, promoting public health, monitoring diseases that can affect island populations, including diseases that spread rapidly in the tropical environment of most SIDs (vector borne diseases and HIV/AIDS). The fourteen independent Pacific Island states which are members of the World Health Organization are expected to have an HIS and to collect and report data on a core set of 100 indicators as a requirement of membership.[12] Few manage to collect data on all 100 indicators, and it is beyond the capacity of many PICS to make effective use of the data they are required to collect.

[11]The WHO is an agency of the United Nations, but countries can be members of the WHO even when they are not member states of the UN itself, e.g. Cook Islands. The non-self-governing territories and dependencies of the Pacific are associate members, through the WPRO.

[12]Data from the Pacific territories is provided by the WHO member country (France, United States, New Zealand) responsible for them.

In most PICs, the data in each HIS is usually based on simple manual data collection, paper forms and reports shipped to the nearest main center (although tablets and mobile phones are beginning to be used for this purpose) and entered into an Excel or Access spreadsheet. Where possible the data is entered into the database as closely as possible to the source, at a regional hospital or large health center and transferred electronically to the central government agency collecting it; however, this can be a function of how dispersed the population is and the quality of telecommunication connections between remote islands. In countries such as Samoa, where the majority of the population is close to the two main urban areas, this is easier than in countries with many islands, where health services are structured around one or two 'referral' hospitals, or several very basic regional hospitals or health centers and where much health care is delivered through isolated rural health clinics staffed by untrained or partially trained health workers with limited access to mobile networks.

For many rural health centers, data collection is often problematic. In rural areas, where not all members of the community attend the health clinic because it may not be open when needed or may be too far away, data is incomplete. Conversely, the collection of data in larger health centers is compromised by internal migration. Since most PICs have no national system of patient registration, as people move from rural to urban centers and vice versa, they may be counted more than once. They may also be listed under variants of their names (with first and last names confused) and the collection of the data itself may be irregularly carried out. Data that must be carried considerable distances on foot may be lost before it is transferred to an online system, but electronic data transmission is affected by infrastructure issues—connectivity and power supply. Apart from the examples of Papua New Guinea and Fiji (see boxes) which have both had some success installing networked health information systems, the collection of data from networked clinic and hospital systems in the Pacific Islands region remains a predominantly unreliable manual process.

There is one exception to this situation. In both Cook Islands and Niue, data for national statistics is collected from Medtech32, which is now used at health centers throughout both countries and their main referral hospitals. Medtech32, which has been installed as part of the NZAID program, (largely because of the constitutional responsibility New Zealand has for Cook Islands and Niue) provides access to patient diagnoses, allergies, test results and medications, as well as scheduling and reminder systems. Medtech32 is a computerized patient management system and clinical decision-support system with clinician alerts and uniform data standards that enable data to be quickly extracted for local or international reporting. Problems with the system in Cook Islands are still reported, most notably in the quality of connections between more remote areas and the capital, Avarua (on Rarotonga) and in the need for initial and ongoing training (Iorangi 2012), but overall Medtech32 has brought considerable benefits to both these small island states with respect to quality of care and quality of health data.

Medtech32 is a well-supported commercial off-the-shelf system, designed for primary and secondary care contexts, which can handle large numbers of patients; this enables it to function as both a patient management system and an HIS in small countries. While costly commercial systems, even if initially supported by development partners, are unsustainable long term, Cook Islands and Niue are clearly exceptions as long as they retain their current status with New Zealand.

However, in some PICs the collection of data for HIS is now impacted by new software designed for developing countries facing issues of poor connectivity and lack of capacity. The leading product globally is the Norwegian-led open source health information software DHIS2, a web-based HIS with advanced GIS-based visualization features that incorporates mobile and geospatial technologies. It is typically used as a national health information system for data management, analysis purposes and health program monitoring and evaluation although it also offers logistics management and functionalities such as mobile monitoring of pregnant mothers in rural communities. Data can be captured on any type of device, including desktops, laptops, tablets, smartphones and feature phones, and its simple user interface is particularly suited to inexpensive, low-end mobile phones. It can also transmit data though SMS messaging and be programmed to send reminders (of appointments or public health advice) to individuals or groups of patients or health workers. DHIS2 can be implemented to work offline (that is, it does not need to be connected to the Internet to function and data can be uploaded once connected), enabling it to be used in locations with poor connectivity.

While DHIS2 can also be used as a patient management system, handling longitudinal individual patient clinical data, appointment schedules, tracking and reminders, in PICs its main use is as a health information management system, extracting anonymized data using the full set of ICD-10 codes (see below). DHIS2 is used in Samoa, Solomon Islands and Vanuatu, and is likely to be used more extensively in the region in coming years. At present, the full functionality of DHIS2 is not used, and to do so would bring additional training issues. In addition, ICD10, the WHO's international classification codes for diseases used for reporting the burden of disease, hospital discharges and deaths in all member states, is an elaborate, controversial and highly sophisticated system. Applying even the basic set of 14,400 codes, with their attached metadata and definitions, is a challenge to rural health workers in developing countries with limited health knowledge and literacy, and many countries report difficulties. In particular, training for doctors in the use of ICD10 to certify death is needed to ensure accurate statistics. (Even in industrial nations with highly trained coders, applying ICD10 is regarded as a specialized task.) Coding, even at the most basic level, remains a significant factor in the poor quality of data in Pacific HIS.

11.4.2 Benefits and Challenges in Implementing ICT-Based HIS

All the issues noted above are highlighted in a case study of the Kiribati health information system (Tabunga 2012) which reports an increase in communicable and non-communicable diseases, infant mortality, poor data quality, problems with coding, poorly trained and unmotivated staff. Backlogs of unanalyzed data, failure to enter critical data on notifiable diseases at the time of outbreaks and mortality data gaps are also reported. Not only is more accurate and timely health information needed (as noted in the Kiribati Health Strategy), training is needed for health planners and decision-makers to enable them to extract, analyze and apply this information to urgent decision-making, as well as ongoing policy. This is a common story for the region and dramatically illustrates the challenges that confront PICs in collecting and using sound health data, despite their use of ICT applications.

More positively, it is clear that ICTs are a powerful tool for managing the collection, analysis and distribution of health data. Not only do they manage the technical aspects of data collection and management and have the potential to significantly improve the quality and timeliness of data if it is entered into the system close to the source, but they also demand some standardization of data, a principle adopted by the Pacific Heads of Planning and Statistics, which is one of the objectives of the *Ten Year Pacific Statistics Strategy 2011–2020.*[13] Standardization of data has two main outcomes—it allows comparisons over time and between similar countries, facilitating 'shared approaches' to some intractable problems, and it helps break down the 'silos' of information that limit the value of health data. As the Director General of the WHO noted when advocating for good data, accountability and transparency and the use of ICTs at the opening of the 2015 *WHO Summit on Measurement and Accountability for Results in Health*, "Our investment has always been in silos":

> We are moving away from a time when investment in information systems were ad hoc and fragmented, with unsustainable pilots, stand-alone surveys and one-off impact evaluations. … we are moving towards a time of shared approaches that align with country priorities and make smart investments in sustainable ways. … to a systematic and rational [approach] anchored in strong country plans, with aligned investments and inclusive accountability mechanisms… . This is a time to develop self-reliance and self-sufficiency as the best exit strategy for dependence. This is a time when countries need a hand-up, not a handout. … Countries need capacity, not charity (Chan 2015).

11.4.3 Regional Approaches to the Management of Health Data

Shared approaches, strong country plans and developing self-reliance and self-sufficiency could sum up the approach taken by the regional health organizations of the Pacific that have sought to build capacity and self-reliance by education and training and to develop a regional approach to data collection, dissemination

[13]This is discussed in more detail in Chap. 8 on Statistics and ICTs in development.

and analysis. Country data is handled in the first instance by the Statistics for Development Division (SDD) of the SPC, and is collected through each country's HIS system as well as routine censuses, household and expenditure surveys and demographic and health surveys.[14] Some of this data is reported by the SDD's National Minimum Development Indicators Database (SPC 2016), while data on the WHO indicators is transferred to the WHO's regional office and observatory at the Western Pacific Regional Office and is available on its web site.[15]

The SPC's Public Health Division hosts the Pacific Public Health Surveillance Network (PPHSN), with the assistance of the WHO and UNICEF. The PPHSN network links in-country Hospital-Based Syndromic (HBAS) systems, collects data and reports regional data on communicable diseases to member countries and to the WHO's Western Pacific Regional Office through the PacNet system. PacNet links all countries in the Pacific and also functions as a forum for communication regarding public health emergencies, including those of international concern. The PPHSN supports several other networks such as PicNet (which links health professionals dealing with infectious diseases in the region), EpiNet (which focuses on building capacity in the area of infection control and response to epidemics) and the Syndromic Surveillance scheme which uses dashboards to summarize syndrome reports and issue alerts for outbreaks of infectious diseases (see Box 11.2 below and Kool et al. (2012) for more information about syndromic surveillance in the Pacific). In areas where diseases such as malaria, dengue fever, Zika and other mosquito or vector-borne diseases can be rife, these Internet-based early warning systems are very valuable aids in dealing with outbreaks.

In addition, the WPRO and the SPC's Statistics for Development Division provide analytical tools, easily accessible online, which help member countries analyze their health data and package it for inclusion in the policy-making process. For example, the WPRO's Health Information and Intelligence Platform (HIIP), which states on its website that "too often health information is hidden in isolated pockets or stored in ways that make it difficult to use and impossible to share", is focused on making data available and user-friendly. It has online tools, dashboards, table-generating tools and map and chart-making tools to help users analyze country data and explore data across any selection of indicators and any number of countries in the region. The PPHSN goals have a strong focus on harmonization of health data across the region, regional networking and the development of computer applications to exchange and analyze data. The PPHSN uses SPC's data visualization tools to map developing epidemics across the region and communication systems to ensure support for those working to respond to epidemics of communicable diseases.

[14]See also Chap. 8.

[15]http://hiip.wpro.who.int/portal/Home.aspx. This data covers the entire region including Australia, New Zealand, China and Japan and includes health financing data; some data is a decade or more out of date for all countries, and does not always match WHO data which is projected forwards using a larger range of sources.

The other great need in the region, as many commentators have noted, is training. The WPRO and SPC provide courses and workshops in the region, and the WHO collaborates with institutions in the region designated as Collaborating Centers for training (for health information systems this is the University of Queensland's School of Public Health) holding as well as its own conferences. The Public Health Division of the SPC also uses the SPC website to promote important information about the diseases which are its chief focus (TB, HIV and other sexually transmitted diseases) and supports PICs in their management of these diseases with expert advice, regional meetings, briefings and online discussion lists. It also disseminates public health messages about healthy lifestyles, often creating multimedia presentations to help reach a youth audience.

11.4.4 The Role of CRVS and Death Data in the Pacific: Contributing to Better Health Statistics

A major contribution to improving the quality of health data in the region is the Pacific Civil Registration and Vital Statistics (CRVS) Project and the Pacific CRVS Plan. CRVS is defined by the SPC Statistics for Development Division website in the following way:

> Civil Registration (CR) is the compulsory, continuous, universal and permanent recording of vital events such as births and deaths. From these records, vital statistics (VS) on births, deaths, causes of death, fertility and mortality (and where migration data is also available – population estimates) can be produced for policy and planning. Where civil registration has a high level of both coverage and completeness, it is generally accepted to be the preferred source of vital statistics due to the ongoing and timely collection of data, and the ability to include cause of death information (SPC Statistics for Development Division 2016).

ThePacific CRVS project is administered by the Brisbane Accord Group, of which the SPC's Statistics for Development Division is a key member along with the WHO, UNICEF, PHIN, the Pacific Civil Registrars Network and several universities in the region. The project also 'sits under' the *Ten Year Pacific Statistics Strategy 2011–2020* of the regional Heads of Planning and Statistics to ensure that CRVS is incorporated into regional statistical priorities; it is part of the Asian and Pacific Civil Registration and Vital Statistics Partnership and its declared 'Asian and Pacific Civil Registration and Vital Statistics Decade, 2015–2024'.[16]

Reliable data on births and deaths is essential for monitoring key health outcomes such as live births per capita, maternal and perinatal mortality and deaths from non-communicable diseases. The importance of death registration systems, and the ability to generate accurate data on trends in "cause-specific mortality for different NCDs", is essential to help Pacific countries respond to the current crisis in NCDs noted above. "Many countries in the Pacific still do not know the real burden

[16]http://www.getinthepicture.org/subregional-initiatives.

of specific components of NCDs as reliable cause-of-death data is often absent" (Aumua and Hodge 2012). The Pacific Vital Statistics Plan, its emphasis on giving countries the right ICT tools and expertise in data extraction and analysis, as well as its focus on "operational and functional HIS systems in Pacific countries that will give planners and decision-makers the information necessary to make decisions around resources. ... and to monitor the impact of NCD programs on disease burden" (p. 191) will help address this lack.

In addition, a robust CRVS provides direct measures for a number of the SDG Goal 3 *Health and Well-being* indicators. It is also an SDG goal in its own right (Goal 16, *Peace, Justice and Strong Institutions*: Indicator 16.9, *By 2030, provide legal identity for all, including birth registration*) (Sustainable Development Knowledge Platform 2016).

The right to be registered at birth and to have one's existence formally recognized is a universal human right (see Chap. 7). Moreover, effective CRVS systems do not only provide essential information for public health policy management and the prevention of disease, but also help ensure an individual's entitlements and access to health services. "Well functioning civil registration and vital statistics (CRVS) systems directly benefit individuals and policy, and are independently and statistically significantly associated with better health outcomes" (Phillips et al. 2015). Improving CRVS systems however, is increasingly dependent on technological innovations to make the collection and analysis of data effective and affordable. Mobile Internet-enabled technologies (both tablets and phones) can be used to enhance birth registration (as noted in Chap. 7). The issuance of birth certificates and the transmission of data to central registries, along with analytical tools that can produce real-time data, and the ability to match data from different HIS components will benefit countries, families and individuals (Abouzhar et al. 2015). CRVS systems "provide improved individual rights, equity and accountability in a way that no other data strategy can do". But these must be supported by coordinated assistance from development partners, attention to the legislative frameworks necessary to protect privacy and enable digital authorization, and they will require "sustained and substantial support from national policy makers and form regional and global development partners." (p. 1381).[17]

11.5 Basic ICT Systems for Healthcare Delivery

While HIS systems dominate the use of ICTs and are the most obvious form of e-government in the health sector in PICs, other technologies that are more directly related to the delivery of health care are also used. Furthermore, there is a considerable overlap between HIS and some systems used for patient management in

[17]A monitoring and evaluation program is an essential feature of the Pacific CVRS program and milestones and outcome indicators have been identified, but no reports are yet available.

primary and secondary healthcare. As we have seen above, Medtech32, used to extract data for an HIS and DHIS2, is also used to manage patient scheduling and some clinical data—in short as an electronic health record or EHR, sometimes referred to as an EMR. In many ways, the fragility of paper records in a tropical climate, the disorder of most hospital records (Rooney 2013) and a patient population that is not oriented towards the world of print suggests that electronic patient record systems would be an obvious solution. That has indeed been argued for many years in the health informatics community (Fraser et al 2005; Kalogriopoulos et al. 2009), and the WPRO has published a useful manual, *Electronic Health Records for Developing Countries* (WPRO 2016). The use of open source software is frequently advocated as a way of getting around the cost of a commercial system (Fraser et al. 2005; Wolfe et al. 2006). But it is far from clear that this is an easy or appropriate solution for a number of reasons. First, open source solutions are not always easy software to implement and incur considerable staff training costs; second, staff implementing and maintaining them are often isolated and unsupported. In addition, clinicians in developed countries have long shown resistance to the introduction of EHRs, with widespread and effective adoption often requiring a considerable “change management approach” (Boonstra and Broekhuis 2010). With a far less qualified and stable workforce, and the need for ancillary staff with limited literacy to access systems, the challenge in PICs would be considerably greater. Furthermore, a combination of technical issues (lack of backup system, poor system security, unstable power supplies and lack of technical support) and user issues (lack of user training, poor initial design, lack of consultation, difficulty of systems for users, lack of support staff training, lack of perceived benefit) have affected attempts to deploy such systems in other developing countries (Fraser et al. 2005).

Apart from the use of Medtech32 in Cook Islands and limited use of DHIS2 as a patient information system, the main application in the region is the PATIS system. PATIS was initially developed for Samoa in 1996 with the assistance of the Australian government as a basic HIS and was subsequently developed with Australian funding for use across the Pacific. It is also used in Fiji (see Box 11.1) Written in MS Access using an SQL server database management system, it is effectively a patient administration system (PAS) with functionality equivalent to that of a commercial system that feeds into an HIS. As such it lists basic patient data only (using an individual patient number and ICD10 codes) and has not been networked in Samoa; data is manually entered from a paper-based health record system that covers only hospitals and some larger health centers (Soar et al. 2012). Samoa is now in the early stages of implementing DHIS2, which offers a full EHR system.

Box 11.1: Fiji’s patient information system PATIS

Fiji adopted PATIS soon after it was implemented in Samoa and has put considerable effort into developing further modules for the system: a Patient Master Index, Admission and A & E Transfer and Discharge systems, General and Specialized Outpatient systems, Public health, Dental,

Pharmacy, Microbiology, Disease Index and Radiology modules. The patient's clinical record however, remained in a paper-based system, accessed on arrival, sent when necessary to another health unit or center. When it was first deployed, it was networked to health facilities around the country where it was used on regional servers, data being transmitted and consolidated overnight. The system soon outgrew the constraints of the equipment and communications network and was redesigned (as PATISPlus) and loaded on more powerful MS SQL server software accessible via the web 24/7/365, with a new user interface based on identified user needs. New training programs were put in place and support staff increased. The system in 2012 was still a hospital-based HIS (using ICD10), with data linked to paper-based records, but able to provide clinical and statistical data for either operational or strategic decision-making.

From 2014 the system has been undergoing further 'remediation', based on more extensive consultation with user groups, which is aimed at eventually making it a full patient administration system able to share patient information between health services. Much of the work has been focused on cleaning up the data (e.g. in several hospitals discharge data is incomplete) and making the user interface more user-friendly (Biscoe and Jacobsen 2014). Even so, subsequent evaluation of the system found that most health facilities in Fiji still worked on a manual system, there was a general shortage of computers in government health facilities and inconsistent practice between hospitals; PATISPlus was found to be slow, few users had access to it and many users shared the same login. Considerable further development is needed to make it a functional system and to incorporate the EHR functions that the country's health workforce wants and needs. Some of these issues are not dissimilar to those experienced in industrial countries introducing networked patient information systems. It appears to face the same challenges as both its developed and its less developed neighbors in achieving this.

Sources: Naidu (2012); Ravindra et al. (2015).

11.6 Supporting the Role of the Clinician

Significant use of ICTs is also being made to support clinicians and health workers—from online clinical guidelines, to access to local information and e-learning resources, to international research and telemedicine to support clinical decision-making.

11.6.1 Information Resources for Clinicians

POHLN, the Pacific Open Learning Health Net, a key program of the WHO and WPRO in partnership with Pacific ministries of health, offers free online courses accessed through learning centers in most Pacific Island states (course resources are also available in print form and CD in these centers). The courses, which use Moodle as the platform, provide continuing professional development (CPD) training for both qualified and unqualified health workers in management, dentistry, nursing, pharmacy, laboratory services, radiology, maternity and children's services, health promotion and the more severe communicable diseases of the region such as TB, HIV/AIDS, leprosy and STIs. Courses in Health Informatics and ICD10 are offered. Also playing a key role in continuing education for health workers are clinical guidelines produced by ministries of health in the larger PICs, such as Papua New Guinea's Diabetes Guidelines (Papua New Guinea. Department of Health 2014) initially prepared for Africa and Samoa and adapted for PNG. Many such guidelines are available online to all health workers in all PICs.

Of particular relevance to physicians are initiatives to give them access to research literature through the standard scholarly channels which might otherwise be unavailable to them. In the past two decades, the availability of scientific research through the Internet has increased exponentially, especially health and medical information through the web-based PubMed database which lists more than 26 million citations to the bio-medical literature, a considerable number of which are downloadable from PubMed Central. But there is still much recent and specialized information behind 'paywalls'. The WHO's Health InterNet Access to Research Initiative (HINARI), established in 2002, offers free or very low cost online access to the major journals and a large number of textbooks in biomedical and related social sciences to local, not-for-profit institutions in developing countries (this includes medical schools, teaching hospitals and health centers, government departments and NGOs). All major scientific bio-medical publishers collaborate in HINARI; users register through their institutional library once it has joined the scheme. (HINARI is part of a suite of programs in the fields of Agriculture (AGORA), the Environment (OARE), and Research and Development (ARDI), under the RESEARCH4LIFE umbrella.) Eleven PICs (but not the Pacific Territories or Tokelau or Tuvalu) participate in the scheme, either free or for an annual subscription of US$1500 per institution, through the University of the South Pacific (which has campuses in all of its 12 constituent countries), or through their own universities or nursing colleges.

11.6.2 Telehealth Initiatives

More direct support to clinicians is provided by several telehealth ventures in the region, which operate both within and between countries. These initiatives can take various forms: the use of Internet connections or mobile telephony to exchange images can be a significant aid in the diagnosis of skin diseases (teledermatology) and also some eye problems. Teleoncology systems exchange X-rays, enabling chemotherapy regimens to be supervised remotely in local hospitals. Telemedicine systems provide support for healthcare workers and midwives, allowing them to call regional facilities or a referral hospital for advice. Telemedicine systems are also used to enable physicians in a referral hospital to consult with clinicians overseas (this service is often linked with visiting specialist schemes which operate in several countries.) It can also include the transmission of pathology test results from an overseas laboratory back to a local clinic, with specialist advice attached and comments on X-rays sent to experts in larger centers. The potential of telemedicine is acknowledged in the Samoa National ICT plan (Samoa 2012).

Successful telehealth initiatives depend greatly on the availability of equipment and good Internet connections for either simultaneous transmission and exchange (usually through some form of video-conference) or through 'store and forward' systems. 'Store and forward' technology depends less on high quality connections but can still provide access to expert clinical advice, a protocol much better suited to 'low-resource settings' (Wootton and Bonnardot 2015) and is a procedure becoming so universally and successfully applied as to become 'almost routine'. Case studies identifying potential benefits in the SIDs environment go back over a decade. For example, Khazei et al. (2005) surveyed physicians working on remote Tanna Island in Vanuatu, who found some value in the use of telehealth, provided physicians consulted understood the local conditions and resources. However, the doctors surveyed believed basic public health needs such as tuberculosis control, clean water and improved infrastructure and communication between hospitals and rural outposts were higher priorities than telehealth. But Adam, writing as a clinician from Kiribati in 2000, viewed telehealth initiatives more favorably (while acknowledging the need for infrastructure to support such a service).

From our own experiences, we have identified these as our most urgent needs:

- For the nurses on the atoll without a doctor to be able to consult with a doctor about medical cases.
- For the doctors to be able to consult with each other on management of cases.
- For the doctors to be able to consult with outside specialists on management of cases (Adam 2000).

A Senior Medical Office in Tonga at the same time noted that although PacNet and the WPRO's HealthNet had helped reduce the isolation of doctors and health workers on some of Tonga's 36 islands, better communication through telehealth systems would further enhance opportunities for continuing education through the sharing of health information, leading to better patient management in remote locations (Kupu 2000). At the turn of the century lack of infrastructure and suitable technology made these visions unachievable. That is now changing. Although more remote areas can only access 2G networks, which limit communications to text and voice, the increasing availability of 3G and 4G networks allowing Internet access, is enhancing communications options for health workers throughout the region. Despite the difficulties of extending mobile access to remotes islands and issues around affordability and small markets, mobile broadband connections are expected to exceed 4 million by 2020 (GSMA 2015, p. 3). (See Chaps. 4 and 5 for more detailed information on mobile use and issues.)

Successful examples of telemedicine include the long-running Pacific Island Health Care project, which provides humanitarian medical referral/consultation from the Tripler Army Medical Center in Honolulu to more than 5,000,000 people in the American associated states, Marshall Islands, Federated States of Micronesia, and Palau and U.S. territories Guam, American Samoa and the Commonwealth of the Northern Mariana Islands. Since the mid-1990s a store-and-forward telemedicine program supported by X-ray machines, CTO and MRI scanners, has enabled the treatment of infectious diseases (including TB), NCDs, obstetric cases, leptospirosis (which no longer needs evacuation to Hawaii) and tropical pyomyositis (an acute infection of the muscle caused by staphylococcus aureus). The use of an asynchronous store-and-forward system has proved to be as effective as videoconferencing and more affordable and sustainable. The program has enabled a large number of Pacific Island residents to be successfully diagnosed and treated at home, with those requiring care in the United States to be identified, reducing the numbers turning up in Hawaii (Person 2014). In addition, island physicians now know how to treat patients with these problematic diseases after diagnosis is made via the PIHCP system, increasing capacity in the islands.

Another example of a sustainable telemedicine initiative involves a group of children in Samoa with cochlear implants who are in need of post-operative care after treatment in Australia. Video-conferencing links and links between machines programming the implants connect a clinic in Samoa with the Sydney Cochlear Implant Centre and the Royal Institute for Deaf and Dumb Children in Melbourne in order to monitor the implants and provide training to local clinicians. After the initial trials in 2010, conducted at the premises of the telecommunications provider, the system is now flexible enough to use mobile-enabled tablets at the children's schools (Eikelboom and Swanepoel 2015). These initiatives, driven by clinicians working together for solutions to provide better care to Pacific Islands patients and using affordable technologies, show the undoubted promise of telemedicine throughout the region.

11.7 Mobile Technologies in Public Health and Health Care

The report of a WHO global survey on *mHealth: new horizons for health through mobile technologies* which gathers data from industrialized as well as medium and low-income countries (although only a few Pacific countries appear to have participated and their data is not reported) defines mobile health as "medical and public health practice supported by mobile devices, such as mobile phones, patient monitoring devices, personal digital assistants (PDAs) and other wireless devices (WHO 2011, p. 6). This includes "complex functionalities" such as 3G and 4G systems, GPS, Bluetooth and other wireless applications. The survey classifies mobile heath initiatives into major uses: communications between individuals and health services, communications between health services and individuals, consultations between health professionals, intersectoral communications in emergencies, health monitoring and surveillance, access to information for health professionals at point of care. These communications enable a range of purposes—emergency situations, health advice or helpline, telemedicine, appointment reminders, vaccination reminders, treatment compliance reminders, health promotion and awareness, public health warnings and advice (pp. 12–13). Many of these communications could be implemented through basic SMS messaging.

In countries with more reliable, affordable and secure connectivity, mobile technology can also be used for accessing electronic health records of patients and for clinical decision support systems (WHO 2011). In PICs, they are almost exclusively confined to services that can make use of SMS, although recent innovations use mobile technologies for the collection of birth and death data (as noted above) and some public health data. Applications that require Internet connections are constrained by limited connectivity and the high cost of Internet connections in the region. In addition, applications must take into account other factors such as an understanding of the local context (e.g. which groups have access to mobile phones and whether at-risk groups of women have unimpeded access), levels of mobile literacy and engagement (which should include issues related to language) and fit with local policy (Umali et al. 2016). A strong argument is made by some health professionals that applications of the technology should start with an assessment of existing health programs and any gaps in delivery, identifying problems for which m-health solutions would be appropriate. Many argue that m-health programs should not take priority over existing programs, which are already under-resourced, but others view the technology as a way of enhancing existing programs and services. And although m-health's potential is not seriously contested, several researchers comment on the lack of empirical research in the region to demonstrate the value of mobile technology for health and lack of a viable business model, including ongoing sources of funding, both of which limit the extent to which mobile technologies can be incorporated into health policies (Umali et al. 2016).

However, in the Pacific Islands region there are many informal, ad hoc trials of mobile health (m-health), an increasing number of reports referring to them, and some evaluations. Cave (2012) refers to a program in Fiji, where 13,000 subscribers were reported to be using a mobile health application to receive alerts about emerging diseases and maintain contact during natural disasters. Subscribers could also use SMS to contact a team of doctors from the Fiji Medical Council and Fiji College of General Practitioners who undertook to reply to medical cases within 24 h via SMS, and when required to provide referrals to the nearest medical facility. The system, known as 'Dr SMS', was expected to disseminate up to 5 million texts in the next year (Cave, p. 14). A similar program was launched in PNG around the same time. In July 2012 Papua New Guinea's first free, national m-health program was launched by Population Services International (PSI), a global health NGO that operates in more than 60 countries. The m-health application, called Haus Lain ('members of the household' in Tok Pisin[18]) aimed to send weekly SMS messages in English and Tok Pisin, on issues such as malaria, the prevention of child and infant deaths, gender-based violence and HIV. The mobile program, which was supported by the Australian government and heavily promoted over radio, attracted almost 30,000 subscribers in its first year, and was later expanded to incorporate more interactive mobile application features in partnership with the National Department of Health and the National AIDS Council Secretariat (PSI 2016). Funding has since been withdrawn, due to lack of evidence that it was having any effect.[19]

Mobile telephones and SMS have had more impact in facilitating better communication between health services staff. An initiative deploying mobile phones in the Western Highlands in Papua New Guinea found that the system was especially useful in enabling rural health workers to book beds and transport for patients requiring transfer to hospital and in supporting rural workers dealing with emergency situations; this has had positive impacts on the delivery of health services and on health outcomes, according to evidence gathered by Yamo (2013). Two more recent projects in Papua New Guinea have been formally evaluated and shown to have identifiable benefits. Interviews with health workers in a rural setting and staff in the labor ward in the provincial hospital in Alotau who participated in the successful introduction of the Childbirth Emergency Phone project, set up to counter the very high maternal mortality rate in Milne Bay Province of PNG, considered the project, which made a toll-free phone number available to the health workers, helped to resolve communication problems between the two groups. The evaluation showed that the project played a critical role in enabling the health workers to address life-threatening childbirth complications and has potential for rollout across PNG 'potentially reducing maternal morbidity and maternal mortality rates' (Watson et al. 2015). This is consistent with a report by Mehl et al. (2014) on the use of mobile health solutions in developing countries across the world, especially in the areas of reproductive, maternal, neonatal and child health.

Box 11.2: Evaluating a mobile phone-based syndromic surveillance system in Papua New Guinea

PNG often suffers from severe outbreaks of infectious diseases. especially in remote areas. The PNG National Health Information System (NHIS) has been monitoring outbreaks of infectious diseases in recent years through a Hospital Based Active Surveillance System (HBAS), based on reports compiled

[18]The local pidjin language used in PNG.

[19]Personal communication from Amanda Watson, 17 August, 2016.

monthly by surveillance officers at provincial hospitals using registers of symptoms; the HBAS has been described as "the cornerstone of surveillance for suspected cases of measles, poliomyelitis and neonatal tetanus." Syndromic surveillance, which uses data that is not diagnostic of a disease but that might indicate early stages of an outbreak (Buehler et al. 2004), is considered a useful adjunct to diagnosis-based disease surveillance in developing countries (Kool et al 2012). Such systems allow outbreaks to be detected early and their magnitude and geographical distribution to be quickly identified.

PNG's telecommunications networks have improved rapidly in recent years, providing mobile phone network coverage to increasingly large areas. In response to concern over delays in detecting recent outbreaks, the PNG health authorities piloted a mobile phone-based syndromic surveillance system (MOPBASS) for timelier detection of outbreaks. The MOPBASS system was tested in two health centers in Port Moresby in 2010 and piloted nationwide for ten weeks in 2011. (Test sites received mobile phones with appropriate tools installed, one-day onsite training, guidelines and sample collection materials.) A report on the evaluation of the nationwide pilot of the system compares data from 10 trial sites with data from the HBAS in terms of average reporting delay, completeness of data and cases of measles reported. Attributes concerning the acceptability, stability, usefulness, portability and system costs of the new system were also assessed in a user survey. The MOPBASS system reported 95% of cases compared with 2% identified in the HBAS system (against NHIS clinician-based assessment), and the "average weekly reporting delay was 2.4 (range 0–52 days) compared with 84 days" for the reports sent through to the NHIS by the HBAS system.

Although data quality "associated with the use of clinical case definitions" improved significantly during the trial, accuracy of data transcription into the mobile phone was high (98% accuracy) throughout the trial. Feedback on the attributes of the system was generally very positive (relative validity for reporting for all diseases was high); the system enhanced collaboration between public health authorities and clinical services and "provided opportunities for nurse clinicians to demonstrate innovation, coordination and leadership"; the cost of the trial was US$45,000, excluding staff costs. Despite acknowledging a number of limitations of the evaluation and other concerns that emerged during the pilot, the authors conclude that "MOPBASS reporting was more timely, complete and sensitive than reporting through existing systems", and it was "simple, effective, reliable and acceptable, and enabled the routine, systematic, and ongoing reporting of syndromes of public health importance from the district level."

Source: Rosewell et al. (2013).

A second trial in PNG was of a health call center model, with nurses employed to work on shifts in a call center environment, on standby to take toll-free calls from the public and health workers in Western Highlands Province. A study showed that the majority of calls were from the public on health-related matters (including some which needed urgent medical attention) and from rural health workers (Watson and Poima 2015). Maternity cases were a significant proportion of both public and health workers' calls. It is hoped that the service, which is dependent on callers' access to mobile phones will help improve and alleviate some of the consequences of the low rates of out-patient attendance at health facilities across the region (Watson and Poima 2015).

Some of the examples above are in the category of communications between health services and patients and vice versa. Some are in the category of communications between health professionals for expert advice, and in this category there is a considerable overlap between telemedicine and m-health. For example, a new mobile application has been developed to provide doctors in the remote island of Kirakira in Solomon Islands[20] to connect with specialists at the Gold Coast Hospital in Queensland, Australia. The application allows relatively inexperienced junior doctors running a hospital servicing 60,000 people, with limited resources and access to information, to oversee patients requiring specialist pre- and post-operative care and avoids the need for patients to make an arduous journey to the nearest referral hospital. As the initiator of the scheme Dr Sophie West notes "isolation seemed to be the hardest part of their practice—[junior doctors] had very limited access to peers or specialist medical advice" (ISN News 2014). An evaluation after the pilot was concluded was expected to assess whether users were benefitting from the application and if the advice given had actually improved patient care. If benefits are realized, the early reports noted, it was hoped that the application would be trialed in Papua New Guinea, Fiji, and Tonga. Systems like this are not at all unusual in industrialized countries and there are many commercial applications available. However, to make a simple application work between Kirakira and the Gold Coast of Australia is a major challenge.

Skepticism about m-health projects rightly highlights the need for clear evidence of their value, and for any change in policy to be based on this evidence, especially if any future projects were to displace existing priorities in the health sector. But as some of the examples show, including the trial of a mobile phone-based syndromic surveillance system in PNG outlined in Box 11.2, m-health can justify the relatively modest costs involved, often only the cost of a small smartphone. This can be as little as US$50 (far less for a simple phone with only SMS messaging) since the specialized health applications involved are often made freely available. These initiatives mirror the predominant pattern of technology adoption in the Pacific Islands region—the rapid uptake of mobile technology, its increasing coverage and

[20]A town of about 3500 people, with electricity supplied by generators and extremely limited cell phone coverage and Internet access. Medical students from Australia and nursing staff from Japan provide the main support to the salaried doctor in charge of the 30-bed hospital.

affordability leading to some innovative applications. Some of these small ad hoc projects have the potential for considerable impact on health outcomes, especially those that provide support to isolated health workers; the gains shown in some of these initiatives need to be taken on board by policy-makers.

11.8 Conclusion

When resources are scarce, good policy based on up-to-date and reliable information becomes even more important to decisions on priorities for funding. For this reason, the Pacific Island states in this study have been strongly encouraged to focus on health information systems, rather than look for ICT applications across the entire field of health care. They are also faced with the considerable demands made on their scarce resources to provide data to the WHO and to the SPC's Statistics for Development Division. The value of this data is not in question. But it is essential that the data is both accurate and timely and is utilized for better planning and resource allocation. The health sector is a domain in which the quality of evidence is constantly being challenged. ICTs, both HIS systems themselves and the more modest use of mobile technologies in data collection, have a role in enhancing the quality of health data in the region.

This is starting to occur with the encouragement of regional organizations. One of the most notable aspects of the health information systems being put in place across the region is the 'shared approaches' being applied. Strong regional cooperation is driven by the WHO's WPRO and the SPC's Statistics for Development Division. The networks they have established and the technologies for data collection, analysis and dissemination, as well as the training and encouragement they offer have resulted in real benefits in the area of surveillance and in the collection and use of critical public health data. Although there is much work to be done, the role of evaluation followed by further capacity building has been established, "a notable example of how evidence gathered through a surveillance system evaluation can improve public health surveillance practice" (Paterson and Durrheim 2015, p. 3). It is hoped that this evidence will continue to flow through to policy and that more such research will reinforce the value of health information systems and encourage PICs to maximize the investment they have already made.

At the same time, there is growing evidence of the very real value of some telemedicine and m-health programs being trialled in PICs, and of the value of online resources for clinicians and health workers, which should not be overlooked. At present the value of these initiatives is not widely acknowledged but they represent an affordable and sustainable contribution of ICTs to health outcomes in the region. The evidence suggests that such programs meet very real needs for information using technology that is readily available and familiar to users (Heeks 2014), whether such users are clinicians, patients or communities. They meet Heeks' criteria for sustainable ICT and for e-government projects to have clear objectives and values, to be within the capacity of and to align with the responsibilities of local staff and be manageable within the resources available. Driven by health professionals meeting their own needs, such programs not only meet these criteria, they also reflect the transformational power of mobile technology and Internet access in Pacific SIDs.

The narrative of health IT in the Pacific Islands region to date can be seen as two separate and contrasting stories: the first being the internationally and regionally supported health information systems, gathering and managing critical public health data; and the second a raft of ad hoc applications, which taken as a whole represent a trend that may eventually become the norm—the use of ICTs to keep clinicians and health workers up-to-date and in contact with each other, to improve the quality of healthcare in the region and to get important public health messages to isolated communities. All of these are potentially sustainable solutions led by clinicians and health workers developing their own capacity in the process. But these are tiny individual projects; unless they can reach a critical mass and are given opportunities for sharing what has been learned more widely, these valuable initiatives in e-health, e-learning, telemedicine and m-health will fall into the familiar pattern of short-term projects with limited long-term benefits for the region. A coordinated regional agenda for e-health that goes far beyond the POHLN education network is required, as well as a regional agenda for HIS. There is a clear gap in regional sharing and support for e-health initiatives that needs to be filled if PICs are to take advantage of the new technologies available. The ITU and others have identified the major contribution that ICT can make, and has made elsewhere, to the delivery of healthcare in SIDs (ITU 2016). Regional leadership is clearly needed, from one of the regional organizations working in the health sector, or from one of the WHO regional Collaborating Centers to ensure that the benefits of ICT in the health sector are realized to the maximum possible in the Pacific Islands region.

References

AbouZahr C, de Savigny D, Mikkelson L, Setel PW et al (2015) Civil registration and vital statistics: progress in the data revolution for counting and accountability. Lancet 386:1373–1385

Adam P (2000) Experience and hopes for telehealth in Tokelau. Pac Health Dialog 7(2):61–62

Anderson I, Ivatts S, Somonathan A, Rolfe B (2014) The challenges of health financing in five South Pacific countries. Crawford School of Economics and Government, ANU. Asian Pac Econ Lit. doi:10.1111/apel.12052

Aumua A, Hodge N (2012) The Pacific in crisis: the urgent need for reliable information to address non-communicable diseases. Pacific Health Dialog 18(1):191

Berg M (2001) Implementing information systems in health care: myths and challenges. Int J Med Informatics 64:143–156

Biscoe G, Jacobson C (2014) High level strategic review [of] Fiji health sector support program. Final report. DFATR, Health Resource Facility, Canberra. http://dfat.gov.au/about-us/publications/Documents/fiji-health-sector-support-program-strategic-review-2014.pdf. Accessed 8 July 2016

Boonstra A, Broekhuis M (2010) Barriers to the acceptance of electronic medical records by physicians form systematic review to taxonomy and interventions BMC Health Services Research 10:231. doi:10.1186/1472-6963-10-231

Buchan J, Connell J, Rumsey M (2011) Recruiting and training health workers in remote areas: Pacific Island case studies. World Health Organization, Geneva

Buehler JW, Hopkins RS, Overhage JM et al (2004) Framework for evaluating public health surveillance systems for early detection of outbreaks. Recommendations from the CDC Working Group. Centers for Disease Control and Prevention, Atlanta, GA

Capelle M (2016) Republic of Nauru hospital infection prevention and control program: a Pacific Island perspective. In: Paper presented at the 5th international conference of the australasian college for infection prevention and control. http://www.acipcconference.com.au/wp-content/uploads/2016/02/The-Republic-of-Nauru-Hospital-Infection-Prevention-and-Control-Program-A-Pacific-Island-Perspective.pdf. Accessed 26 July 2016

Carrick D (2014) Solomon Islands HC judge speaks out. Law report. Australian Broadcasting Corporation. http://www.abc.net.au/radionational/programs/lawreport/solomon-islands-hc-judge-speaks-out-v2/5912956. Accessed 16 Feb 2016

Cave D (2012). Digital islands: how the Pacific's ICT revolution is transforming the region. Lowy Institute, Sydney. http://www.lowyinstitute.org/publications/digital-islands-how-pacifics-ict-revolution-transforming-region. Accessed 18 June 2016

Chan M (2015) WHO Director-General's speech at the summit on measurement and accountability for results in health. WHO, Geneva. http://www.who.int/dg/speeches/2015/health-measurement-summit/en/. Accessed 4 July 2016

Curtis M (2004) The obesity epidemic in the Pacific Islands. J Dev Soc Transform 1:37–42

Eikelboom RH, Swanepoel DW (2015) Teleaudiology. In: Eren H, Webster JG (eds) Telemedicine and electronic medicine. CRC Press, Boca Raton, pp 539–560

Fiji. Ministry of Health (2010) Shaping Fiji's health strategic 2011-15. Suva. http://www.wpro.who.int/health_services/fiji_nationalhealthplan.pdf?ua=1. Accessed 1 July 2016

Fraser HS, Biondich P, Moodley D, Choi S et al (2005) Implementing electronic medical record systems in developing countries. Inform Prim Care. 13(2):83–95. http://hijournal.bcs.org/index.php/jhi/article/viewFile/585/597. Accessed 8 July 2016

GSMA (2015) The mobile economy: Pacific Islands 2015. https://gsmaintelligence.com/research/?file=23485245295f02524925b2bd3aeec6de&download. Accessed 20 Aug 2016

Hayes G (2009) Population ageing in the Pacific: a situation analysis. Suva: UNFPA Pacific Sub-Regional Office. http://countryoffice.unfpa.org/pacific/drive/Ageingpopulation20.10.10.pdf. Accessed 30 June 2016

Heeks R (2006) Health information systems: failure, success and improvisation. Int J Med Inform 75:125–137

Heeks R (2014) A Development 2.0 research agenda. ICTs for development. 25 April 2014. https://ict4dblog.wordpress.com/2014/04/. Accessed 3 Feb 2016

Hetzel MW, Page-Sharp M, Bala N, Pulford J et al (2014) Quality of antimalarial drugs and antibiotics in Papua New Guinea: a survey of the health facility supply chain. PLoS ONE. doi:10.1371/journal.pone.0096810

Hodge N (2012) Quality for health information. What does it mean, why does it matter and what can be done? Pacific Health Dialog 18(1):120–124

Industrial Safety News (2014) Mobile app to connect Pacific Island doctors with Gold Coast specialists. http://isn.co.nz/node/947

International Telecommunications Union (2016) ICTs for a sustainable world. Goal 3. Health and well-being. http://www.itu.int/en/sustainable-world/Pages/goal3.aspx. Accessed 23 Aug 2016

Iorangi T (2012) Issues and challenges for enhancing statistical capacity: Cook Islands perspective Pacific Health Dialog 18(1):155–157

Johnson G (2015) Corruption undermining sustainable development. Online commentary. Port Vila: Pacific Institute of Public Policy. http://pacificpolicy.org/2015/11/corruption-undermines-sustainable-development/. Accessed 30 June 2016

Kalogriopoulos NA, Baran J, Nimunkar AJ, Webster JG (2009) Electronic medical record systems for developing countries: review. Ann Int Conf IEEE Eng Med Biol Soc 1730–1733. doi:10.1109/IEMBS.2009.5333561

Kaplan B, Harris-Salamone K-D (2009) HealthIT success and failure: recommendations from literature and an AMIA workshop. J Am Med Inform Assoc 16(3):291–299

Khazei A, Jarvis-Selinger S, Ho K, Lee A (2005) An assessment of the telehealth needs and healthcare priorities of Tanna Island: a remote, under-serviced and vulnerable population. J Telemed Telecare 11(1):35–40

Kool Jl, Paterson B, Pavlin BI et al (2012) Pacific-wide simplified syndromic surveillance for early warning of outbreaks. Glob Public Health 7(7):670–681. http://dx.doi.org/10.1080/17441692.2012.699536. Accessed 1 July 2016

Kupu S (2000) Health and telecast: a milestone for Tonga into the 21st century Pacific Health Dialog 7(2):54–56

Lal BV, Fortune K (2000) The Pacific Islands: an encyclopedia. University of Hawaii Press, Honolulu

Landry M, Novak J, Whittaker M (2012) Why HIS? Pacific Health Dialog 18(1):9–13

Mehl G, Vesudevan l, Gonsalves L, Berg M et al (2014) Harnessing mHealth in low-resource settings to overcome health system constraints and achieve universal access to health. In: Marsch LA, Lord SE, Dallery J (eds) Transforming behavioral health care with technology: the state of the science. Oxford University Press, Oxford

Minges M, Stork C (2015) Economic and social impact of ICT in the Pacific. Pacific Region Infrastructure Facility (PRIF), Sydney. http://www.theprif.org/index.php/resources/document-library/121-prif-ict-study-report-2015. Accessed 3 July 2016

Naidu S (2012) Developing a patient information system in Fiji. Pacific Health Dialog. 18(1):158

Palau. Bureau of Public Health (2008) Palau public health strategic plan 2008–2013. Koror, Palau. http://www.wpro.who.int/health_services/palau_nationalhealthplan.pdf?ua=1. Accessed 1 July 2016

Papua New Guinea. Department of Health (2010) National Health Plan 2011-2020. Government of Papua new Guinea, Port Moresby. Available at: http://www.wpro.who.int/health_services/papua_new_guinea_nationalhealthplan. Accessed 10 Oct 2016

Papua New Guinea. Department of Health (2012) Diabetes: clinical practice guidelines for Papua New Guinea, July 2012. Port Moresby. https://spmt.files.wordpress.com/2013/03/guidelines-png-diabetes-final.pdf. Accessed 8 July 2016

Parry J (2010) Pacific islanders pay heavy price for abandoning traditional diet. Bull World Health Organ 88(7):481–560

Paterson BJ, Durrheim DN (2015) Surveillance system evaluations provide evidence to improve public health practice. Online J Public Health Inform 7(1):e202. doi:10.5210/ojphi.v7i1.5959

Person RA (2014) The Pacific Health Care Project. Front Public Health 2:175. doi:10.3389/fpubh.2014.00175

Phillips DE, AbouZahr C, Lopez AD, Mikkelsen L et al (2015) Are well functioning civil registration and vital statistics systems associated with better health outcomes? Lancet 386:1386–1394

PHIN (2011) Health information systems. http://www.phinnetwork.org/Resources/HIS.aspx. Accessed 4 July 2016

Population Services International. Papua New Guinea (2016). Improving the quality of life for people in PNG. Port Moresby: PSI PNG. http://www.psipng.org/about-us. Accessed 13 July 2016

Ravindra S, Chandra R, Dhenesh VS (2015). A study of the management of electronic medical records in Fiji hospitals. arXiv preprint. arXiv:1507.03659

Rooney PJ (2013) Health record systems in Papua New Guinea. Papua New Guinea. Health [online discussion list]. https://papuanewguineahealth.com/2013/01/28/health-record-systems-in-papua-new-guinea/. Accessed 8 July 2016

Rosewell A, Ropa B, Randall H et al (2013) Mobile phone-based syndromic surveillance system, Papua New Guinea. Emerg Infect Dis 19(11). http://www.doc88.com/p-3109735296023.html

Samoa. Ministry of Communications, Information and Technology (2012) National ICT policy 2012-2017. The Ministry, Apia. Available at: http://www.mcit.gov.ws/images/mcit/NICTPOLICY2012-2017.pdf. Accessed 10 Oct 2016

Schäferhoff M, Fewer S, Kraus J et al (2015) How much donor financing for health is channelled to global versus country-specific aid functions? Lancet 386. http://dx.doi.org/10.1016/S0140-6736(15)61161-8

Soar J, Gow J, Caniogo V (2012) Sustainability of health information systems in developing countries: the case of Fiji Health Information Management Journal 41(3):13–19

SPC (2016) National Minimum Development Indicators. SPC Statistics for Development Division, Noumea. https://www.spc.int/nmdi/. Accessed 10 October 2016

SPC Statistics for Development Division (2016) Pacific CRVS. SPOC SDD, Noumea. http://sdd.spc.int/en/sdd-programmes/pacific-crvs. Accessed 10 October 2016

Tabunga T (2012) Issues and challenges for HIS in a small island nation. Pacific health Dialog 18 (1):25–28

Tuvalu (2009) Strategic health plan 2009-2018. http://www.mcit.gov.ws/images/mcit/NICTPOLICY2012-2017.pdf. Ministry of Health, Funafuti. http://www.wpro.who.int/health_services/tuvalu_nationalhealthplan.pdf?ua=1. Accessed 1 July 2016

Umali E, McCool J, Whittaker R (2016) Possibilities and expectations for mHealth in the Pacific Islands: insights from key informants. JMIR Mhealth 4(1). doi:10.2196/mhealth.4626.e9

UNFPA Pacific Subregional Office (2014) Women should not die giving life—an article on maternal mortality. http://countryoffice.unfpa.org/pacific/2014/05/12/9686/women_should_not_die_giving_life_an_article_on_maternal_mortality/

United Nations (2016) Sustainable development knowledge platform. https://sustainabledevelopment.un.org/sdgs. Accessed 8 July 2016

Watson A, Poima R (2015) Healthy phoning. Papua New Guinea's first ever health call centre. Australas Psychiatry 23(6 Supplement):48–51

Watson A, Sabumei G, Mola G, Iedema R (2015) Maternal health phone line: saving women in Papua New Guinea. J Pers Med 5:120–139. http://www.mdpi.com/2075-4426/5/2/120

Wolfe BA, Burke W, Mamlin MD, Tierney WM (2006) The OpenMRS system: collaborating toward an open source EMR for developing countries. In: AMIA annual symposium proceedings 2006: 1146. PMCID: PMC18393

Wootton R, Bonnardot L (2015) Telemedicine in low resource settings (Editorial) (Article 3). Front Public Health 3

World Bank (2013) The economic costs of non-communicable diseases in the Pacific islands: a rapid stocktake of the situation in Samoa, Tonga and Vanuatu. World Bank Group, Washington, DC. http://www.worldbank.org/en/news/feature/2013/01/16/the-economic-costs-of-noncommunicable-diseases-in-the-pacific-islands.pdf

World Bank (2016) GDP per capita, PPP (current international $). Overview per country. World Bank Group, Washington. http://data.worldbank.org/indicator/NY.GDP.PCAP.PP.CD?view=chart

World Health Organization (2008) Health information systems: toolkit on measuring health systems strengthening. WHO, Geneva. http://www.who.int/healthinfo/statistics/toolkit_hss/EN_PDF_Toolkit_HSS_InformationSystems.pdf. Accessed 26 July 2016

World Health Organization (2011) mHealth; new horizons for health through mobile technologies: second global survey on eHealth. WHO, Geneva. http://www.who.int/goe/publications/goe_mhealth_web.pdf. Accessed 26 July 2016

World Health Organization (2014) Achieving the millennium development goals in the Western Pacific Region. WHO, Geneva. http://www.wpro.who.int/health_information_evidence/documents/mdg2014leaflet_final.pdf?ua=1. Accessed 26 July 2016

World Health Organization (2016) Health and Development. http://www.who.int/hdp/en/. Accessed 10 October 2016

WPRO (2016) Electronic health records: a manual for developing countries. World Health Organization Regional Office for the Western Pacific, Manila. http://www.wpro.who.int/publications/docs/EHRmanual.pdf. Accessed 30 June 2016

WPRO (2016a) Human resources for health country profiles. World Health Organization Regional Office for the Western Pacific. http://www.wpro.who.int/hrh/en/. Accessed 30 June 2016

WPRO (2016b) About the health information, evidence and research unit. Issues, challenges and action. Manila. http://www.wpro.who.int/health_information_evidence/strategy/en/. Accessed 4 July 2016

Yamo H (2013) Mobile phones in rural Papua New Guinea: a transformation in health communication and delivery services in Western Highlands Province. Master of Communication Studies (thesis). Auckland University of Technology

Chapter 12
ICT in Education in Small Island Developing States of the Pacific

Ioana Chan Mow, Emma Kruse Vaai, Ian Thomson and Katalina Pasiale Taloka

Abstract This chapter outlines how Information Communication Technology (ICT) in education supports and facilitates the processes of e-government. ICT education, its use in teaching and learning and for enhancing education administration and governance, are pivotal components in regional and national development for the small island developing states of the Pacific. The identification and discussion of various issues and challenges which have hindered the effective utilization of ICT in education in the Pacific region provides a reality check. At the same time, it suggests a way forward for major initiatives grounded in the context of small islands where a growing dependence on the knowledge economy is increasingly facilitated by ICT capability. Progress in different countries of the Pacific region and the reasons underlying varying rates of development particular to each country are also noted together with outside support from various sources. Sound planning, appropriate education and training, together with enabling and well informed partners and policies can make a difference to progress and the sustainability of such developments.

12.1 Introduction

This chapter outlines how Information and Communication Technology (ICT) in education supports and facilitates the processes of e-government. The chapter begins by introducing the key components of ICT. We argue that the use of ICT in education for enhancing administration and governance, for teaching and learning and for ICT education itself are pivotal components in regional and national

I. Chan Mow (✉) · E. K. Vaai
National University of Samoa, Apia, Samoa
e-mail: i.chanmow@nus.edu.ws

I. Thomson
University of the South Pacific, Suva, Fiji

K. Pasiale Taloka
Ministry of Education, Funafuti, Tuvalu

R. Cullen and G. Hassall (eds.), *Achieving Sustainable E-Government in Pacific Island States*, Public Administration and Information Technology 27,
DOI 10.1007/978-3-319-50972-3_12

development for the small island developing states (SIDS) of the Pacific. We then provide a background section introducing the Pacific's education systems and ICT infrastructure. The importance and impact of regional and national ICT policies, plans and strategies in education is then discussed and followed by a consideration of the main uses and impacts of ICT in education: in administration, education, national and regional initiatives and in teaching and learning. We then identify and discuss various issues and challenges that have hindered effective utilization of ICT in education and the things that can facilitate and support the processes of e-government in the Pacific's SIDS. This assessment provides a reality check for our discussion of future directions via major initiatives in light of global ICT developments. We take account of the immense diversity in size, populations, economic development and cultural and physical resources in Pacific Island countries (PICs) as we suggest practical steps that may enable them to participate in the ICT-facilitated knowledge economy. The chapter then concludes with a summary of common themes, reflections and recommendations.

Information Communication Technologies are seen as potentially powerful enabling tools for expanding access to education and improving the quality of education (Daniel 2009; Kanwar et al. 2014; Whelan 2008). ICT innovations have revolutionized and enhanced the learning and teaching process, opening up new learning spaces, opportunities and better access to educational resources. Integration of ICT into schools, colleges, universities and other educational institutions has provided novel ways to access resources, implement information gathering and analysis; it also provides the means to improve the administrative and management capacity of educational institutions. In the development context ICT plays a critical role in providing an enabling environment for local development and people's empowerment. It can provide employment opportunities, improve people's access to basic services, create networks, share information, build knowledge and increase the transparency, accountability and effectiveness of development actors (Koshy and Mataki 2005).

Many definitions of e-government distinguish between developed and developing countries. However, the World Bank's definition used in this book states:

> E-Government refers to the use by government agencies of information technologies (such as Wide Area Networks, the Internet and mobile computing) that have the ability to transform relations with citizens, businesses and other arms of government. These technologies can serve a variety of different ends: better delivery of government services to citizens, improved interactions with business and industry, citizen empowerment through access to information, or more efficient government management… (World Bank 2015)

E-government success demands levels of literacy and ICT competence in society for which the education system must prepare young people. Because ICT in education refers to its use for enhancing skills in administration and management, for ICT education and for teaching and learning at all levels, several important considerations follow.

The first major consideration is how ICT in education can support and facilitate e-government activities in government to government, government to citizen and government to business (often referred to as G2G, G2C and G2B). The 2015 Pacific Region Facility's Report (PRIF 2015) states that the PICs' main uses of ICT in education and e-government are administrative. ICT in education enhances administration and management by establishing systems leading to improved efficiency, effectiveness and productivity. This also results in greater transparency and accountability with stakeholders. Evidence of the importance of ICT in administration and management is the establishment of a Pacific regional Electronic Management Information System (EMIS) in education (see Sect. 12.4 below).

Second, one of the keys to good governance lies in the education system. E-governance is becoming ubiquitous as countries become more and more ICT-oriented in their communication strategies. ICT education and training provide the technological skills and digital literacy needed to enable people to engage as active citizens in local and global governance in a world that is increasingly informed by and communicates via various forms of electronic media. In addition, ICT education and computer literacy provide citizens with technological and communication skills not only for the workplace, which is increasingly dominated by technology, but even more so for their social interactions with peers and other groups. ICT education provides the training to enable a skilled workforce to use technology for better delivery of government services to citizens, improved interactions with business and industry, citizen empowerment through access to information and more efficient government management. This need for ICT capacity building in Pacific Island countries is well reflected in both national (AMFIE 2013) and regional (SPC 2010; PIFS 2010; Samoa Pathway 2014) plans and policies. These initiatives are described later in the chapter.

Third, the use of ICT in education for teaching and learning provides further avenues for access to education as well as educational resources; enables efficiency and effectiveness in administration and overall improvement in the quality and delivery of teaching and learning. Open Educational Resources (OER) provide an excellent example of democratizing education by making high quality educational resources available at low cost. Not only are they freely accessible online, they can also be changed and adapted to suit the needs of the user. The creation of large repositories of materials and resources for use without charge significantly reduces the cost of researching and developing course content by educators and their institutions. Moreover, when resources are developed collaboratively they lead to the creation of online social and professional networks. A good example of this kind of international collaboration is the Virtual University of the Small States of the Commonwealth (VUSSC). Under the umbrella of the Commonwealth of Learning, VUSSC has enabled small states to embrace ICTs through education and ensured they would "not be left behind". VUSSC enables ICT skills training for online collaboration in content development for post-secondary education that is appropriate for adaptation and use either face-to-face in the classroom, in distance mode or in blended mode. By expanding access to learning VUSSC also enables small states to contribute to global development and become practitioners and leaders in educational reform through the innovative use of ICTs. Furthermore, as well as being members of the Commonwealth, as most PICs are, they are also members of other international entities which mandate compliance with internationally agreed goals such as the Millennium Development Goals and Education For All. The use of ICTs in education became necessary to advocate for and realize these goals.

The right to education and the government's responsibility to provide for it has also been facilitated through the use of ICT. One of the best Pacific examples was the creation of the regional University of the South Pacific (USP) in 1968 (Whelan 2008). With its member countries far flung throughout the Pacific Ocean, the use of ICT for distance mode delivery provided access to higher education for many Pacific peoples. National universities have been established more recently in several PICs and many are now also gravitating toward distance learning modes in their own countries to provide greater access. Although distances are not particularly great in some of these small island states, transport systems are not well developed, so on-campus attendance can be quite irregular. The use of ICT is pivotal to advances in distance learning (Vaa 2015).

While government is the predominant provider of education in PICs, other providers such as churches and private schools offer school level education. Although these providers have not made much use of distance learning, the inclusion of computer studies in the national school curriculum ensures students learn how to use and appreciate the importance of ICT in their lives. In most PICs, all education providers are subject to education policies and legislation for the development of well-educated and productive citizens, and increasingly governments are coming to realize that ICT education and computer literacy contribute to and are as important as the traditional three Rs (reading,'riting,'rithmetic). Student access to computers is therefore essential, which means computer labs must be a key resource in every school, although at present this is not always possible due to budget constraints.

12.2 Background

The Pacific region consists of thousands of islands in more than 40 island groups which can be divided into three sub-regions based on their people's cultural and physical attributes: Polynesia to the east, including Samoa, Tonga and Tuvalu; Melanesia in the west, including Fiji, Papua New Guinea (PNG), Solomon Islands and Vanuatu; and Micronesia in the north-west, including Kiribati and Nauru. These island groups vary in size, population and, significantly, level of economic development, cultural and physical resources. This diversity and the vast distances involved in some island groups pose some challenges in communication and access to education.

12.2.1 Education in the Pacific

Pre-missionary education in the Pacific islands was undertaken in villages by older village members imparting by word of mouth and demonstration values, skills and knowledge to younger generations on a daily basis in practical situations. The concept of "formal schooling" was introduced with the arrival of Christian missionaries in the early 1800s and the creation of mission schools which had their greatest impact not only in evangelization but in teaching people how to read and write. In some islands (e.g. Samoa) the translation of the Bible into the vernacular raised literacy rates to almost 100% in the early 1900s. With the advent of colonization of island groups by various Western powers, there was a further move towards formal schooling through the establishment of government schools that operated side by side with mission schools. Before independence, government involvement in education in many of the islands occurred in two phases. In the first phase the focus of the governing colonial powers was on training clerks and public servants for effective administration. In the second phase the focus was on education systems preparing citizens for political independence. Today government and mission schools exist side by side, providing the backbone of the education system in most of the PICs (Baba 1985). The different island groups which emerged as independent island nations in the early 1960s largely retained the education systems and language of their former colonial administrators alongside their own. Early childhood, primary and secondary education, as well as tertiary education of some form are now provided in all Pacific countries, although not all children have access to education due to cost or distance.

An important consideration in the use of ICT in education is the diversity in the education needs and systems across the region. At one end of the spectrum is Papua New Guinea with a huge population dispersed over many islands and with several hundred languages, in contrast to Samoa with its moderate population, occupying a few islands, with one language and a strong missionary education system. In the Melanesian countries in particular, a very large number of languages are spoken. This creates a problem when it comes to using technology in education. For example, Vanuatu has three official languages (English, French, Bislama) and over a hundred indigenous dialects. Samoa's official languages are English and Samoan; and Fijian, Hindi and English are Fiji's official languages. All may be used at some level of education (although not all indigenous languages in the Pacific islands have written form.) The creation of educational materials that can take advantage of ICT must deal with the range of languages used in schools.

Most PICs have established Technical and Vocational Education and Training (TVET) facilities in each island group and some form of community training. However, university education is largely provided by the University of the South Pacific (USP) (which has campuses in all twelve member countries) and the national universities established by some (e.g. Samoa, Solomon Islands, Fiji and PNG) to enable more of their own populations to access higher education. Scholarships to universities in New Zealand, Australia or other countries are available to top students, but many other equally capable students are unable to continue on to tertiary education due to limited scholarship opportunities and distance. Large numbers of students in some PICs finish their schooling at the end of primary school.

With the dotcom boom at the turn of the century Pacific ministers of education together with other small states of the Commonwealth were concerned that their countries should engage with the online world and sought to increase access to tertiary education in their countries. They doubted that they had the critical mass of either expertise or equipment to engage with technology and virtual learning on their own. They saw their countries as being on the wrong side of the digital divide and lacking the capacity to benefit from developments in technology that would help develop tertiary institutions in the region. The Virtual University for Small States of the Commonwealth (VUSSC) was created to develop a continuous cadre of people with the skills and expertise to implement virtual education systems, strengthen existing tertiary institutions and enable the delivery of courses in a variety of formats. VUSSC started its activities by strengthening the capacity of institutions and educators in open and distance learning, using appropriate technologies while at the same time developing OER free content (Kanwar 2015). At present the Pacific region's average participation rate in tertiary education is less than 5%. Even Fiji, the country with the highest rate, has only 15% of its young people in tertiary education. While many PICs still lack adequate ICT infrastructure, the challenge is to enable them to make use of ICT for greater access into tertiary education in the Pacific via the University of the South Pacific, local universities and others offering distance learning opportunities.

12.2.2 Infrastructure

A robust basic telecommunication infrastructure is vital to use learning technologies effectively. Without it PICs are seriously limited and their access and penetration rates will continue to lag behind the rest of the world. As the Framework for Action on ICT Development in the Pacific (FAIDP 2010) review report shows there is a vast disparity in penetration rates among the nine South Pacific Island countries. Fiji, Tonga, Vanuatu and Papua New Guinea have developed fairly rapidly, while others have yet to experience the full range of technologies (Vaa 2015). While mobile phone use has grown exponentially across the Pacific region, a common issue in all countries has been the high cost of Internet connectivity and long distance calls. However, in recent years most countries have invested in infrastructure, such as satellite or submarine fiber optic cable to improve the telecommunication infrastructure (Vaa 2015).[1]

[1]A more detailed exploration of this issue is found in Chaps. 4 and 5.

12.3 National and Regional Policy Frameworks, Plans and Strategies for ICT in Education

Another important consideration is the role and impact of national and regional policy frameworks and plans in the promotion and realization of the benefits of ICT in education, which is seen in the region important to ensure "quality education" that will enhance the capacity and sustainability of SIDS on the global stage.

12.3.1 National Policies

By 2002 many PICs were developing national ICT policy frameworks (Thomson 2015). These national policies included key capacity building to enable citizens to participate and take advantage of the full benefits of ICT in national development. For example, the Papua New Guinea (PNG) National Vision 2050 included strategies for the promotion and use of ICT for sustainable education. The PNG National Education Plan 2005–2014 (Papua New Guinea Department of Education 2005) proposed that "information communications technology systems and process be developed to support the management of educational programs". The plan also included the use of radio and television as suitable technologies for "delivering teaching programs and supporting learning and assessment programs through a distance mode when it becomes available and is affordable and sustainable" (Papua New Guinea Department of Education 2005: 94). The National Media Policy for Papua New Guinea also articulated strategies for promoting ICT for education (Papua New Guinea Department of Education 2010).

In Samoa, the National ICT Strategic Plans of 2003, 2008 and 2012–2017 all affirm the importance of ICT in education for building capacity. The Ministry of Education, Sports and Culture (MESC) ICT plans of 2002–2003 and 2005 and ensuing MESC strategic and corporate plans emphasize ICT development. Objective 4 in the MESC Corporate Plan for 2012–2015 prioritizes the improvement of ICT standards. Objective 5 includes the improvement of ICT services for the Ministry and all schools by strengthening ICT Human Resource capacity, utilizing appropriate ICT for asset management, management and maintenance of MESC's website, upgrading data centers and infrastructure and extending ICT centers to all schools (Samoa Ministry of Education, Sports and Culture 2012). (see also Box 12.3) These improvements in standards, infrastructure and capacity were intended to provide a highly competent, skilled, ICT-literate workforce and, ultimately, better ICT services in the country.

Other countries (Solomon Islands and Tuvalu) developed national ICT policies later; for example, the ICT Policy and Strategy for Tuvalu was drafted and finalized in 2014 through consultations with relevant stakeholders including the Education Department. Currently all PICs, with the exception of Kiribati and Solomon Islands, now have ICT in education policies in place. Of these, however, only Fiji and

Tuvalu have a provision for OER in their policies (Vaa 2015). These national ICT policies and plans also have universally addressed the need for infrastructure and more accessible and affordable ICTs. This has been done through a range of strategies (discussed in more detail in Chap. 4), but the net effect of regulating the cost and use of ICT and removing monopolistic pricing has reduced the cost of access, particularly the cost of mobile calls in all PICs. This reduction in costs is also evident in the availability of affordable tablets, allowing mass deployment of these devices for use in education.

12.3.2 Regional Policies

The key role of ICT in education was well reflected in a Pacific Island Forum Education Ministers meeting in 2010 that culminated in the establishment of a regional framework for education in the Pacific (PIFS 2009), updated in 2012 (PIFS 2012). In fact, the 2007 Pacific Plan (reviewed in 2010), Asia-Pacific Ministerial Forum on ICT in Education (AMFIE 2013), the Framework for Action on ICT for Development in the Pacific (FAIDP 2010), Review of Pacific Regional Digital Strategy Part A: Technological Capacity (PIFS 2010), the Pacific Plan (2010) and the more recent Samoa Pathway (2014) all included strategies for promoting ICT in education. In 2014, at a Pacific Island Ministers of Education Forum in the Cook Islands, delegates underscored the importance of ICT as a tool to improve access, quality and inclusion across the region. This was further progressed at the Pacific Island Forum Ministers of ICT meeting in Tonga in 2015 that endorsed an update of the FAIDP; more recent progress is seen in the Pacific Regional ICT Strategic Action Plan (PRISAP) for 2016–2020 which will follow on from FAIDP (ITU 2015).

Regional strategies can also be used to target specific educational needs and identify ICT solutions. At the 2015 Conference of Commonwealth Education Ministers (CCEM) the critical importance for small states to safeguard and sustainably develop their marine resources was noted. Education Ministers requested Commonwealth of Learning to develop the OER content for courses on the 'blue economy' through VUSSC. Educational resources on climate change and climate change adaptation were also requested, a major educational need in the region, climate change being one of the biggest challenges facing small and vulnerable island nations.

However, notwithstanding the increasing awareness of ICT's importance and its inclusion in regional and national planning, progress has been slow and below expectations (Pacific Region Infrastructure Facility 2015; Thomson 2015). Despite early awareness of the issue and its inclusion in regional and national policy and planning frameworks, several challenges have hindered effective implementation of ICT in education. Poor supporting infrastructure, the high cost of technology, lack of finance for curriculum development, high turnover of ICT teachers, lack of ICT experts and teachers, lack of a planned ICT curriculum, poor project coordination and lack of capacity of implementers on the ground are some of the factors preventing successful implementation.

Strategies and solutions that have been suggested focus on the need for policy and planning to be supported by follow-up activities, such as targeted budget allocations, the alignment of school development strategies with national education development objectives and the necessary organizational adjustments (distribution of roles and responsibilities) in government and other education stakeholders (Strigel et al. 2007). Although these strategies are not yet fully implemented, there has been progress, as reflected in the 2013 review of FAIDP 2010, which revealed increased ICT usage in education in all the countries. This has been attributed mainly to the continuous efforts of governments to upgrade infrastructure and communication tools coupled with efforts by institutions to take advantage of new and more affordable communications and technologies to improve teaching and learning (Vaa 2015).

12.4 ICT in Administration and Management

According to the PRIF report on the economic and social impacts of ICT, ICT in the education sector has mostly focused on its use for administration (PRIF 2015). The focus on administrative systems does not necessarily include capacity building; there is little focus on training to develop adequate ICT skills and knowledge. For example, the Solomon Islands National Education Plan 2013–2015 has as its outputs for information and communication an improved Electronic Management Information System (SIEMIS), the establishment of an asset management system and an ICT plan for the HR division to ensure two-way communication internally and with provincial authorities (Cullen and Hassall 2016a). Resources for ongoing capacity building are not included. The experience of Papua New Guinea implementing an EMIS (Box 12.1) illustrates other challenges of developing management information systems in the Pacific due to factors such as geographical dispersion and disparity of data-handling, as well as budgeting. That state of progress in other countries is noted below.

Box 12.1: Papua New Guinea EMIS: the challenges of developing EMIS in the Pacific

The education system in Papua New Guinea includes nearly 13,000 schools, 55,000 teachers and close to 2 million students. The system includes Elementary schools covering grades 1–2 (teaching in the local language), Primary schools grades 3–8, (teaching also in English), Secondary schools (grades 9-12) and Technical and Vocational Education and Training (TVET) institutions. Pupils unable to attend a school in their area or who need to repeat a Grade examination to progress further may study through one of the Flexible Open and Distance Education (FODE) provincial centres.

The education system administers national, provincial and schools: Provincial Education Boards manage Elementary, Primary and Secondary Schools (about 50% of elementary and primary schools are administered by church agencies), while FODE Centres, national high schools, technical and business colleges are administered at the national level by the National Education Board (NEB), the Secretary for Education and the Teacher Services Commission (TSC). The PNG Department of Education (referred to as the National Department of Education or NDoE to distinguish it from provincial administrations) manages payroll for teachers while the Teacher Services Commission (TSC) manages all other affairs related to HR and appointments. The NDoE also manages the school registration process, the national curriculum, quality control of teaching and administration and measurement and assessment for the whole student body. It also manages the Tuition Fee Free system, a subsidy disbursed to schools (especially in remote areas) to encourage enrolment and ensure that no fees are charged. The TFF scheme has seen significant improvement in school enrolments in recent years, but puts considerable pressure on teachers and classrooms as student numbers expand.

The NDoE's Education Management Information System (EMIS), based on an Oracle database (supported by the Australian government since 2007) manages data for the education system; this data covers location and infrastructure of schools, budgets and enrolments (including grade, gender and age), and forms the basis of the TFF disbursement system. The EMIS, for which data was initially collected manually and transported by road, was inaccurate due to manual entry and the fact that schools are geographically dispersed across the country with rugged terrain and limited road access. In 2011 a project to automate data collection was initiated and a gradual rollout of the system to provincial centres was undertaken. This has already improved the quality of data and the new network allows schools to access the EMIS directly, staff to access the government payroll system (and get their payslips online) and documents to be submitted to the NDoE by email. Under development are systems related to teacher employment, better HR management (including a clean-up of employee data), access for the TSC to teacher data and a full student record system (national student assessment records are still paper-based, copies sent manually to the NDoE). It is hoped that when complete this centralised database will better support the education system and provide "a more complete picture of education in Papua New Guinea" even though the EMIS has many challenges still to overcome.

Sources: Kamrooz Khademazad (personal contribution) and UNESCO (2015)

In Samoa, linked administrative and information systems allow the effectiveness, efficiency and productivity of the education sector to be monitored and managed. Significant investments have been made by Ministry of Education, Sports and Culture in developing and integrating its internal systems with government-wide systems (such as finance and HR), made possible by various donor-funded projects. The systems include:

(i) Pelican—the Ministry's personnel information system;
(ii) Manumea—the system that records and compiles the Schools Census information;
(iii) ATLAS 2000—the system used for National Assessments;
(iv) SNAP—the Special Needs Assessment Program database which records information on those with special needs;
(v) FinanceONE—the Government of Samoa Financial and Analytical Reporting system; and
(vi) The Public Service Recruitment and Selection system (MESC Strategic Plan 2006–2015).

The Tuvalu Education Management Information System (TEMIS) has a school-based implementation called the School Information Management System (SIMS) for school administration. Current modules mainly focus on school statistics; however, scheduled upgrades include a wider variety of school administration features, including a class timetabling system.

The Vanuatu Education Management Information System (VEMIS) contains information collected from the annual education census and includes information from early childhood, primary, secondary and post-secondary and training sectors. VEMIS records examination results and includes results of the Vanuatu Standardised Test of Achievement (VANSTA) as well as financial information (Cullen and Hassall 2016b); work is on-going to align school income with expenditure through improved quality and coverage of financial data.

Tonga's Education Ministry is developing an Education Management Information System to improve record keeping and school administration that will replace an earlier system provided by UNESCO. A major hindrance in tailoring the new system to the local context is the lack of technical capacity, but there are now plans to move to a cloud-based system (PRIF).

In the Cook Islands, the EU-funded EduNet system, launched in 2003, was intended to connect all schools to provide them with access to curriculum materials. However, the main use of EduNet so far has been to enhance school administration and communication (PRIF 2015).

Fiji has also had a major focus on EMIS, developing an in-house system (FEMIS) that is becoming a regional benchmark. All schools are now connected to it, enter data and obtain reports to help them provide better school management (PRIF 2015; Vaa 2015).

12.5 ICT Education

The status of ICT education in the Pacific is summed up by Thomson (2015) who states that most Polynesian schools have computer labs and teach ICT, whereas Melanesia and Micronesia lag behind (Thomson 2015). For example, in Vanuatu inclusion of training in ICT use has not been part of the educational program at secondary or tertiary level (Cullen and Hassall 2016b) although this is now changing with a number of schools participating in the One Laptop per Child (in this case One Laptop per Pacific Child) program and the Commonwealth of Learning project *Classrooms without Walls*, both of which are described below.

In Samoa, quite a few schools offer computer training from Year 9 to Year 13, and in 2005 a national Computer Studies curriculum was established for Years 12 and 13. However, as the literature points out, curriculum development should be accompanied by teacher training (Chan Mow 2010); and indeed, this has resulted in the revision of teacher training programs at the National University of Samoa to include relevant courses.

In Fiji, a Computer Science curriculum in schools was developed as early as 1996, but by 2003 only 55% of secondary schools had implemented it (PRIF 2015). However, in recent years there has been progress—nearly all schools now have computer labs and teach computer science (Thomsen 2015). The three universities in Fiji (USP, Fiji National University and the University of Fiji) all teach IT courses. USP extends these IT programs to 11 other member states in the South Pacific through its USPNet for distance learning (Vaa 2015).

Papua New Guinea has included computer basics and word processing skills (in its lower secondary curriculum for over decade, as well as database, spreadsheets and Internet as more advanced options in its curriculum for some years; computer studies (focusing on information literacy, computer-based communication and computer technology and solutions) has been available as part of the curriculum for the two senior years of high school since 2008 (Papua New Guinea Department of Education 2008).

In Tonga, computer science is included in the curriculum in secondary schools. At tertiary level, USP and the Tonga Institute of Higher Education (TIHE) also offer programs in computer science. Diploma programs are offered by church schools and some private companies. In addition, the Tonga Chamber of Commerce offers two-week training programs in office applications, coordinated through the Tonga Business Enterprise Centre with assistance from the New Zealand government to build up business skills in the country (PRIF 2015).

Some universities in the Pacific Region include ICT education and training as a major part of pre-service teacher education. At the National University Samoa, for example, all pre-service teacher education students take an introductory course in computer basics and computer applications. Pre-service bachelor and diploma programs include minors and majors in computing for those intending to teach computer studies in secondary schools (Chan Mow 2010). Pedagogy courses in computing are also a compulsory component for diploma and bachelor of education students preparing for secondary teaching, and a new computer literacy course is also incorporated into the primary pre-service teacher education program as of 2016. However, much work remains to be done for the professional development of a large number of teachers who are already in the service and often reluctant to learn new skills and methods. In other PICs teacher training is not necessarily provided by a university and the training does not seem to include ICT education to the same extent. A recent report on education in Papua New Guinea, for example, commented that teachers' colleges included basic IT skills but that this did not include the Internet, so teachers were left unable to use the Internet for teaching materials or use ICT in teaching and learning, let alone teach the subject (UNESCO 2015).

However, the private sector is making a contribution in this area. Basic Microsoft technical courses and Cisco courses are available in most countries, provided by the private sector and often funded by employers. Universities are also starting to offer technical networking skills; two Cisco academies have been set up in Samoa, one by the National University of Samoa another by the University of the South Pacific campus at Alafua. This is a significant area for USP to be involved in; Cisco academies provide much needed training to provide the technical skill base to support some of the national ICT initiatives outlined above.

12.6 ICT in Teaching and Learning

The use of ICT for improving teaching and learning is still relatively undeveloped in the Pacific region. As noted in *Education for All* (UNESCO 2015) although policy managers and education administrators may be aware of the benefits of ICT in education, lack of infrastructure, unreliable electricity supply and lack of capacity to maintain ICT in schools (to which we could add lack of trained teachers) often frustrates attempts to make effective use of ICT in PICs schools; this problem is more severe in remote rural schools. Despite pockets of progress (e.g. One Laptop per Child, PFNet and USPNet, which are discussed below) proven technologies already in use are often the best solution, while the challenges noted above are addressed. (And of these three examples only USPNet has been shown to be truly sustainable.)

Radio is still therefore the predominant technology in education in most PICs. For example, early use of the technology in Samoa was for delivering lessons to primary schools in 1940. The Samoan Educational Broadcasting Unit has undergone several upgrades through a variety of donor projects with broadcasts supplemented by CD-ROMs, printed materials and Internet resources to enrich course content. The advantages of radio broadcasting lie in the fact that it is sustainable and can be maintained at minimal cost; hence it is economically viable in a developing country (Chan Mow 2010). The fact that most, if not all, families in Samoa can afford and need a radio for daily general information, entertainment and particularly weather forecasts (including cyclone warnings) means that it can also be utilized for educational programs without extra personal or network costs. It is less effective in countries where several languages are spoken, although local pidgin can be used as a broadcasting medium

Radio Pasifik Nauru is a community-based educational radio station launched in 2007 to help students overcome isolation, frequent power cuts and the scarcity of transportation and fuel. Using a solar-powered, 30-watt FM "radio in a suitcase" obtained through Commonwealth of Learning (COL), the station broadcasts a range of programs, including lectures and tutorials recorded weekly at USP in Fiji.

In the Solomon Islands, high frequency (HF) radio is used to broadcast information about education and educational opportunities to officials and teachers in remote locations. However, it is not used for broadcasting educational content (Cullen and Hassall 2016a).

Radio is gradually being replaced by e-learning tools. The One Laptop per Child (OLPC) project was piloted in eight Pacific Island countries including Samoa, Fiji, Vanuatu and Solomon Islands from 2008 to 2010. However, the project could not be scaled up from pilot to full implementation in all these countries due to a variety of factors—in particular the likelihood some pilot projects may have been initiated before there was adequate connectivity. There were also issues about the contextual relevance of OLPC's bundled software and its support and maintenance (PRIF 2015). In Vanuatu, for example, involvement in the OLPC project, has been quite successful. But Vanuatu is starting to manage its own more sustainable e-learning opportunities. As part of its Universal Access Policy, Vanuatu successfully launched 22 community tele-center sites, 15 of which are co-located in schools that provide access to ICT and Internet to local villagers after school hours (PRIF 2015).

In Tuvalu, each school is equipped with a digital projector for in-house professional development as well as class activities such as student presentations. A few schools are equipped with computer labs with expectations for national coverage in the near future. Fiji has also launched several projects that provide computers and Internet access to schools. One such is the Share, Engage and Educate (SEE) project providing over 200 computers as well as robotic kits, digital cameras and data projectors to schools. Equipment for the project was donated by individuals and the Queensland University of Technology in Australia (PRIF 2015). The Fiji Department of Communications established a tele-center program providing after-school computer access for the community. Fiji also benefited from an OLPC initiative in which 1500 laptops which were distributed to primary schools and 5000 tablets from India were delivered to rural and remote schools in late 2015.

The Commonwealth of Learning has also been involved in introducing technology into schools in the region. Its APTUS project is a low cost, offline virtual classroom. This 'classroom without walls' which facilitates content sharing and learning interaction between schools using a Wi-Fi network that is not linked to the Internet is being trialed in several Pacific countries. This is a successful OER initiative for creating and sharing educational resources through collaboration which helps reduce the costs (time and money) of obtaining such resources. It has also enabled a growing sense of community and cooperation among people from different countries facing similar challenges. To date APTUS has proven to be relevant and robust in that it can be solar powered, start up easily without Internet and connect easily to many devices such as tablets and mobile phones (Balasubramanian 2012). However, in the Pacific region, most of the initiatives to put computers and technology into schools have been donor driven and funded (Thomson 2015) and risk coming to an end when funding is no longer available.

12.6.1 Regional Initiatives and International Collaborations

With the growth in the use of ICTs linking institutions and learners around the world and the slowly increasing affordability of Internet access, there are many opportunities for regional and international institutions to enhance their ability to support both education and ICT for development in the Pacific. USP and the Commonwealth of Learning have initiated programs of considerable significance in the region.

Box 12.2: USP playing a leading role in ICT education in the Pacific
The University of the Pacific, which delivers courses to all 12 of its member countries, leads the way in the Pacific Island tertiary sector with USPNet connecting all its satellite campuses. Through flexible and online learning for distance education USP is the sole provider of tertiary education through online learning in most of the smaller island countries in the region. The University provides Preliminary and Foundation programs through distance learning that cater for the last year of high school and the preparatory year for degree studies in all member countries dof the university; for some countries, these are the only senior level programs available before university studies. In most of the smaller island countries it also offers degree level courses, some of which are offered only on satellite campuses (e.g. the Law Faculty in Port Vila). However, USP is the sole provider of online tertiary education.

Today around half of USP students are studying via distance education using applications such as audio and video conferencing and accessing Internet resources for some or all of their courses. Regionally, the Pacific e-learning observatory was established in 2006 at USP to facilitate monitoring of ICT education in 12 USP member countries. USP leads the way within the region in trialling ICT-based learning with the innovative use of SMS, iPads and tablets to support student learning (Vaa 2015). An upgrade of USPNet in 2010 allowed for interactive and synchronous audio and visual conferencing for the delivery of open and distance learning. This opened the way for further innovations, the use of OERs, MOOCs (Massive Open Online Courses) and smart classrooms. The Pacific e-learning Observatory was established in 2006 at USP to facilitate monitoring of ICT in education in all 12 USP member countries (Thomson 2015).

12.6.1.1 Commonwealth of Learning (COL)

Also playing a significant role in the promotion of ICT in education, is the Commonwealth of Learning which, in addition to its 'classroom without walls' initiative APTUS, spearheads initiatives in e-learning policy, training in online course development and creation of OERs. Projects spearheaded by COL in the Pacific include the VUSSC, FLOSS4Edu, and Learn4Content.

12.6.1.2 Virtual University of the Small States of the Commonwealth (VUSSC)

Of the Commonwealth's 53 member states, 31 are classified as small states in Asia, the Pacific, Africa and the Caribbean. These small states make up the membership of VUSSC (Chan Mow 2010). Content developed by VUSSC members can be shared under a Creative Commons license. Specifically, VUSSC countries have chosen to focus on the development of postsecondary, skills-related courses in areas such as tourism, entrepreneurship, use of information and communications technologies, life skills, teacher training, maritime and disaster management. The VUSSC course materials are non-proprietary and readily adaptable to the specific context of each country. They can be used in the offering of credit-bearing qualifications as well as strengthening educational capacity and access in member countries.

12.6.1.3 Free/Libre and Open Source Software for Education (FLOSS4Edu)

Another COL initiative is the promotion of open source software via the FLOSS4Edu, along with Open Office and Linux (Chan Mow 2010). FLOSS4Edu is a rapidly-growing Commonwealth initiative to tackle the dire shortage of free knowledge for education and to build capacity in the use of free software tools for educators. WikiEducator's FLOSS4Edu project strives to build a thriving community of teachers sharing knowledge to widen access to quality learning opportunities.

12.6.1.4 WikiPasifika

In relation to the FLOSS4Edu project, a workshop held by the Commonwealth of Learning in in Wellington in August 2007 saw the launch of WikiPasifika, the Pacific chapter of FLOSS4Edu. The workshop, entitled WikiPasifika, was intended to build capacity by focusing on the skills required to develop free content on WikiEducator of relevance to ICT educators in the Pacific. Participants were encouraged to share resources and strategies to enable them to return home and conduct local workshops that would introduce educators in their own country to the range of resources and projects on WikiEducator which are freely available for use in facilitating capacity building workshops in country. The workshop also hoped to build a strategy for growth and sustainability of the Pacific Chapter of the FLOSS4Edu project (Chan Mow 2010).

The Commonwealth of Learning also conducted a series of Learn4Content workshops in which teachers were given lessons in authoring tools such as WikiEducator in exchange for a free lesson that could be shared as part of OER.

12.6.1.5 Pacific Rural Internet Connectivity System (PacRICS)

The PacRICS initiative attempts to address some of the major telecommunication challenges facing rural and remote communities in the Pacific Islands region, including lack of access to communications technology, high costs of ICT and its access. In Kiribati, PacRICS has provided Internet access to several secondary and tertiary institutions (Vaa 2015).

12.7 Other Country-Specific ICT Initiatives

12.7.1 PFNet

In Solomon Islands, the People First Network (PFNet) delivered distance learning in 2000 using High Frequency (HF) radio for a basic email communication system. This was followed in 2004 by establishing distance learning centers in nine provinces, funded by the European Union and NZAID (Cullen and Hassall 2016a). Each center was provided with a short-wave radio, a VSAT, six computers, a printer, a scanner and solar panels. Like many such projects, however, it is now not as active as it was, suffering the same fate as many externally funded projects.

12.7.2 OUM Blended Learning

Another noteworthy e-learning initiative in Samoa is the Oceania University of Medicine's (OUM) blended learning system. Making use of problem-based learning and Moodle to deliver courses for onsite and distance students since 2002, OUM's integrated, state-of-the-art approach also uses Elluminate Live to build virtual classrooms in which students attend online lectures, study groups and tutorial sessions. After delivery, online lectures can be saved as podcasts, either as audio only or full video/audio format. The National Hospital of Samoa also serves as a teaching hospital, providing clinical observations and supervision by qualified doctors who contribute in face-to-face learning sessions.

12.7.3 SMS Story in Papua New Guinea

SMS Story is a research project funded by the Australian government under its Economic and Public Sector program (Kaleebu et al. 2013) which evaluated whether daily mobile phone text message stories and lesson plans would improve children's reading in Papua New Guinea elementary schools. The project was an attempt to improve English reading in primary schools where at the outset students had limited reading ability and limited access to English reading books. The controlled trial had half of the teachers receiving text messages and lesson plans to improve children's reading whilst the other half of the teachers did not. Findings of the study showed that on average, across both sets of schools, children's reading did improve over the two terms with children at SMS Story schools improving significantly more.

Box 12.3: SchoolNet Samoa: A success story in ICT in learning but with some challenges

The Samoa SchoolNet project was initially piloted in 2006–2007. The project aimed at piloting an appropriate model for introducing ICT in schools and also sought to extend this exposure to ICT to school communities to improve their knowledge and capacity to become productive members of their local communities (Chan Mow 2010). The project was implemented from March 2005 until December 2006. Activities in School-Net included:

(i) the provision of computer labs, Community Learning Centers (CLC), fully equipped with hardware and software (including pcs, servers, video camera, printers, fax, web camera, photocopier, Internet access);
(ii) creation of a School-Net portal for hosting learning objects and e-resources;
(iii) provision of a training program in basic ICT skills, use of ICT hardware for ICT administrators;
(iv) provision of training materials on basic applications, managing access through policies and regulations, business planning for managing the centers; and
(v) provision of one year's maintenance of SchoolNet equipment by a local vendor.

The pilot project was scaled up to include 37 schools and each school received ICT equipment (for the Community Learning Center) and an e-library, a repository of e-resources covering all grades and content areas for the following priority subjects: Science (Agricultural Science, Biology,

Chemistry and Physics), Mathematics and English. SchoolNet is implemented by the Ministry of Education, Sports and Culture (MESC) and funded with a grant from the Asian Development Bank (ADB). Considered as an example of a successful implementation of ICT in learning in the Pacific, the main challenge for SchoolNet is reluctance of the teachers to use the SchoolNet resources, probably due to lack of familiarity and confidence in the use of ICT. As Vaa (2015) reported, "Even when provided with a library of first-class interactive resources, teachers hardly know how to integrate them into their lessons". This has prompted MESC to plan more training and mentoring for teachers.

Sources: Chan Mow 2010, ADB 2013, Vaa 2015

12.7.4 e-Learning Project Tuvalu

The e-Learning Pilot Project that commenced in the last quarter of 2014 was scheduled for completion (including deployment, training and initial monitoring and evaluation) in 2016. The pilot included the compilation of an initial monitoring and evaluation report and a reflective proposal for national-level implementation of a suitable eLearning System. The e-Learning System mainly includes a server housing two virtual intranet web servers, NComputing labs as access devices and a wired and wireless network infrastructure to allow for connectivity within each school to all hosted resources. The two intranet web servers will host a digital library based on WorldPossible's RACHEL package and Moodle as a means of linking hosted OERs to specific learning outcomes. In combination with continuous capacity-building for teachers, increased efficiency and improved teaching and learning is expected by shifting from traditional textbook procurement to dissemination of OER with prescribed links to the curriculum. This change will also allow for the implementation of basic research activities for students in a learning environment that can enable independent learning to flourish.

The eLearning system features various redundancies to enhance resilience especially for the remote outer islands that at best have fortnightly boat visits. The web servers' host hardware is partially and/or wholly interchangeable with a primary backup server at the Education Department, and a secondary backup server that also provides for continuous research and development. As standard protection against power failures, uninterruptable power supply (UPS) units are paired with each server. The NComputing lab environment (in addition to its cost effectiveness) also includes redundant high-end servers that will automatically re-assign users connected to a failed server within 10 s of failure.

To ensure the success of the eLearning pilot project, its follow-up nation-wide implementation and other education ICT activities, it was determined that a dedicated staff position must be established to spearhead such activities. This position is especially critical in carrying out collaborative efforts with other education units as well as with other sectors. Collaboration with the Curriculum Development Unit (CDU) at the Ministry was expected to play a key role in prescribing existing OERs and developing contextualized OERs as the mainstream resources for teaching and learning in place of traditional textbooks. Collaboration with Sonoma State University (SSU) in the United States to develop teacher training resources is also an important initiative to enhance the teacher training process. In place of traditional and costly face-to-face teacher training, a change to a more flexible environment, where possible, will allow for the use of teacher training resources, such as video capacity building sessions for asynchronous and flexible training while minimizing or even eliminating travel, daily subsistence and other training costs.

12.8 Key Issues and Challenges in ICT in Education

Despite the early awareness and commitment to ICT in education in the Pacific Islands region (evident by its inclusion among key goals in both national and regional policy and planning frameworks), progress has been somewhat slow. The gap between planning and reality is due to multiple issues, challenges and factors hindering implementation.

Perhaps most inimical to realizing national ICT goals has been the lack of coordination between projects (PRIF 2015; Thomson 2015), in spite of their clear articulation in national and sector plans. In Papua New Guinea, for example, the ad hoc setting up of projects was seen to have "resulted in incompatible appliances and applications being installed, as well as different ICT approaches on the part of many donors" (Vaa 2003). In Fiji, extemporized initiatives to put computers into schools included the Telecommunications Authority of Fiji (TAF) and the Department of Communications tele-center program as well as the Share, Engage and Educate (SEE) project, outlined in Sect. 12.6 above, which has been implemented in many countries, including Solomon Islands, and installs second hand computers donated by the Queensland University of Technology. The lack of coordination and absence of a clear strategy for integration of ICT in education pervades all countries in the Pacific region.

A second major problem identified across all PICs is the lack of ability in ground-level implementers (mainly teachers, but also teacher educators) to use the technology to enhance education (AMFIE 2013). The root of this problem lies in the lack of thorough teacher training in new technology when it is being introduced for use in education. For example, a critique of the OLPC project is that it should have begun with one laptop per teacher. And, as mentioned above, most school computers in many PICs have been gifted (often as second hand equipment) or received from donors. In Tonga, most secondary and primary schools receive computers from benefactors, such as a local bakery owner who donated 100 PCs. Although these efforts in supporting schools and ICT education are commendable, they pose maintenance challenges due to the lack of standardization, given the variety of models and software versions (PRIF 2015).

The resource intensiveness and the dynamic nature of ICT leads to critical problems and challenges of sustainability and maintenance. Factors that inhibit the successful realization of the potential benefits of ICT in education, common to all SIDS, include lack of infrastructure; the high cost of technology; lack of finance for curriculum development; high turnover of ICT teachers; lack of ICT experts and teachers; lack of a planned curriculum; inability to integrate ICT into curriculum; lack of awareness and support for development and access to ICT (AMFIE 2013; ICT in Secondary Education in the Pacific Report 2005).There are additional problems (e.g. of size and geographical dispersion as well as lack of capacity) confronting the introduction of technology to educational administration, as the case of the Papua New Guinea EMIS shows (see box 12.1 above). Integration of systems institution-wide is a problem for most PICs—certainly in the case of the Solomon Island Educational Management Information system (SIEMIS). And all PICs face the challenges of rolling out systems to remote government offices and schools although, as noted above, these are exacerbated in Melanesian and Micronesian countries where Internet access for many schools is either non-existent or very expensive, making the use of ICTs for administration inefficient.

The shortage of competent maintenance service providers for all aspects of technology poses issues on maintenance service level agreements. The costs of equipment maintenance are also very high and add to the problem. Cisco academies at NUS and USP were set up to provide this much-needed training in technical hardware and networking skills. But there are still many gaps. Samoa's education ministry (MESC) lacks sufficient staff in the ICT unit to deliver adequate services to the whole Ministry.

All these factors affect sustainability. This was exemplified by some examples from Samoa, but it is affecting all PICs and some even more than Samoa. The need to revise and upgrade systems within limited ministry budgets poses a constant challenge (Chan Mow 2010). Sustainability is affected by many things—from funding coming to an end to lack of capacity and lack of affordable access to the Internet. It is also reliant on reliable infrastructure and adequate government ICT infrastructure, in particular. As the example of the Samoa education ministry shows, the government cannot always support the ICT initiatives that various ministries wish for. Furthermore, sustainability is impacted by the problem all PICs face in the demands of technology for upgrades and greater bandwidth. With limited budgets, most education ministries cannot keep upgrading to more modern technologies which demand Internet access, so until resources permit, there is a need to focus on low-cost sustainable technologies such as OERs, CD-ROMs and radio broadcasting. Some of these issues are alleviated by the fact that many PICs are setting up powerful government LANS and making major infrastructural investment (as is the case with the Samoa National Broadband Highway launched in 2014). Such initiatives will result in enhanced access and connectivity but, more importantly, are enabling precursors to more sustainable and achievable e-government initiatives. This should make any planned ICT initiatives in e-learning more achievable.

New technologies may also make the use of ICT more sustainable and affordable. Mobile technology is now readily available but still underutilized in the sector. Despite the pervasiveness of mobile phones in most countries in the Pacific region, very few projects have aimed at utilizing mobile phones in any of the three areas of ICT in education discussed in this chapter—administration, ICT in education and ICT education itself. Apart from the PNG SMS project noted above (Kaleebu et al. 2013) the most recent innovation using mobile technology is the APTUS device, whereby students connect offline via computers and their mobile phones. Their delight in extending their mobile phones from a device for talking, information and entertainment into a learning tool was noted in an initial trial of APTUS at the National University of Samoa.

However, in teaching and learning, it must be emphasized that new teaching approaches and methodologies go beyond computers, incorporating new communication technologies and multimedia for learning. The use of English, the language of computer communication and often a second or third language in most Pacific countries, is also pivotal as the language of instruction for tertiary education. As evident in various recent initiatives, the integration of ICT into general education affects curriculum development and delivery, pedagogical practices of teachers and learning needs of students. Yet, as identified in the FAIDP report, curriculum development is not accompanied by a systematic teacher-training program for in-service teachers. As in the rest of the Pacific, there is lack of human capacity in the ICT sector to deal with the challenges and opportunities that technology brings (FAIDP 2010).

Moreover, very few studies in PICs are available to identify the educational outcomes of ICT. A few have been conducted at the regional level by UNESCO and ITU. A few studies have been conducted in Fiji, Samoa and Papua New Guinea. However most of these are anecdotal and none employ rigorous methodology (PRIF 2015).

12.9 Major Initiatives and Future Directions

It must be noted that interest in using ICTs has recently increased significantly, such that most ministries of education and key donors in the region are now either planning or deploying various ICT initiatives to address some of the key issues. The major difference between these and previous efforts is that today's approaches are more strategically driven rather than purely technology driven and are more holistic in their approach. It seems that the (sometimes difficult) lessons learned from the past are being applied today (although not always perfectly). It is refreshing to see ICTs considered as tools to help solve education problems rather than being the reason for the initiative. But problems of sustainability (both financial and technical) are still not fully addressed. As mentioned earlier, the key issue of using ICTs to improve teaching and learning is being addressed fully in only one or two countries and should receive more wide-spread attention in the future.

12.9.1 Major Initiatives

Many believe that the region's education sector is not getting full benefits from the ICT infrastructure and regulatory reform that has occurred in recent years. This is reflected by the PIFS agreement to include ICTs as one of the five critical themes for development. At the time of writing, the details are being worked though; however, there is a strong belief that the many stakeholders in ICT need better coordination and focus in supporting economic and social development and better delivery of government education services.

It seems the best way to look at major initiatives is initially to take account of major global ICT-related developments that will affect the region. Three such developments show great promise for Pacific education:

- Development of Open Education Resources (OERs)—high-quality teaching and learning digital multimedia resources that are available free with copyright conditions that encourage free use and adaption
- ICT enabled distance and flexible learning—new models of training, professional and personal development
- Rapidly reducing cost of devices—e.g. tablets/phones and microcomputers that cost around US$50, that can access the Internet and provide learning "apps" and store hundreds of megabytes of educational resources

The region is now in the early stages of trialing these developments.

12.9.2 Future Directions

Given the rapid development of ICTs in education in nearly all developed countries and the associated transformation of education, the Pacific is coming from behind and our children are being disadvantaged by not developing the skills required in today's knowledge economy. The first future direction must be a stronger sense of urgency to make progress. The challenge seems to be a lack of in-country capacity to plan, implement and sustain holistic ICT initiatives.

The recent regional initiative of DFAT/Aus Aid and SPC to support the Pacific EMIS throughout the region perhaps exemplifies how to overcome the capacity limitation through focused regional initiatives. Such an approach that addresses the three previously identified major global initiatives by adapting and implementing them in the region could be of great assistance to many ministries of education in their decision making.

A good example would be the use of OER. The increasing availability of these resources promises to take Pacific schools from being education resource-poor to resource-rich affordably. This is particularly relevant in the trend away from teacher-centered to learner-centered education in the region. The driver behind the OER movement has been free access to resources for students and teachers to help bring about a change in the traditional superior/inferior relationship between teacher and student. The use of technology enables the teacher to facilitate student access to information rather than to be the sole font of knowledge. Student access to a range of high quality digital resources would assist in achieving this difficult change.

Establishing a regional facility for the development of Pacific OER packages for primary, secondary and TVET, similar to the SPC EMIS facility could be a similar initiative, saving costs in producing and distributing teaching and learning resources for all Pacific schools. Models of this way of working exist in other developing regions of the world and could be adapted for the Pacific.

The MOOC model of ICT-enabled collaborative learning shows great potential when a course can be delivered on a USB key to a head teacher or principal to run an "in school, collaborative professional development" program, complete with a rich set of training videos, reference material and exercises. Courses could include planning segments on how to implement the learning outcomes in the classroom.

Finally, OERs and the decreasing cost of ICT devices should lead some PICs to develop offline ICT solutions for schools that cannot yet afford to connect to the Internet. Such approaches could include highly portable and affordable packages of microcomputers and tablets (powered by solar), pre-loaded with appropriate educational resources that are accessed as if they were on the Internet. The packages could be delivered to poor schools and shared around classes, developing ICT and related new skills. Packages could include appropriate professional development training in the use of the package.

This is a key issue affecting the use of ICTs in education–the need to train teachers in these new approaches, particularly with in-service professional development. Distance and flexible learning approaches are being developed, enabled by ICTs, which can facilitate not only the understanding of their use in school but also other teachers' professional development issues. For example, in Samoa's Science Teachers Accelerated Programme (STAP Samoa) over 60 science, mathematics and IT teachers are in their final year of an upgrade that has used Distance and Flexible Learning and blended modes of learning enabling teachers to study "without leaving their classrooms" (Vaa 2015).

12.10 Summary of Common Themes, Reflections and Recommendations

In summary, this chapter shows that since early 2000 there has been growing awareness within PICs of the importance of ICT in education as reflected by its inclusion in national goals and policy frameworks. However, despite this awareness, progress has been slow and somewhat uncoordinated. Most ICT initiatives in education have been donor driven projects. USP is leading the way in online and distance learning; other recently-established universities in the region are expected to follow suit. Factors hindering progress include inadequate supporting infrastructure, the high cost of technology, lack of finances for curriculum development, high turnover of ICT teachers, lack of ICT experts and teachers, lack of a planned ICT curriculum, inability to integrate ICT into curriculum, lack of awareness and support for ICT development and access (PRIF 2015).

From an examination of ICT in education and from selected examples, some common themes, reflections and recommendations have emerged. The need for a planned, holistic, coordinated approach is crucial and is identified as the main underlying cause of failure for most failed ICT initiatives in all countries in the region. Innovative ways of using ICT to improve both access and quality of teaching and learning need to be examined. Examples include tapping into pervasiveness of mobile technology for use in learning, promotion of Open Educational Resources and co-sharing of expensive resources such as infrastructure. Sustainability should be a key consideration in the choice of technology suitable for Pacific Island countries, which are all small developing states. It is also important to consider sustainable older technology such as radio and CD-ROMs. Similarly, open source software needs to be assessed for its potential to offset licensing obligations. The importance of research needs re-emphasis (Thomson 2015) in the form of a regional research agenda and council such as the Pacific Island Universities Research Network (PIURN).

This chapter has demonstrated the valuable role ICT in education plays in the context of supporting and facilitating e-government in the Pacific Region. However, despite awareness of the transformational nature of ICT and the commitment of Pacific Island countries to ICT in education, various challenges have hindered the realization of its effective use, which has a follow-on effect on the processes of e-government in the Pacific region. Initiatives and recommendations for the way forward lie primarily in the training of teachers in ICT and the availability of appropriate resources. Strategic and appropriate use of ICT in education provides expanded access to quality education empowering educated citizens to influence and shape the course of national development. The key to good governance lies in innovative and robust education systems which, through the use of ICT, enable access to quality learning at low cost. Quality education enables small states to participate in a globalized economy while strengthening their local economies, cultures and sense of community. Small PICs will attract investment in their economies and their own people will be able to direct the course of their development on their own terms if well informed, well connected and well educated.

References

Asian Development Bank ADB (2013) Pacific ICT Investment Planning and Capacity Development Facility. Project Number: 47114. Regional–Capacity Development Technical Assistance (R–CDTA) December 2013. Manila: ADB. Available at: https://www.adb.org/sites/default/files/project-document/79686/47114-001-tar.pdf Accessed 12 Oct 2016

Asia-Pacific Ministerial Forum on ICT in Education (AMFIE) (2013) Outcome Document UNESCO Bangkok. Available at: http://www.unescobkk.org/fileadmin/user_upload/ict/Workshops/amfie2013/report/AMFIE_2013_Outcome_Document.pdf. Accessed 15 June 2015

Baba TL (1985) Education in the Pacific Islands. In: Kiste RC, Herr RA (eds). The Pacific islands in the year 2000. Working Paper series [1985]. Honolulu, Hawaii: Pacific Islands Studies Program, Center for Asian and Pacific Studies, University of Hawaii at Manoa, pp 125–150

Balasubramanian K (2012) Bridging the Technology gap: Mobile phones for Education and Learning in presentation Slides from the Ministerial Roundtable, 18th Conference of Commonwealth Education Ministers (18CCEM), Mauritius, 30 Aug 2012

Chan Mow IT (2010) Technology Enabled Education in Samoa: Issues, Challenges, Strategies and Recommendations. In: Mukerji S, Tripathi P (eds). Cases on interactive technology environments and transnational collaboration: concerns and perspectives. New York: IGI Global Hershey, pp 193–210

Cullen R, Hassall G (2016a) e-Government in the Pacific island states: ICT policy and implementation in small island developing states. Solomon Islands Country Report, October 2016. Available at: http://pippr.victoria.ac.nz/handle/123456789/35. Accessed 10 Aug 2016

Cullen R, Hassall G (2016b) e-Government in the Pacific island states: ICT policy and implementation in small island developing states. Vanuatu Country Report, 2016. Available at: http://pippr.victoria.ac.nz/handle/123456789/30. Accessed 10 Aug 2016

Daniel J (2009) The Virtual University for the Small States of the Commonwealth. What is your vision for the future? In: Open and distance learning in a changing world. Selected speeches of Sir John Daniel and Colleagues 2007–8. Available in http://hdl.handle.net/11599/131

Framework for Action on ICT for Development in the Pacific (FAIDP) (2010) SPC, Noumea. Available at: http://www.spc.int/edd/images/stories/ictpapers/Final%20Pacific%20ICT%20Framework.pdf. Accessed 12 October 2016

ICT in secondary education in the Pacific region: status trends and prospects (2005) edited by ICT Capacity Building at USP Project. University of the South Pacific, Suva. Available at: http://www.usp.ac.fj/jica/ict_research/documents/pdf_files/ICT%20in%20Secondary%20Education%20in%20the%20Pacific%20Region.pdf. Accessed 12 October 2016

ITU (2015) Concept note, Pacific ICT ministerial and officials' meetings 2015, Nukualofa, Tonga. Available at: https://www.itu.int/en/ITU-D/Regional-Presence/AsiaPacific/Documents/Events/2015/June-Pacific-Ministerial-Meeting/DOC-01-Concept Note_Pacific_ICT_Ministerial_Meeting_v6.2.pdf. Accessed 12 Oct 2016

Kanwar A (2015) Virtual university for small states of the commonwealth: its journey, evolution and future. Presented at the ministerial pre-conference meeting, Nassau, The Bahamas, June 2015. Burnaby, BC: Commonwealth of Learning. Available at: http://oasis.col.org/handle/11599/883. Accessed 12 Oct 2016

Kanwar A, Aguti J, Balasubramanian K (2014) How ICT can support teacher training. Burnaby, BC: Commonwealth of Learning. Available at: http://hdl.handle.net/11599/757. Accessed 31 Jan 2016

Kaleebu N, Gee A, Maybanks N, Jones R, Jauk M, Watson A (2013) SMS story: early results of an innovative education trial. Contemporary PNG Studies: DWU Research Journal 19 November, pp 50–62. http://www.dwu.ac.pg/en/images/Research_Journal/2013_Vol_19/5_Kaleebu_Gee_Maybanks_Jones_Jauk__Warson_SMS_Story.pdf. Accessed 12 Oct 2016

Koshy K, Mataki M (2005) Sustainable development and Pacific island countries. Suva: Pacific centre for environment and sustainable development. The University of the South Pacific

Pacific Islands Forum Secretariat (PIFS) (2009) Pacific education development framework 2009–2015. http://www.forumsec.org/resources/uploads/attachments/documents/Pacific%20Education%20Development%20Framework%202009-2015.pdf. Accessed 23 Jul 2015

Pacific Islands Forum Secretariat (PIFS) (2010) Review of the Pacific regional digital strategy (2010). Final report for the Pacific Islands Forum Secretariat (PIFS). http://www.spc.int/edd/images/stories/ictpapers/Digital%20Strategy%20Review-%20Final.Pdf. Accessed 20 Aug 2015

Pacific Islands Forum Secretariat (PIFS) (2012) Regional framework for ICT in education in the Pacific. Presented at the Forum Education Ministers Meeting Port Vila, Vanuatu, May 2012

Pacific Region Infrastructure Facility (PRIF) (2015) Economic and social impact of ICT in the Pacific. Available at: http://www.theprif.org/components/com_jomcomdev/files/2015/10/40/124-PRIF%20Pacific%20ICT%20Report%202015.pdf. Accessed 20 Aug 2015

Papua New Guinea. Department of Education (2005) Achieving a better future. A national education plan 2005–2014. Port Moresby: Department of Education. Available at: http://www.education.gov.pg/quicklinks/documents/edu-plans/national-education-plan-2005-2014.pdf. Accessed 10 Oct 2016

Papua New Guinea. Department of Education (2008) Computer studies. Upper secondary syllabus. Port Moresby: Department of Education, Papua New Guinea. Available at: http://www.educationpng.gov.pg/Teachers/secondary/upper/syllabus-upper-secondary-computer-studies.pdf. Accessed 10 Oct 2016

Papua New Guinea. Department of Education (2010) The national education media policy for Papua New Guinea. Port Moresby: Department of Education, Papua New Guinea. Available at: http://nemc.education.gov.pg/Documents/NEMP_final.pdf. Accessed 10 Oct 2016

Samoa. Ministry of Education, Sport and Culture (2012) Corporate plan July 2012–June 2015. Apia: Ministry of Education, Sport and Culture. http://www.mesc.gov.ws/images/downloads/CP_2012_2015_Eng.pdf. Accessed 15 Aug 2015

Samoa Pathway (2014) SIDS accelerated modalities of action outcome statement. https://sustainabledevelopment.un.org/samoapathway.html. Accessed 23 Jul 2015

Secretariat of the Pacific Community (SPC) (2010) Framework for action on ICT for Development in the Pacific. http://www.spc.int/edd/images/stories/ictpapers/Final%20Pacific%20ICT%20Framework.pdf. Accessed 20 Jun 2015

Strigel C, Chan Mow IT, Vaa R (2007) Teachers and technology in education in Samoa. In: UNESCO ICT in teacher education: case studies from the Asia Pacific Region, pp 116–130

Thomson I (2015) Overview of ICTs in Pacific education. Presentation to Pacific ICT officials, 17 Jun 2015

UNESCO (2015) Education for all review 2015 report: Papua New Guinea. Papers prepared for the World Education Forum, Incheon, Korea, May 2015. http://unesdoc.unesco.org/images/0023/002316/231679e.pdf

Vaa, R. (2003) Samoa: ICT use in education. In: Farrell G, Wachholz C (eds). Meta-survey on the use of technologies in Education in Asia and the Pacific. Bangkok: UNESCO

Vaa R (2015) A baseline study on technology-enabled learning in the Commonwealth Pacific Island countries: Report. Burnaby, BC: Commonwealth of Learning. http://hdl.handle.net/11599/1738

Whelan R (2008) Use of ICT in education in the South Pacific: findings of the Pacific eLearning Observatory. Distance Educ 29(1):53

World Bank (2015) e-Gov research and resources. http://www.worldbank.org/en/topic/ict/brief/e-gov-resources. Accessed 10 Dec 2015

Chapter 13
E-Government and Civil Society: Exploring New Relationships in Pacific SIDs

Graham Hassall

Abstract This chapter examines the extent to which Pacific Island governments are using ICTs to engage with civil society, through the initiation of service provision, or some form of policy dialogue. It finds that whereas moderate progress has been made with the establishment of key websites, in only a few instances have these been designed as "portals" through which the public can link to any government department or agency, or through which these departments can communicate with citizens and civil society organizations online. There has been, on the other hand, considerable growth in government activity on social media sites, which are easy to initiate and manage. More active than government-initiated sites are the social media sites established by individuals, civil society groups, think tanks and academia, for the express purpose of commenting on government policies and activities. Communicative practices between state/civil society/citizenry are still being negotiated, particularly in states where government has traditionally been the voice of authority and government information has been protected by Official Information Acts. Government and civil society alike have to work more collaboratively to maximise citizen engagement in public governance. Whilst such situations do not fit neatly into the 'transformative' models of e-Public Services and e-Democracy articulated by the World Bank and the UN, they do constitute a beginning to online interaction between citizen and state.

13.1 E-Government and Civil Society: The Pacific Context

Pacific people and communities have traditionally identified with villages, clans and religious organizations, and many now also identify with the "modern" spaces in which contemporary civil society is formed. In the Pacific context, the term *civil society* includes not only non-profit community focused entities like CSOs (Civil

G. Hassall (✉)
School of Government, Victoria University of Wellington,
Wellington, New Zealand
e-mail: Graham.Hassall@vuw.ac.nz

R. Cullen and G. Hassall (eds.), *Achieving Sustainable E-Government in Pacific Island States*, Public Administration and Information Technology 27,
DOI 10.1007/978-3-319-50972-3_13

Society Organizations, also known as NGOs—Non-Governmental Organizations), social and community service organizations, charitable groupings, churches, faith-based organizations and sporting bodies, but the more traditional community-family clan groupings and village-based voluntary formal and informal groups and associations.

The NGO is a relatively new entity in Pacific SIDs, whose communities have traditionally organized themselves and their collective interests in other ways. CSO methods of advocacy are often more direct and adversarial, as CSOs contend that this new approach is necessitated by the manner in which modern government imposes governance systems that threaten and compete with traditional governance by self-governing communities. Governments are accountable to citizens for their services and there is need for activism by citizens. Civil society organizations across the Pacific advocate for good governance, transparency and rule of law. They monitor usage and exploitation of the environment and natural resources, and human rights abuses, including violence against women and view their methods as promoting good governance and holding leaders to account for the types of public policies developed, their use of public funds and legal powers. Some Pacific Island governments (particularly Papua New Guinea and Fiji) have begun to require CSOs to not only register with government, but to seek permission to hold public events. Pacific civil society activists took note of the role that social media played in the "Arab spring" which resulted in the removal of several undemocratic regimes; government leaders also observed that phenomenon and are all the more wary of activist voices. Nevertheless, social media has become a popular medium of advocacy for Pacific civil society and use of democratic space.

13.2 Civil Society, "Communication for Development" and E-Participation

Communication and participation are viewed by civil society as key pillars of development. In the early stages of "communication for development" (C4D), communication between institutions and their stakeholders, including for CSOs, was almost always one way, providing information from government to the public or from an organization to its audience. The 2006 World Congress on Communication for Development defined C4D as "…a social process based on dialogue using a broad range of tools and methods. It is also about seeking change at different levels including listening, building trust, sharing knowledge and skills, building policies, debating and learning for sustained and meaningful change" (The Communication Initiative, Food and Agriculture Organization of the United Nations (FAO) and The World Bank 2007, p. xxxii). C4D gradually broadened to incorporate interpersonal communication, i.e. face-to-face communication that can either be one-on-one or in small groups. This came alongside the general push for more participatory approaches to development and greater representation of voices from the Global South.[1]

While mass media allows for the learning of new ideas, interpersonal networks encourage the shift from knowledge to continued practice. Communication for development, especially for CSOs, has thus come to be seen as a way to amplify voice, facilitate meaningful participation and foster social change. ICTs have now advanced to a stage where communications can easily facilitate such meaningful participation, whether the transactions be public hearings, debates, deliberations and stakeholder consultations, participatory radio and video, or web forums.[2] McIver has defined such "community informatics" as "…an emerging, interdisciplinary field concerned with the development, deployment and management of information systems designed with and by communities to solve their own problems." (McIver 2003, p. 33).

At global scale, development thinking now recognizes the critical role civil society plays in demanding good governance and inclusive development. This includes defending human rights and ensuring that citizens enjoy the rights that are inherently theirs that governments have the duty to deliver. The development models espoused by the United Nations and the World Bank, discussed below, contribute to a wider movement toward deliberative democracy, which, as Gimmler points out, has always been the reason for establishing a "public sphere" (Gimmler 2001). The Busan Partnership for Effective Development Cooperation[3] gives global recognition and legitimacy to the role of CSOs. Paragraph 22 of the Busan Partnership Agreement states:

> CSOs play a vital role in enabling people to claim their rights, in promoting rights based approaches, in shaping development policies and partnerships, and in overseeing their implementation. They also provide services in areas that are complementary to those provided by states. Recognising this, we will (a) Implement fully our respective commitments to enable CSOs to exercise their roles as independent development actors, with a particular focus on an enabling environment, consistent with agreed international rights, that maximizes the contributions of CSOs to development. (b) Encourage CSOs to implement practices that strengthen their accountability and their contribution to development effectiveness, guided by the Istanbul Principles and the International Framework for CSO Development Effectiveness. (4th High Level Forum on Aid Effectiveness 2011)

International development agencies now see e-Democracy and e-Participation as essential components of a successful e-Government rollout. The UN's concept of e-Government, for instance, includes "eDemocracy", measured by an e-Participation index that combines evidence on a country-by-country basis for e-Information, e-Consultation and e-Decision-making.[4] Of course, understanding what constitutes

---

[1]The Global South refers to least developed and developed countries, which are for the most part former colonies, and very many of which are in the Southern hemisphere.

[2]http://www.worldbank.org/en/news/press-release/2012/07/25/world-bank-group-announces-new-focus-on-using-ict-for-greater-development-impact.

[3]www.oecd.org/development/effectiveness.

[4]The model by Lee (2014, p. 16) is discussed in chapter one.

these requires some consideration, as these are difficult enough to implement without also being mediated through ICTs. Enabling citizens to complete transactions online certainly increases service provision, but has fewer implications for government than enabling citizens to offer their views on issues, or to contribute to decision-making. The work in this field by UNDP has contributed new understandings not only of the term "governance", but of governance when adapted for electronic processes:

> E-governance is a multifaceted concept that refers to the use of ICTs for improving collective governance. This definition covers a wide range of activities that can be grouped in four clusters of ICT-enabled reform objectives: I. Facilitating access to political information and improving the means for political expression, discourse, mobilization and advocacy; II. Enhancing the democratic quality of the political process and public administrations through more transparency, accountability, participation and disintermediation; III. Making the internal working of public administrations more effective and efficient; and, IV. Enhancing the range and utility of public services on offer and making their delivery to citizens and business more accessible, efficient and responsive to the needs of all groups of clients. (UNDP 2004)

Other theoretic contributions have come from academia and from practitioners. Waseda University's e-Government ranking indicators include the criterion "EParticipation/Digital inclusion", with sub indicators being E-information mechanisms, consultation and decision-making. (Obi 2015, p. 4) The communications spectrum of the International Association for Public Participation has been adapted to an "e-spectrum" (Macintosh and Tambouris 2009). There is, too, as Aur has pointed out, a relationship between social media and phases in the public policy cycle, from information gathering through to "decision to implement", implementation and evaluation—stages reflective of the classic analysis by Harold Laswell (Auer 2011).

These concepts and others, when actualized, help create the "voice and accountability" in democracies and indicate the extent to which a country's citizens are able to participate in selecting their government and in expressing their views freely, including through the Internet. In 2012 the World Bank announced a new focus on the development of ICT capability, reliant on the principles themes: *Transformation, Connectivity, and Innovation*:

- *Transformation*: Making development more open and accountable and improving service delivery, for instance by facilitating citizen feedback to governments and service providers;
- *Connectivity*: Scaling up affordable access to broadband, including for women, disabled citizens, disadvantaged communities and people living in remote and rural areas;
- *Innovation*: Developing competitive IT-based service industries and fostering ICT innovation across the economy with a focus on job creation, especially for women and youth.[5]

[5]http://www.worldbank.org/en/news/press-release/2012/07/25/world-bank-group-announces-new-focus-on-using-ict-for-greater-development-impact.

While these three processes may proceed simultaneously, the priority of Pacific Island governments—as noted in earlier chapters—has been connectivity. The implications of transformation in Internet-mediated government-citizen interactions are yet to be worked through and realized. This chapter therefore considers these interactions in two sections. The first examines the expansion of e-Information, which includes the ability to complete transactions online, and the second examines e-Democracy, which introduces "citizen voice" on matters of government policy and action.

In broad terms, Pacific Islands civil society organizations have participated in policy dialogue about the Internet, its governance and use, at global, regional and national levels. The Pacific Islands Association of Non-Governmental Organisations (PIANGO) and the Pacific Islands News Association (PINA), for instance, contributed reports to the preparatory phases of WSIS, participated in a survey conducted by UNESCO (Zwimpfer Communications 2002) for that event and attended the Tokyo Asia Pacific conference (Guild and Taufao 2003). The Pacific Islands Forum, too, has acknowledged the role of ICTs in engaging civil society voices in policy dialogue. In 2002, for example, it stated:

> Leadership from governments and partnerships with businesses, non-government organisations (NGOs), religious groups and the community at large are required to facilitate participation in the knowledge society and to make their countries part of the global knowledge economy. All need to work closely to develop a connected population and to use information and communications technology to maximise the potential of the region and its people. Governments will also need to provide special attention to protect the social, cultural and ethnic diversity of the Pacific Community. By committing to individual and collective action partners will increase the use of ICT to benefit the people of the Pacific Community. (Pacific Islands Forum Secretariat 2002, p. 1)

The 2002 Pacific Islands Information and Communication Technologies Policy and Strategic Plan envisaged working with community groups, NGOs and the media, to promote the goal of promoting awareness of ICTs. It envisaged holding an annual event to celebrate and highlight the significance of ICT such as ITU's World Telecommunications Day, and even declaring a Year of Pacific ICT (Pacific Islands Forum Secretariat 2002).

In 2011 SPC hosted the Pacific Internet Governance Forum (Secretariat of the Pacific Community 2011). However, such high-level discussions do not equate with the expansion of online government-citizen dialogue within countries and whereas research for this chapter found a considerable amount of CSO and citizen voice *about* government, it found very little CSO engagement *with* government. There has been virtually no CSO input for instance—whether online or offline—about individual countries' ICT and e-Government policies and plans and few civil society organizations (apart from IT professionals) have contributed ideas to the formulation of national ICT plans.

13.3 E-Public Service

Although models of e-Government refer to the goals of "Government to Government" (G2G), Government to Business" (G2B) and "Government to Citizens" (G2C), Pacific Island governments have to date prioritized joining up government departments (G2G) and assisting business processes (G2B) over transforming engagement with citizens. But the sparse physical presence of government services in many PICs (i.e., most offices are in the national capital, with a smaller number in some provincial centers) means that the smallest of innovations to put government information and services online makes a substantial difference to citizens. Rural dwellers, in particular, have faced difficulties when seeking information, registering births, deaths and marriages, checking land ownership records, applying for passports, seeking permits, claiming welfare vouchers, or correcting administrative errors, etc. As expressed by Sao:

> The provision of government information to citizens online has considerable impact on the lives of Pacific Islanders. Prior to such access, the smallest query, or the need to lodge any type of application, necessitated travel to the nearest government centre. (Sao 2012)

13.3.1 Government Central Websites

Approximately half of all PIC governments have established websites to act as central hubs for government ministries, departments, agencies and services. They include Fiji, Kiribati, FSM, Nauru, Niue, Tonga and Vanuatu. In the case of Solomon Islands this function is taken up by the Ministry of Commerce, Industry, Labour and Migration. In other cases, a web presence is coordinated through the office of the Prime Minister, as in Papua New Guinea, or even through the Public Service Commission, as in Cook Islands. Government entities that are constitutionally separate to government, such as parliaments, leadership commissions, ombudsman, etc., often have a separate website.

Fiji's website (at http://www.fiji.gov.fj/), one of the more comprehensive and largest (with links to almost sixty ministries, departments and other organizations), has not kept pace with government restructures, nor with link verification. In Papua New Guinea, the Department of Communication and Information "launched" the site www.gopng.gov.pg in 2015, but it is not yet operational in 2016.[6] Kiribati's Ministry of Finance and Economic Development site http://www.mfed.gov.ki includes links to sixteen other government websites, including Ministries, Parliament, Public Service and Audit Office; and a Kiribati Development hub that includes key policy and program documents on everything from aid effectiveness, climate change and governance to health, poverty, private sector, tourism, food

[6]http://www.emtv.com.pg/news/2015/08/government-portal-launched/.

nutrition and energy. It is also possible to manage taxes online https://etax.tax.gov.ki/.

Palau's government site http://palaugov.pw/, updated in 2016, provides links to all branches of government, to the ministries and to the individual states, as well as to the offices of statistics, tax and passports. The latest government documents are made available for download.

Samoa's site http://www.samoagovt.ws has comprehensive links to government, as well as to government communications channels using RSS, Twitter, Facebook and YouTube. Tonga's government site http://www.tongaportal.gov.to/, launched in 2016 by the Secretary General of the International Telecommunication Union (ITU), Mr. Houlin Zhao, is possibly the best conceived of all PIC government portals at the current time. It provides links to all government ministries, agencies and public enterprises, and has dedicated sections for government, citizens, business and investors and visitors. It also has links on Twitter and Facebook.

The existence of a main website demonstrates a government's capacity for coherent communications across its ministries, departments and dependent agencies. Without such, government websites have an orphan quality that risks divergence in corporate identity, in key government messaging and with data accuracy. Without connection to a central hub they may be more difficult for citizens and other stakeholders to find and their effectiveness is reduced.

Although each of the current PIC government websites includes (some to greater extent than others) links to legislation, policy, government structure, vacancies and even tender opportunities, very few actively seek public feedback or active engagement on proposed plans, policies, laws, or other government actions. Notwithstanding such current limitations these websites are an essential step toward enhanced citizen engagement.

Box 13.1: Pacific Islands Government Main Websites

Cook Islands—http://portal.cook-islands.gov.ck/
Fiji—http://www.fiji.gov.fj/
Kiribati—http://www.mfed.gov.ki/
Marshall Islands—none
Federated States of Micronesia—http://www.fsmgov.org/
Nauru—http://www.naurugov.nr/
Niue—www.gov.nu
Palau—http://palaugov.pw/
Papua New Guinea—none
Samoa—http://www.samoagovt.ws/
Solomon Islands—http://www.commerce.gov.sb/
Tonga—http://www.tongaportal.gov.to/
Tuvalu—http://www.tuvaluislands.com/gov
Vanuatu—http://governmentofvanuatu.gov.vu/

13.3.2 Government and Social Media

Pacific government departments have more active interaction with the public via social media than via websites. Social media has been defined as "… a group of Internet-based applications that build on the ideological and technological foundations of Web 2.0, and that allow the creation and exchange of user-generated content" (Kaplan and Haenlein 2010, p. 61). Tang and Liu have articulated no less than six social media formats—blogs (e.g. Blogger, Wordpress), forums (e.g. Yahoo! Answers), media sharing (e.g. YouTube), microblogs (e.g. Twitter) and social networks (e.g. Facebook)—while also acknowledging that such platforms are dynamically changing and evolving (Tang and Li 2010, p. 1). Facebook, for example, has a variety of genres, including *pages*, which are open for all to see and interact with, and *groups*, which can either be open, or closed (discussion viewable by members only), or secret (not advertised). Indeed, it could be the case that websites, rather than social media sites, are more suitable for government publication because they use permanent urls and because interaction between government and citizens take place on sites designed for such inter-activity.

13.3.2.1 Freedom of Speech

From a governmental perspective, social media is the most fluid and volatile means communication, while for civil society it has become the most popular mode for freedom of expression. Citizens have taken the lead in initiating ad hoc and unofficial "e-Information" and "e-Participation" which has, through such public queries, complaints, assertions and challenges, provoked responses from government officials and political leaders. Content can be disseminated globally in an instant. Governments have expressed concern about the ability of social media users to hide their real identity when posting material or making allegations. Individuals can post a document, image, or comment online, with little regard for its veracity, accuracy, authorship, authenticity, or legitimacy. From individuals' perspectives, on the other hand, such instantaneous reach frees them from the burden of establishing and maintaining their own website, or cultivating their own readership and, for many, from being victimized. Where government information is poorly disseminated, social media facilitates access to valuable information about the activities of parliaments, parliamentarians, government agencies and officials.

Government response to the emergence of social media varies not only across the PICs, but within countries. On the one hand, there are threats to freedom of expression coming from leaders (usually members of parliament but sometimes also from government and the public service), while there are also genuine efforts by government departments within those same countries to reach particular audiences via social media.

Vanuatu provides a case in point. Most government departments and much of the private sector cooperate in celebrating World Telecommunications Day with a well-planned program of education about the Internet, yet freedom of speech remains under threat from potential law reforms and from public service directives (discussed below).

In Papua New Guinea Transparency International has expressed concern about the potential impact of the Social Media Bill and Cyber-Crime Act[7]—the latter threatening free speech at the same time that most government agencies are increasing their engagement with the public via social media. The mass media support the new laws to ensure that online bloggers have the same responsibilities before the law as do traditional reporters (Anon 2014). As one Minister expressed the point of view on the floor of Parliament:

> Mr. Speaker, today some of us are facing allegations of corruption, and because we are holding public office, we are liable and subjected for public scrutiny and that comes with the territory and the jurisdiction of public office holder … On many occasions, there are many people who hide behind pen names and continue to rattle on about how corrupt this country is. I ask the Minister to look into how the ICT Sector can be regulated so that whilst we allow freedom of speech, we also take responsibility in what is being expressed. There must be names and addresses included so that we allow for responsible writing. (Marape 2013)

In Nauru, the government blocked access to Facebook in 2015, arguing that it sought to block pornography, but this was not considered to be a strong argument since the site already has a policy guarding against this; the reason was more likely an attempt to stifle discussion of the country's political situation (Radio New Zealand International 2015).

In Fiji, which was under military control between 2006 and 2014 and where the military commander of that period is now the parliamentary Prime Minister, the social media's numerous sites (both government and non-government) invariably express entrenched and divergent views on issues of democracy, race, the constitution and religion. When the military ruled by decree and the right to speech was most severely curtailed, bloggers established anonymous sites on which to voice their opposition and the regime responded with corresponding efforts to unmask them. With the promulgation of a new constitution in 2013 and an ostensible return to democracy, online discussion of government in Fiji has become even more expansive. The government's view is that the freedom of speech enshrined in the constitution comes with responsibilities (Bainimarama 2016). Although official censorship ended in 2012 with the lifting of the 2009 Public Emergency Regulations, a Media Industry Development Authority was established, with the power to impose penalties on work deemed to be against the "public interest or public order". While it has expressed concern about social media's unregulated content, the authority understands that blocking access would compromise the democratic values that the post-coup government claims to uphold (Tarai 2015).

Uneven policy approaches to the phenomena of social media suggests that these countries are still negotiating the boundaries between control and transparency. The

---

[7]http://www.pireport.org/articles/2016/08/14/concerns-raised-about-new-png-cyber-crime-laws.

conservative view has to be appreciated: social media provides individuals with the ability to comment on government business with little accountability for the accuracy or comprehensiveness of the information presented, without verification of their true identity, across fluid communications networks that flow far beyond the boundaries of state control. But so, too, should the merits of social media be considered by government: the immediacy of delivery/receipt of a message, the ease with which the ordinary citizen can participate, the equalization of voice that the Internet gives to young as well as old and irrespective of gender, and the advantage of anonymity when required (e.g. when "whistleblowing", which government leaders should appreciate). Social media interactions can also provide rapid appraisal of the strength of public concern, for example, by noting the number of comments contributed and the number of "views", "likes", "shares" and "re-tweets". When the legitimate concerns of government are addressed and the public attains a certain command of "digital literacy", social media can more adequately contribute to the e-Public Service and e-Democracy that governments have pledged to deliver. These issues and others have been explored in a useful policy primer on government use of social media published by the OCED (Mickoleit 2014). In the Fijian context, the European Union in 2016 funded a program "Valuing Voices: Social Media Masterclass", a "3-day masterclass on digital strategy" that aims to "...contribute to democratic process by enabling greater breadth of voice in mainstream media and social media through creative arts."[8]

13.3.2.2 Facebook

Government departments from virtually all PICs have initiated pages on social media sites—on Facebook in particular. At least five ministries in the Cook Islands have established Facebook pages. The Department of Agriculture established its page in September 2014 and in April 2015 its staff received Web 2.0 training from the Netherlands-based Technical Centre for Agricultural and Rural Cooperation in collaboration with SPC.[9] The Ministry of Health page, which also commenced in September 2014, promotes health awareness campaigns (e.g. concerning AIDS, obesity, cervical cancer and smoking), distributes health surveys, promotes healthy lifestyle information, advertises upcoming conferences and seminars and covers the activities of the Minister and health sector leaders. In March 2015, the Ministry posted two draft policies on its website[10] and used its Facebook page to invite public comment.

[8]https://www.facebook.com/fijimediawatch/posts, accessed 29 September 2016.

[9]http://www.cta.int/en/article/2015-04-16/building-capacity-in-using-social-media-for-agricultural-development-in-cook-islands.html.

[10]Ministry of Health Patient Referral Policy 2015, and Ministry of Health National Health Service fees schedule 2015.

The Cook Islands' Ministry of Foreign Affairs and Immigration page was created in January 2015, motivated in part to promote awareness of activities celebrating 50 years of self-government. The Page's first message read:

> Kia Orana to you all and welcome …Through this page you will be introduced to the mechanics of our Ministry and the work we do including details about the Cook Islands diplomatic relationships, the Treaties and Conventions the Cook Islands have signed up to, Cooperation Programs across our various partners including trade initiatives, the Cook Islands involvement and interest in multilateral groupings as well as how we as a Ministry operate to advance our interests on the global stage. We will update you on the Current Issues being addressed as well as an insight into upcoming Meetings and Engagements the Cook Islands have with our various partners and groupings around the world as well as in country. We will also provide you updates of developments in our immigration services as we strive to improve quality and deliver through strengthened systems and wider policy. We hope the content of this page will encourage robust discussions within the community as well as ownership to our countries future aspirations.[11]

The page presents the history of Cook Islands foreign relations, provides updates on diplomatic visits and staff movements, advertises scholarship opportunities and records participation in UN events—including UNGA, UNESCO, ILO, IMO, IPCC and WHO. In August 2015, the page announced the final draft of the Cook Islands National Trade Policy Framework and invited public feedback before finalizing its submission for cabinet. Several other Cook Island Ministries on Facebook, including Education, Infrastructure and Planning, Internal Affairs and Marine Resources similarly post draft legislation and report on events and staff movements. The "Climate Change Cook Islands" site was created by the Office of the Prime Minister to promote the Unit's mission to "… strengthen the ability of all Cook Island communities, and the public service, to make informed decisions and manage anticipated climate change driven pressures in a community based, proactive, integrated and strategic manner."[12]

Ministries and departments of the Government of Fiji are extensively represented on Facebook. Ministries with pages include Foreign Affairs, Youth and Sports; Agriculture; Education, Health and Medical Services; Infrastructure and Transport; Fisheries and Forests; Rural and Maritime Development and National Disaster Management; Public Service Commission and iTaukei Affairs.[13] The page "The Fiji National Development Plan 2015" was created in June 2015 to advertise consultations around the country as part of formulating a 20-year development plan.[14] The Fiji Parliament's page provides links to parliamentary schedules and documents as these are posted to the parliament's website, plus photos of parliamentary sessions and other events. The Public Service Commission page

---

[11] https://www.facebook.com/mfaicookislands/?ref=br_rs, accessed 28 September 2016.

[12] https://www.facebook.com/sriccookislands/about/?entry_point=page_nav_about_item&tab=page_info, accessed 28 September 2016.

[13] Indigenous Fijian Affairs.

[14] https://www.facebook.com/Fiji-National-Development-Plan-2015-1584845028434612/.

established in 2014 notifies of positions vacant, scholarships, staff movements, events, driving readers to the PSC's website.[15]

Smaller PICs such as Kiribati also have a presence on Facebook. Official pages include Kiribati and Climate Change, Kiribati Environment and Conservation Division, Kiribati Embassy in Taiwan, Kiribati Meteorological Division, Kiribati Housing Corporation and the Communications Commission of Kiribati.

Although (as noted above) the Papua New Guinea government currently lacks a web portal, individual ministries, departments and other agencies have developed Facebook pages that assist in driving traffic to their websites. These include the Ministry of Tourism, Arts and Culture PNG, the Coastal Fisheries Development Agency of Papua New Guinea, the Internal Revenue Commission, National Capital District, National Parliament (and also some of its committees), National Research Institute, National Statistical Office, the PNG Science & Technology Council and the Departments of National Planning and Monitoring, Labour & Industrial Relations, Provincial and Local Government Affairs, and Public Enterprise & State Investment. The last of these used its Facebook page in 2015 to solicit views on the country's draft *National Energy Policy 2016–2020*. In October 2015 Water PNG's page directed readers to a new Online Water Bill Payment Scheme at http://www.waterpng.com.pg/site/. NICTA, PNG's "converged Licensing and regulatory authority for ICT services", refer readers to new uploads on its website (www.nicta.gov.pg).

The Investment Promotion Authority created its page in October 2013 in time to promote an Online Registry System, which became operational a month later, and to drive traffic to its website www.ipa.gov.pg. The Facebook page of the state's commercial arm, Kumul Holdings' (formerly IPBC) was launched "to coincide with the historic signing of the Charter of Good Governance, that took place on Thursday May 22, 2014 at the State Function Room, Parliament Haus". Kumul Holdings' announcement the following July that the government-owned mobile operator bmobile was entering a business partnership with Vodafone provoked one of the largest ever online responses, the website posting on July 12th:

> Thank you to all who responded to our post regarding the strategic partnership between bmobile and Vodafone. 1,300 "likes" in 24 h is quite a response! We also appreciate those who offered congratulatory messages and/or constructive comments. Your voice has been heard by bmobile senior management. To keep informed of further developments please visit www.bmobile.com.pg".[16]

The "Papua New Guinea Tax Review" page was created in September 2013 "… to engage with Papua New Guineans active on social media to have a say in the design of Papua New Guinea's future taxation system",[17] and directed readers to the Tax Review website at http://taxreview.gov.pg/. The Ombudsman Commission

---

[15]www.psc.gov.fj.

[16]https://www.facebook.com/IPBC.PNG.1/posts, accessed 30 September 2016.

[17]https://www.facebook.com/Papua-New-Guinea-Tax-Review-239167382953463/about/?entry_point=page_nav_about_item&tab=overview, accessed on 29th September 2016.

of Papua New Guinea established its Facebook page in December 2014 when seeking public input to a review of its enabling legislation. Its page directed readers to its website http://www.ombudsman.gov.pg/ (created at the same time), where a discussion paper was downloadable and where the public could access an online complaint form, greatly increasing the Commission's reach into the community. A newly established government body, PNG NID (National Identity Card), established its Facebook page in June 2015 and immediately began receiving suggestions from the public that application forms be made available online. The Regional Security Liaison Office explained in its first post in January 2012 the functions of its field offices in Vanimo, Aitape and Wutung in researching and understanding issues concerning the security of PNGs' land border with Indonesia:

> Traditional Security concerns were Border Security primarily monitoring and control and state of domestic law and good order in the province, but due to the international trends in globalization, revolution in technology complimented with the PNG multi-billion kina projects, Sandaun and its principle centers of Vanimo, Aitape and Wutung also have newer security developments. These new trends in security development have come about due to the boom in commercial natural resource extraction operations throughout the province and border trade.[18]

All such pages contribute to the emergence of e-Government in the Pacific. They facilitate access to political information and improve opportunities for expression; they render political processes and public administration more transparent and accountable; and they improve efficiency and effectiveness by expanding the range of public services available online for citizens and businesses.

13.4 E-Democracy

The first section of this chapter has explored the extent of PIC governments' online presence through website and social media presence, the vast majority of which focuses on improving access to information and increasing transparency and efficiency. But, as noted at the outset, such one-way information sharing is less of a challenge than is than more extensive online interaction and deliberation. The following section consequently examines online government-civil society engagement in policy dialogue and decision-making.

A considerable number of government entities have entered the social media space. In addition, the many actors within civil society are rapidly expanding their online presence, whether as think tanks, professional organizations, individual activists and commentators, consultants, academics, or NGOs. They are using email lists, blogs and social media. There are at least three "fluid" qualities to these online spaces and discourses. First, there is no essential demarcation between government and non-government actors, in the sense that public servants sometimes express

[18]https://www.facebook.com/Regional-Security-Liaison-Office-351634404864469/.

personal views online which diverge from government policy; second, public servants wear an official title in the workplace but may participate in online policy dialogues in their personal professional capacity, or even as NGO activists; and third, these sites tend to blur the traditional distinction between public and private spheres to an extent that deters some actors, yet for the same reason attracts others. Taken together, these factors can render online policy dialogue effective but unacknowledged, influential but not official.

13.4.1 Government Initiatives

The largest public ICT event in the region, Vanuatu's annual celebration of World Telecommunications Day, demonstrates the fluid inter-connections between state and non-state actors. Since 2012 the Office of the Chief Information Officer has convened a multi-day program in Port Vila's main park each May 17th. The event commences with a parade down the main street, announced by a marching band. The procession includes ICT officials from government, the regulator's office, and the private sector followed by floats celebrating Internet access: service providers, super funds, banks, government agencies and educators. Speeches are given by the Prime Minister, the head of the OGCIO, the Mayor of Port Vila—and even by Miss Vanuatu!—all extolling the benefits of ICTs to ordinary citizens.[19] Government and private sector displays are open to the public. The proceedings are transmitted live on YouTube.

The structure and operation of the Papua New Guinea Resource Governance Coalition (PNGRGC) provide a second example of these characteristics of fluidity. The coalition comprises representatives of government, the extractive industries and civil society, and has been established as a requirement of the government's bid to join the Norway-based Extractive Industries Transparency Initiative (EITI). It receives government funds to coordinate civil society's participation in the country's efforts to improve governance and financial transparency of in the resources sector. In 2015 the PNGRGC established a Facebook page after the first meeting of the group in Port Moresby on March 26th. The page reports on the Coalition's public awareness outreach across the country, international fact-finding tours, participation in EITI global meetings and the launch in early 2016 of PNG's First EITI Report (including providing the Report's link[20]). The findings of the report presented a critical moment for PNGRGC, as they were highly critical of practices within the sector, and called on the industry as well as government to make rapid improvements (Extractive Industries Transparency Initiative 2013).

Elsewhere in the region government agencies are similarly initiating consultative processes aligned with their respective needs. In Papua New Guinea NITCA—the

[19]http://events.gov.vu/ictday2014/ and http://ictdays.gov.vu/index.php.

[20]http://www.pngeiti.org.pg/download/pngeiti-report-2013/.

National Information and Communications Technology Authority of Papua New Guinea—includes "public inquiries" and "consultative papers" on its webpage.[21] One specific consultation in 2015 exercise concerned soliciting proposals from "… interested stakeholders including the general public and ICT operators and service providers" for projects to be funded under the Universal Access requirements of the *National ICT Act* of 2009.

Mention has been made of the PNG Tax Review website at http://taxreview.gov.pg/, which recorded the titles (but not the text) of all submissions received as part of its review exercise. In 2016 Vanuatu's tax office similarly established a website www.revenuereview.gov.vu to inform the public of a Revenue Review Team's proposals concerning the introduction of personal income tax to the country. Questions and their answers are posted on its Facebook page.[22]

13.4.2 Think Tanks

Significant think tanks in the Pacific Islands include the National Research Institute and the Institute of National Affairs in Papua New Guinea, the Pacific Institute of Public Policy in Vanuatu and the Hawaii-based Pacific Islands Development Programme. The research activities and outputs of the University of the South Pacific can also be included in this category.[23] Think-tanks are, suggests McGann, "…public policy research, analysis and engagement institutions that generate policy-oriented research, analysis and advice on domestic and international issues that in turn, enable both policymakers and the public at large to make informed decisions about public policy issues" (McGann 2010, p. 10). Each of the Pacific think tanks has an active publication program, including reports available for download.

The Pacific Institute of Public Policy (PIPP), established in Vanuatu in 2007, has undertaken innovative surveys of ICT use in the Pacific (O'Connor et al. 2011; Pacific Institute of Public Policy 2012) and has also used ICTs to conduct its own public policy civic education exercises. One of these was a first-ever telephone poll across four Melanesian countries to gather public opinion about the sub-region in the lead up to a meeting of the Melanesian Spearhead Group (Pacific Institute of Public Policy 2011). In 2016 the Institute rebranded itself as PolicyPacific.Network, explaining:

---

[21]http://www.nicta.gov.pg/public-inquiries.

[22]https://www.facebook.com/Vanuatu-Revenue-Review-604092133102460/.

[23]This chapter does not review the activities of think tanks and universities outside the region that have a focus on PICs. Most recently, the Royal Oceania Institute was established in Tonga "… to support and pursue Pacific-centered, ethical, fact-based research and analysis" and to "… facilitate open dialogue on topics relevant to community, sustainable development, and regional relations": https://royaloceaniainstitute.org, accessed 30th September 2016.

> The Pacificpolicy.networkbuilds on our rich history of engaging with the public policy community, bridging gaps between technical knowledge and policy-making, and using story-telling to inform a broad audience. Aided by the rapid improvements in online connectivity, we are forging new ways of public problem solving to better connect citizen need to government response.[24]
>
> The social media explosion has given rise to the democratisation of information. We see a future where the people of the Pacific further shift from being informed to be active participants in the policy decisions that affect their lives…"[25]

In Papua New Guinea, the National Research Institute (NRI) occupies an interesting position in the country's public policy dialogue space. Although established with government funding, NRI maintains a fiercely independent voice. It has championed the development of e-Government in Papua New Guinea and has close contact with the regulatory authority NICTA. NRI advocated opening the telecommunications market to competition in the face of some Government Ministers' opposition. In 2014 it pushed for registration of customary land and for an electronic voter ID card. When NRI learned of government's plans to digitize the common electoral role it sought a role in the process through an intervention to the Minister for National Planning and consequently chaired a committee responsible for designing the process for collection of voter data.[26] In October 2016 the National Research Institute launched an updated website (https://pngnri.org).

The Institute of National Affairs (INA), established in Papua New Guinea in 1979 with funding from the private sector, has published on policy issues for four decades, with publications in recent years increasingly made available online (http://www.inapng.com). The INA works in collaboration with the Consultation and Implementation Monitoring Council (CIMC), a body established in 1998 by the government, to assist civil society to provide government with independent feedback on its performance in implementing its programs in accordance with law and policy. Among other projects, CIMC convenes an annual National Development Forum and makes available the proceedings online.[27]

These think tanks have earned reputations for publishing independent analyses and for contributing substantially to public policy debates. Their reports have also, on occasion, met with disapproval from government departments and agencies, resulting in exchanges online and in the media, which have produced additional clarification of the issues at hand. The directors of all four think tanks comment actively online and in the media.[28]

---

[24]http://www.pacificpolicy.network/#re-thinking, accessed 30 September 2016.

[25]http://www.pacificpolicy.network/#future, accessed 30th September 2016.

[26]Charles Yala interview, Port Moresby, 14 November, 2013.

[27]http://www.cimcpng.net/index.php/forums/national-development-forum, accessed 30th September 2016.

[28]Thomas Webster then Charles Yala at NRI; Paul Baker at INA, Jerry Finin at PIDP, and Derek Brien at PIPP.

13.4.3 Non-governmental Organizations

Non-Governmental Organizations that promote democratic governance operate throughout the Pacific Islands. Individually they advocate of a wide range of causes, which the strategic plan of their regional representative organization, PIANGO, groups as promoting human rights and social justice; gender equality and equity; people's empowerment, democratic ownership and participation; environmental sustainability; transparency and accountability; knowledge sharing; and realizing positive sustainable change (Pacific Islands Association of Non-Governmental Organisations 2013, p. 5).

Although NGOs across the Pacific use the web to promote their values, ideas and activities, they have not yet focused on e-Government as a means to accomplish their objectives. They have not, for example, monitored their governments' progress toward universal access, or toward participation in open government initiatives and transparency initiatives. However, this situation will no doubt change as PIANGO and other organizations begin to use the Internet in more sophisticated ways. PIANGO already participates in global policy dialogue online (with UN agencies and with CIVICUS) but much of its advocacy in the region remains "face-to-face." Pacific cultures place much emphasis on the value of personal relations and these are difficult to replicate online. However, online consultation with five Pacific partners (Fiji, Samoa, Solomon Islands, PNG and Kiribati) during the "Beyond 2015" project in the lead up to the UN's formulation of the Sustainable Development Goals marked a beginning; but an ongoing challenge is the variation in the quality of Internet access available to each CSO.

13.4.4 Pacific Internet Society

The Pacific Internet Society (PICISOC; http://www.picisoc.org/), a sub-branch of the Internet Society[29] established in 1999, is the organization most focused on issues of Internet governance and policy in the region. It exemplifies the "fluid" civil society organization in that some office bearers (and many members) are senior public servants. Furthermore, it seeks involvement in government-sponsored events relating to the Internet and it involves government officials in its conferences. Since 2002 PICISOC's annual conference (PACINET) has focused on Internet regulation, Internet security, the digital divide, access, gender, as well as a range of technical subjects, including geographic information systems (GIS), remote sensing (RS), the move to IPv6, PacCERT, routing leading to the establishment of PacNOG (Pacific

[29]The Internet Society http://www.internetsociety.org was established in 1992 as a global organization open to all who are interested in issues of Internet policy, technology, and capacity building.

Network Operators group,[30] FOSS (Free and Open Source Software), and the formation of a Pacific IGF (Internet Governance Forum). PICISOC has established interest groups promoting ICTs in Pacific Education, FOSS, GIS/RS and Internet freedom, and has made policy recommendations on Internet governance to WSIS and to the Pacific Islands Forum Secretariat. PICISOC supports the work of Internet organizations in each PIC, which include the Internet Users Society, Niue, iPNG Clique; the Software Foundation of Fiji; the South Pacific Computer Society; the Vanuatu Information Technology Users Society and the Cook Islands Internet Action Group.

There are indeed technology groups in most of the Pacific countries which, following the PICISOC model, include government and non-government members as members and office bearers. These include Samoa Information Technology Society[31] established in 2003, with office bearers who are senior public servants in the Ministry of Finance, together with others from private sector companies, the Samoa National Provident Fund, Samoatel, Computer Services Limited, Polynesian airlines, SPREP, and with Prime Minister Tuilaepa Aiono Sailele Malielegaoi as Patron.

13.4.5 Academia

Academic institutions are also interestingly situated between the institutions of government and civil society. University programs undertake the capacity building that e-Government transformation requires. Yet whereas most universities in the Pacific Islands are publicly funded to produce graduates, they are also expected to comment on matters of public interest, including the performance of government, and to this extent they constitute a well-resourced component of civil society. Some aspects of USP's activities toward developing ICT capacity in the region have been noted in Chap. 3. At different times the University has maintained institutes with think-tank missions, including some with a focus on government functioning—but not with specific focus on e-Government. Online publications such as the *Journal of Pacific Studies*[32] make a valuable contribution to knowledge of the region's social, economic and political conditions. In Papua New Guinea, the only Pacific country that is a member of the Asia Pacific Economic Community (APEC), APEC Digital Opportunity Centres were established at the University of Technology (2006) and the University of Papua New Guinea (2011) to "help APEC member economies enhance their information and communication technology (ICT) application capabilities and to transform the digital divide into digital opportunities"—although unrest on both campuses has hindered significant progress. One external academic

---

[30]https://www.pacnog.org/.

[31]https://www.facebook.com/Samoa-Information-Technology-Society-SITS-1125975097417973/, accessed 30th September 2016.

[32]http://www.usp.ac.fj/.

program, run by the Diplo Foundation, a joint initiative of the Maltese and Swiss governments that specializes in building the capacity of diplomacy in small island states, has run a capacity development program for Pacific Islands states (Diplo 2013).

13.4.6 *Other NGOs*

A considerable number of Pacific Island-based NGOs are actively engaging online on issues of government functioning. They include media and media watchers, women's organizations, religious communities and churches, consumer groups, political parties, political and social movements and organizations advocating indigenous rights and culture, integrity in government, environmental stewardship and peace.

Fiji Youth for Democracy, "… a movement by Fijian youth to promote Democracy, Human Rights and the Rule of Law" has an active discussion page on Facebook"[33] and a website on WordPress.[34] The *Fiji Native and Tribal Congress*, an organization established by the Paramount Chiefs of Fiji's tribal confederacies in 2011 to "… advance, protect and maintain the group rights of indigenous Fijians" in accordance with the 2007 UN Declaration of the rights of Indigenous People, established a Facebook page in 2015 to promote its cause.[35] In Solomon Islands, Transparency International initiated a petition on the website Change.org "Reject the 'Constituency Development Funds Bill 2013'", through concern that the bill, if enacted, would shift service delivery to constituents from machinery of government to MPs and place public funds directly in the hands of MPs. The petition only received 54 votes, but was handed to parliament.[36]

13.4.7 *Individuals*

It is quite possible that websites and pages established by individuals not only outnumber those established by organizations but have been more effective in gaining the attention of governments. In Fiji, a military coup in December 2006, which provoked many ordinary citizens to express their views, occurred just as useable platforms such as Blogspot (1999) and Wordpress (2005) became

[33]https://www.facebook.com/FijiYouthForDemocracy/about/?entry_point=page_nav_about_item&tab=page_info.

[34]https://fijiyouth.wordpress.com/.

[35]https://www.facebook.com/Fiji-Native-and-Tribal-Congress-1722175551343168/about/?entry_point=page_nav_about_item&tab=page_info.

[36]https://www.change.org/p/the-national-parliament-of-solomon-islands-reject-the-constituency-development-funds-bill-2013-2, accessed October 7th, 2016.

accessible. Some sites had a short life and are no longer available: *Fiji Democracy Now*, *Nadina Fiji truth*, *Pacific Freedom Forum* and *Resist Frank's Coup*. Others were only active in 2007 or 2008 but remain available online: *Fiji Coup name and shame list*, *Fiji Blog Dump*, *Free my Fiji*, *Illegal coup 2006*, *Kaiviti Freedom* (all 2007); *Tard Watch Fiji* (2008) and *Fiji Uncensored* (2009). The pro-coup site *IG-Fiji* only operated Jan–April 2009. Sites with a slightly longer life included *Digital Fiji* (2006–2008), *Fiji Democracy and Freedom Movement* (2010–2012), *Good men and women doing something* (2007–2012), *Soli Vakasama* (2007–2010) and *Tears for Fiji* (2008–2010). A few have continued from the time of the coup to the present day: *Stuck in Fiji M.U.D.* (from 2005), *Discombobulated Bubu* (from 2007), *Fiji Girl's Weblog* (from 2008), *Luvei Viti* (from 2008), *Namuamua journal* (from March 2008), *Fiji CoupFourpointfive*, *Fiji Exposed* (from 2010) and *Intelligentsya*. The sites *Rawfijinews* and *Realnewsfiji* can only be read with the permission of their webmasters.

Whereas each of the sites mentioned above operated anonymously, USP academic Waden Narsey chose to speak openly, with his WordPress blogsite *Wadan Narsey on Fiji—fighting censorship in Fiji* remaining unsparing in its criticism of the Fijian government—even after he lost his position at the university and emigrated. Other academics who have chosen to follow the Fijian situation include Croz Walsh (*Fiji: The Way it Was, Is and Can Be*) and Victor Lal (*Musings on Fiji*).

Papua New Guinea's "blogosphere" is equally active, spurred not by the overthrow of democratic institutions by the military but by the threat to democracy by evidence of corruption at the heart of government. *Malumnalu,*[37] *which records more than 2,700,000 views since it commenced in 2006*, was awarded the UNESCO/Divine Word University Award for Communication and Development in 2011, is archived by the National Library of Australia, and claims to be the "No. 1 Blog in Papua New Guinea". Together with sites like *PNGexposed* and the *Masalai Blog*, Malumnalu provides evidence of corrupt practices of political leaders, senior public servants, their business associates and their legal teams in blushing detail. Sites like *ActNow* focus on specific campaigns, such as examining the impact of development models on Papua New Guinea's economy and society, public sector corruption, seabed mining and the impact of the SABL land regime. Much discussion also takes place on old-style discussion lists such as *Papua New Guinea Online Community*[38] and *PNGGossip* (established August 2008). Newer sites, such as *PNGmirror*,[39] established in 2014, has the backing of government bodies to focus on initiatives in natural resource development, tourism, infrastructure and small business that are contributing to positive change in the country.

Political leaders are among the active bloggers and webmasters. Former PNG Prime Minister Sir Mekere Moratu commenced the site *Papua New Guinea Observer* because of his concern

[37]http://malumnalu.blogspot.co.nz.

[38]pngscape.net.

[39]pngmirror.com.

…about the state of the nation - the increasingly undemocratic actions of the O'Neill Government, its economic and financial mismanagement, and the destruction of the institutions of state and weakening of the structures and processes of decision-making, laying the ground for high levels of systemic and systematic corruption. These factors have also resulted in rapidly declining development indicators, attacks on freedom of expression and association, and social division and inequality. Above all I was concerned about the lack of informed public debate on a wide variety of issues, and the Government's intimidatory suppression of such debate. This blog is my way of spurring public debate and providing facts on which people can base their arguments, rather than on spin and hyperbole."[40]

Other MPs focus on particular discussion groups to share their viewpoints: Gary Juffa, Charles Abel, Sam Basil and National Capital District Governor Powes Parkop in Papua New Guinea; Ralph Reganvanu in Vanuatu; Derrick Manuari in Solomon Islands. In some instances, sites that focus on the conduct of government are created by individuals intending to stand as candidates, such the founder of the WordPress site *PNGanticorrruption*.[41]

13.4.8 Facebook Groups and Pages

Some Facebook groups established by individuals have had significant impact on public discussion of government and have influenced government behavior. *Sharp Talk*, created by Douveri Henao in April 2012 out of frustration about the lack of information and dialogue on topical issues concerning Papua New Guinea, reaches 27,000 members. Its ongoing discussions about domestic violence, foreign ownership of businesses and employment in local positions, foreign relations, the cost of banking services, the quality of customer service by government departments, the conduct of leaders on and off duty, the situation in West Papua, the status of government revenue, parliamentary affairs, the social and environmental impacts of mining and corrupt transactions by leaders—and many other topics—have raised public awareness of what is working and what is not in their country and provided a means of their mobilization.

In one very specific episode at the time of writing, motorists in Port Moresby whose cars were rammed by a police vehicle with an intoxicated police inspector at the wheel and then were physically intimidated by him immediately posted photos of the incident on *Sharp talk* and multiple witnesses "replied" with their corroborating evidence.

"Do not blindly accept anything you read on Niugini Outlook or anywhere else", the editors of the Facebook page *Niugini Outlook* wrote on 7 October 2016,

… Think about what you read and question and reach your own conclusions… We put out the allegations we receive in the hopes of attracting more information about the topic so we all come closer to the truth. Thanks to the outstanding PNG coconut wireless and relatively

[40]http://www.mekeremorauta.net/about, accessed 14 September 2016.

[41]https://pnganticorruption.wordpress.com.

> small PNG elite population, our postings seem to always become known by the public figures mentioned in our postings. They have every opportunity to deny anything they believe to be untrue. We NEVER censor anyone who replies to our postings with alternative viewpoints or to deny the stories. We leave those comments on as long as the postings remain.[42]

In Solomon Islands, the Facebook group Forum Solomon Islands International (FSII—political system change for Solomon Islands)[43] has had similar impact as Sharp Talk has had in Papua New Guinea. In this instance the page was created by Solomon Islanders living abroad, introducing a new factor to the merits of online dialogue, namely the role of expatriate Pacific Islanders. The page, specifically created to expose government corruption and hold the Solomon Islands Government to account, was registered as a Civil Society Organization to provide it with some solidity (Finau et al. 2014, p. 5).

An example of FSPI's influence occurred when details of the cost of a tour to Indonesia by then Prime Minister Gordon Darcy Lilo and his delegation was leaked to the press. Whereas the Prime Minister assured the public that the costs were borne by Indonesia, contributions to the FSII site proved that 99% of the trip was funded by Solomon Islands government funds—a revelation that caused considerable political controversy (Finau et al. 2014, pp. 5–6).

In Vanuatu, two online initiatives have revolutionized public discussion of government activities: *Yumi toktok stret* and *Vanuatu Daily Digest*. *Yumi toktok stret*, with some 20,000 members including MPs, was created to provide a forum to discuss political, social and economic issues affecting Ni-Vanuatu. As explained by Finau et al., "The page is primarily used by citizens but government officials as well also use this forum to defend themselves and sometimes proactively to promote themselves" (Finau et al. 2014, p. 5). One incident, recalled by a student at USP, demonstrates the site's effectiveness:

> The [Vanuatu] government had been very late with sending our allowances. A student complaining about delayed allowances on the Facebook page posted a picture of three loaves of bread with the captions: "Breakfast", "Lunch" and "Dinner" on Yumi Tok Tok Stret. The post made front page on the Daily Post [Vanuatu national newspaper]. The government sent us the allowances the following week. (Finau et al. 2014, p. 5)

The site is a "model of e-Democracy":

> Not only is it used by citizens it is also used by politicians. Constructive debates are held on this page between politicians, citizens and even a group named—Vanuatu Opposition Party. The page has become so popular that posts on this page regularly appear as stories on Vanuatu's national newspaper, the Daily Post. (Finau et al. 2014, p. 7)

---

[42]https://www.facebook.com/niuginioutlook/posts/330469603964227:0, accessed 8 October 2016.

[43]https://www.facebook.com/groups/324585784285971/.

A second site, *Vanuatu daily digest*, promotes "… media freedom, open democracy, transparency in government, better delivery of services and greater accountability from our elected representatives…"[44] In 2013 the site covered the intriguing but ultimately scandalous story of the luxury yacht *Phocea* that had appeared in Port Vila Harbour in 2011 only to be, apparently, abandoned. The commissioner of police announced two years later the discovery aboard the vessel of the biggest ever stash of cocaine to be found in the South Pacific. Questions then turned to who owned the yacht, who had processed it into the harbor and allowed the crew ashore and why Australian Federal Police who had been looking into the matter had been expelled from the country in May 2012. The owner turned out to be one of the country's international representatives, and the ship had been met upon arrival by the then Foreign Minister: "Only the *Vanuatu Daily Post* and the *Vanuatu Daily Digest* persisted with the story."[45]

The *Vanuatu Daily Digest* established in 2012 a Facebook page *Daily Digester*, which in September 2016 announced that the Government of Indonesia had put a request to Facebook to close it down, presumably as a consequence of the site's support for self-determination for West Papua. On June 28th 2016 the site published a facsimile of a memo from the Office of the Public Service Commission[46] dated 27th June, which banned the use of social media sites by all public servants, unless written permission was obtained from the Office of the Government Chief Information Officer (OGCIO). This of course, would be a difficult order to implement, as it contravened the mission of the OCGIO, which is to "to encourage the spread of ICTs (information and communications technologies) in society to efficiently and effectively achieve an educated, healthy and wealthy Vanuatu", and "to encourage the right to information (RTI) in Vanuatu". It also goes against the freedom of expression provisions of Vanuatu's Constitution and would be a challenge to those government departments that have developed a presence on Facebook and Youtube.

Vanuatu pages devoted to discussion of public policy are emerging rapidly. *Public Policy Toktok*,[47] for instance, was created by a public servant in 2016 with the explanation: "This group is dedicated to critical discussions on public policy issues and challenges relevant to Vanuatu. As a learning platform, it intends to raise understanding of policy and regulation issues. Our Freedom of expression is guaranteed by the Constitution of the Republic of Vanuatu Article 5(1)(g) and the Universal Declaration of Human Right Article 19."

---

[44]https://vanuatudaily.wordpress.com/about-site, accessed 30th September 2016.

[45]Van Trease (2014).

[46]https://vanuatudaily.wordpress.com/2016/06/28/blanket-censorship-of-social-media-announced-for-all-government-workers/#more-4618.

[47]https://www.facebook.com/groups/173629386363100/.

13.5 Conclusion

This chapter has examined the current and potential role of civil society in e-Government in Pacific Island countries. It has suggested that ICTs are valuable tools for civil society's role in protecting and defending human rights and in promoting access to information and participation in decision making/government—all for the purpose of ensuring good governance through inclusive policy development. However, this "opening up" of information, consultation and participation to non-state actors requires changes in political culture which take time to achieve. E-government models advocated by the World Bank and the United Nations envisage the emergence of "ePublic Services" and "eDemocracy" in ways that will benefit advantaged and disadvantaged citizens alike. Government departments have made great progress toward sharing information with the public, although not all are doing so through a "whole of government" approach. However, from Pacific Island civil society's standpoint, engagement, involvement and inclusion of CSOs in spaces making decisions has commenced physically, but hasn't yet started virtually in the Pacific, although this has opened up immensely at the global level especially with UN.

The Internet has revolutionized the notion of democratic space and enhanced the role and contribution of civil society in promoting good governance and democratic governance. But transformations in state-civil society discourse are not occurring easily, since new notions of citizen engagement imply a questioning of prevailing state-society power relations. Communicative practices between state, civil society and citizenry that the new ICTs enable and facilitate are thus still being negotiated, particularly in states where government has traditionally been the voice of authority and government information has been protected by Official Secrets Acts. Without changes in political culture in such contexts, ideas of open government, freedom of information, sharing of data on social media and other forms of online public discussion of government activities, tend to be viewed as challenges to government authority rather than as lawful queries by legitimate stakeholders. Increased access to information has enhanced civil society's ability for evidence-based advocacy. The technology now exists to facilitate online dialogue between government and civil society organizations, and governments could now make serious efforts to engage their citizens in meaningful policy dialogue both face-to-face and online.

References

4th High Level Forum on Aid Effectiveness (2011) Busan partnership for effective development co-operation. Fourth high level forum on aid effectiveness, Busan, Republic of Korea, 29 Nov–1 Dec 2011 http://www.oecd.org/development/effectiveness/49650173.pdf

Anon (2014) All media carry a great responsibility. The National (11 July). http://www.thenational.com.pg/all-media-carry-a-great-responsibility/

Auer MR (2011) The policy sciences of social media. Policy Stud J 39(4):709–736

Bainimarama V (2016) PM warns young people of social media danger. Fiji Sun Online (30 July). http://fijisun.com.fj/2016/07/30/pm-warns-young-people-of-social-media-danger/

Diplo (2013) CD Pacific: capacity development programme in multilateral diplomacy for Pacific Island States. https://issuu.com/diplo/docs/cd_pacific

Extractive Industries Transparency Initiative (2013) PNG EITI Report 2013. http://www.mra.gov.pg/Portals/2/docs/PNG%20EITI%20Report%202013_PNGEIETI%20Website.pdf

Finau G, Prasad A, Kant R, Tarai J, Logan S, Cox J (2014) Social media and e-Democracy in Fiji, Solomon Islands and Vanuatu. Paper presented at the twentieth americas conference on information systems, Savannah. http://aisel.aisnet.org/cgi/viewcontent.cgi?article=1232&context=amcis2014

Gimmler A (2001) Deliberative democracy, the public sphere and the internet. Philos Soc Criticism 27(4):21–39

Guild R, Taufao S (2003) Pacific Islands and the WSIS. In: Report of conference and side events: opportunities for the Pacific Islands region. Paper presented at the World Summit on the Information Society. Asian regional conference, 13–15 Jan, Suva, FIji

Kaplan AM, Haenlein M (2010) Users of the world, unite! The challenges and opportunities of social media. Bus Horiz 53(1):59–68

Lee YB (2014) E-government for promoting sustainable development in Small Island Developing States (SIDS): background paper for the special event of the 69th Session of the United Nations General Assembly Second Committee. http://workspace.unpan.org/sites/Internet/Documents/UNPAN93784.pdf

Macintosh A, Tambouris E (eds) (2009) In: Electronic participation: first international conference, ePart 2009 Linz, Austria, 31 Aug–4 Sept 2009 proceedings. Springer, Berlin

Marape J (2013) Communication and information technology—communication sector—ministerial statement—paper noted. Resumption of debate from 13 Nov 2013. Hansard, Port Moresby

McGann JG (2010) The think tanks and civil societies program 2009: the global "Go-To Think Tanks". http://repository.upenn.edu/cgi/viewcontent.cgi?article=1001&context=think_tanks

McIver WJ (2003) A community informatics for the information society. In: Girard B, O'Siochru S (eds) Communicating in the information society. United Nations Research Institute for Social Development, Geneva, pp 33–64

Mickoleit A (2014) Social media use by governments. A policy primer to discuss trends, identify policy opportunities and guide decision makers. Retrieved from Parris: http://www.oecd.org/gov/digital-government/government-and-social-media.htm

O'Connor S, Naemon A, Sijapati-Basnett B (2011) Net effects: social and economic impacts of telecommunications and internet in Vanuatu Research findings report 2011–2012. http://www.pacificpolicy.org/wp-content/uploads/2012/05/PiPP-Net-Effects-2011-Vanuatu-Telecomms-report1.pdf

Obi T (ed) (2015) 2015 Waseda—IAC international E-government ranking survey. Waseda University Institute of e-Government, Tokyo

Pacific Institute of Public Policy (2011) Melanesia poll results March 2011. http://pacificpolicy.org/wp-content/blogs.dir/2/files/2015/02/Melanesian-Poll-.pdf

Pacific Institute of Public Policy (2012) Net benefits. Upgrading the coconut wireless: internet uptake in the Pacific. http://www.internetsociety.org/net-benefits-%E2%80%93-upgrading-coconut-wireless-internet-uptake-pacific

Pacific Islands Association of Non-Governmental Organisations (2013) Strategic Plan 2013-2015. CSO leaders rethinking development in the Pacific Islands. Retrieved from Piango, Suva. http://www.piango.org/PIANGO/strategic-plan.html. Accessed 24 Oct 2016

Pacific Islands Forum Secretariat (2002) Policy and strategic plan. Information and communication technologies for every Pacific Islander. http://www.forumsec.org/resources/uploads/attachments/documents/Pacific%20Islands%20ICT%20Policy%20and%20Strategic%20Plan%20(PIIPP)%202002.pdf

Radio New Zealand International (2015) Nauru opposition MP says new laws about stifling criticism. RadioNZ, Wellington. http://www.radionz.co.nz/international/pacific-news/273656/nauru-opposition-mp-says-new-laws-about-stifling-criticism, 19 Oct 2016

Sao FM (2012) Information and communication technology for development in Samoa: exploring the interaction between the local traditions and information found on computer technology. (Master of Business (MBus)), Auckland University of Technology

Secretariat of the Pacific Community (2011) Pacific internet governance forum "Developing the future together": report for sponsors and supporters of the inaugural pacific internet governance forum (Noumea, New Caledonia, 9–10 Apr 2011). www.intgovforum.org/cms/documents/publications/175-developing-the-future-together/file

Tang L, Li H (2010) Community detection and mining in social media. Morgan & Claypool, Williston, VT

Tarai J (2015) To regulate or not: Fiji's social media. State Soc Gov Melanesia 53(2)

The Communication Initiative, Food and Agriculture Organization of the United Nations & The World Bank (2007) World congress on communication for development. Lessons, challenges, and the way forward. Retrieved from Washington DC

UNDP. Evaluation Office (2004), E-governance, Essentials No. 15, New York.

Van Trease H (2014) Vanuatu. Contemp Pac 26(2):524–551

Zwimpfer Communications (2002) Internet infrastructure and e-governance in Pacific Island Countries. A survey on the development and use of the internet. UNESCO, Wellington

Chapter 14
E-Government for Development in Pacific Island States: Achievements and Future Directions

Rowena Cullen and Graham Hassall

Abstract E-government can facilitate more efficient, transparent government and better communication between government and citizens; and in this volume we have been exploring the extent to which it has done so in Pacific Island countries. The chapter initially returns to the objectives of the volume and some of the questions posed in Chap. 1. It summarises the key findings of Chaps. 2–13, accentuating the evidence of critical factors in e-government success that can be identified from these findings. This includes a discussion of the attributes of sustainability that are identified in the various chapters, and the five criteria of sustainability proposed in Chap. 1. The theoretical frameworks used in the study (Heeks ICT4D 2.0 Manifesto, his 'design-reality gap' and his work on E-Government for Development, and the 'information ecology' construct applied to e-government by Bekkers and Homberg) are revisited, and the extent to which these contribute to the analysis of e-government in the Pacific Islands examined. In the final sections of the chapter the focus shifts to e-government policy processes and the question of how well policy processes have contributed to some of the successful e-government initiatives outlined in various chapters in the book is addressed. The chapter, and the volume, conclude with some final reflections: acknowledging the achievements of PICs in ensuring that the basic technical, legislative, policy and regulatory infrastructure for e-government are in place; noting the challenges they have faced with their small economies, lack of resources, and the tensions created in balancing development with traditional lifestyles and forms of governance; emphasizing the need for PICs to build on their achievements to date by focusing on sustainable interactive e-government services that meet local contexts and needs, and using technology to enhance communication with citizens. The lessons learned in the study are applicable to all countries seeking to benefit from the application of ICT to make government more efficient, inclusive and accountable and enhance good governance.

R. Cullen (✉) · G. Hassall
School of Government, Victoria University of Wellington, Wellington, New Zealand
e-mail: Rowena.cullen@vuw.ac.nz

R. Cullen and G. Hassall (eds.), *Achieving Sustainable E-Government in Pacific Island States*, Public Administration and Information Technology 27,
DOI 10.1007/978-3-319-50972-3_14

14.1 Introduction

This book has investigated a range of e-government initiatives across Pacific Island countries (PICs) in an endeavor to identify how ICTs are being used to bring better government to the people of the Pacific Islands region. It has presented a substantial body of evidence of international investment to bring e-government to the region. It has analyzed the forums and processes leading to policy frameworks that have underpinned regional collaboration on some important e-government initiatives. It has provided examples of significant activity and some successful e-government applications in many PICs. Overall the book has noted considerable progress in connectivity in PICs and the use of ICT to make governments more efficient, more transparent and better connected with citizens. In all three of the dimensions that are sometimes used, G2G, G2B, G2C, ICT has improved government processes and service delivery. Although this progress has been uneven, and in many cases not sustained, as we noted in Chap. 1, this could be also said of many e-government projects in larger industrialized countries. Accordingly, in summarizing the various chapters and what conclusions we can draw, we have not sought to document failures, but to focus on what has worked, and the factors that have led to this success. In most PICs the fundamentals of e-government are now in place, especially connectivity and basic policy, as are in most cases the necessary legal and regulatory frameworks. As we noted in Chap. 3 "While government in the Pacific Island Countries has not yet been 'transformed', it has in some sectors experienced incremental improvement in little more than a decade and this, given the region's challenges with capacity, is good progress." What PICs need to consider next is how best to capitalise on this achievement, and how to ensure that the next phase of e-government in the Pacific Islands is sustainable and contributes to development.

Setting out on this journey, we asked a number of questions. What benefits does the use of ICT in government bring? What challenges do countries face in implementing such technologies? How well are e-government initiatives aligned with local contexts and local needs? How sustainable are they? Who is driving these initiatives and how are they funded? In asking these questions we sought to identify factors critical to the successful implementation of e-government in Pacific Island states and the factors that lead to failure. We also noted that the resources, environment and political context of Pacific Island countries (PICs) are very different from their larger neighbors and that given these differences, e-government solutions that apply to industrialized countries might not necessarily be appropriate for PICs. We wanted to know what form successful and sustainable e-government initiatives would look like in this environment and to test our suggested criteria for sustainability.

In order to understand more fully the significance of the cultures and the systems of governance in PICs, and the impact of these on ICT adoption in various government sectors, we adopted an 'information ecology' approach as proposed by Bekkers and Homberg (2005). We used this perspective, which focuses on stakeholders, their motivations, values and practices as well as the cultural, political, intellectual and economic environment within which they interact, avoiding the techno-centric model that underpins most e-government research. The concept of an information ecology also underpins the other theoretical framework that informs this study, Heeks' ICT4D 2.0 Manifesto (Heeks 2009), which highlights the factors that influence how ICT is applied in government in developing countries and the relationships that impact on the sustainability of both ICT projects and policies. Heeks' proposition that a 'design-reality gap' underlies the failure of many e-government applications and the dimensions used to analyze the reasons behind it (information; technology; processes; objectives and values; staffing and skills; management systems and structures; and other resources) have proved useful in the analysis of success and failure in several chapters.

But focusing on an e-government perspective rather than just ICT use within government implies that there is in such projects a more deliberate attempt to transform business process to ensure greater efficiency and accountability; add value and enhance access to government information; and make government services more readily available. It is, in short, a more conscious effort to harness the transformational power of e-government. This is a matter of government policy, and brings into question how policy is made and implemented in the Pacific Islands context. Throughout the volume, and particularly in Chaps. 2 and 3, the role of aid agencies and development partners emerges as one of the most powerful drivers behind e-government initiatives. While it is important to note that PICs and their regional organizations determine regional policy (to some extent influenced by the views and experience of the industrialized nations that are also members of the same organizations)[1] when it comes to implementation at country level, the majority of e-government projects, once approved, are driven as much by development partners and external consultants as by local staff. And although some of the major development agencies such as the World Bank recently acknowledged the need for 'country ownership' to ensure project success and sustainability (see Chap. 6) and the need to work alongside local staff, the development of e-government policy in the local context is not always achieved through dialogue, consultation and coordination, to the extent that Bridgman and Davis (1998, 2003) advocate.

So in this final chapter we must consider whether a robust policy process is discernible in e-government implementation in PICs, and the extent to which policy is developed with due regard to local circumstances, rather than externally imposed. When development partners advocate for policies and projects that have been implemented elsewhere, is there evidence that Pacific governments are active participants in the policy process? To what extent are policies and projects developed in the context of each country's own development plans and what impact does this have on their sustainability?

[1]Australia and New Zealand are members of the Pacific Islands Forum. Regional arrangements are discussed in more detail in Chaps. 1 and 3.

14.1.1 Can the Benefits of E-Government in PICs Be Identified?

It could be argued that the focus in this final chapter, and indeed some earlier chapters, should be on an evaluation of the benefits of these projects. This would indeed have considerable value to all countries implementing e-government projects, and a major contribution to the e-government literature. But apart from some valuable in-country research on the benefits of mobile technology, reported in Chap. 5 in particular and also in Chap. 9, this is not an aspect of e-government research that has been well explored, and there are few tools that can assist (apart from Heeks' design-reality gap). While the World Bank and IMF have resources to investigate the effectiveness of financial management systems, particularly in strengthening public institutions (as noted in Chap. 5), there is little research on e-government applications across government of the kind that is found in the broader public administration and information systems literature relating to industrialized countries. More general research on e-government in SIDs, cited elsewhere in this volume, is all that is available. For example, Lees' analysis of the relationship between the three indicators (EGDIs) of the 2014 E-government Survey (ICT Infrastructure, Human Capital, and e-Services), treating them as independent variables, found that all three significantly affect the variables of sustainable development (Lee 2014), and influence, respectively, the indicators of good governance, transparency, and government effectiveness and regulatory quality. From this, Lee argues, good governance can be shown to have an impact on sustainable development. In addition, from the literature, Lee identified seven "key success factors" of e-government implementation (political commitment, the establishment of legal frameworks for ICT and e-government, a coordinating body responsible for e-government, sufficient budget, a national e-government policy, creation of awareness in user groups, and necessary government reform); also essential are adequate technological infrastructure before e-government implementation and the contribution of international partnerships and cooperation.[2] Achievement on all these criteria was found to lead to higher economic and social development, and sustainable development; these factors are discussed below along with our own criteria for sustainable e-government.

A few other studies have attempted to verify benefits (see for example the *E-government Benefits Study* 2003); some have been focused on costs and benefits (Gupta and Jana 2003; Kertesz 2003). But, like Lee's analysis above, which uses global measures which are to some extent weighted against SIDS, these methodologies do not translate well to the context of Pacific islands, and the limitations that this imposes. Independent research on ICT/e-government initiatives in the region is rare, and the few examples that exist seem to be confined to the mobile and health

[2]In addition to a literature review, Lee surveyed senior ICT and e-government officials in SIDs but primarily to determine what policy and legislative frameworks were in place, and what activities any e-government initiatives covered.

sectors. It has not been possible to survey government officials, citizens, or users of government services to provide the kind of empirical data that is usually found in reports on the adoption and effectiveness of ICT use in government. Even the reports of donor agencies on e-government projects tend to be limited to the fact that a project has been concluded and the budget disbursed. What evidence of benefits that can be found has been presented here; the rest of our data is drawn from accounts of participants, (individuals and organizations) annual reports of government and inter-government agencies, news reports and press releases, and commissioned expert reports on the status of e-government in specific fields (usually commissioned by the World Bank, IMF and other such agencies). Together they present a fairly full picture of what is occurring. But more research, and in particular applied research, is needed to help contribute to the formation of policy, and to a better understanding both locally and globally of how e-government initiatives can aid development.

14.2 Summary of Chapter Findings

Part 1 of the volume concentrates on the contextual factors and the global, regional and national policy frameworks and processes that have affected the ability of PICs to exploit the opportunities for development that the Internet and ICT afford. Reflecting the fact that the Internet is a global communication system essential to many e-government applications, Chap. 2 discusses the significance to PICs of policy made in the global forums that determine the governance of the Internet and the parameters of e-government interventions in PICs by international development agencies such as the UNDP and other United Nations agencies, the World Bank and the OECD. The contributions of key forums to global policy are identified. These include the biennial World Summit for the Information Society, the United Nations 2030 Agenda for Sustainable Development and the Third SIDS Conference in Apia in 2014—the latter in particular underlines the need for Pacific SIDS to embrace ICT for e-government and for development, and to promote the need to build e-government capacity. The chapter explains what the leading inter-government and non-government international agencies have contributed to e-government policy affecting PICs and outlines some of their projects in individual countries. Despite the difficulty small states experience in participating in this level of global discourse, the participation of PICs as development partners in the global e-government policy debate has increased in recent years. This participation has resulted in greater recognition of the needs of SIDs, including those in the Pacific Islands, in the development and implementation of e-government policy.

Chapter 3 investigates how the formulation of policy at regional and national levels assists PICs and their development partners to translate some of these global statements into workable e-government solutions that take account of each country's unique circumstances. The chapter notes the institutions and networks engaged in setting e-government goals and policy at regional level, and how regional policy has developed over the past three decades. At the national level the importance of policy, legislative frameworks and technical infrastructure are highlighted, and the chapter identifies where responsibility for ICT and e-government policy lies in each PIC, and what policy exists. The chapter also lists a number of e-government projects financed by international agencies such as the ITU and the ADB. It concludes that while regional policy dialogue identifies needs and assists the coordination of capacity building and development assistance, the roll-out of e-government in PICs depends to a great extent on leadership and capacity in individual countries.

The central theme of Chap. 4 is the importance of a reliable and affordable telecommunications market for successful e-government implementation. The chapter identifies the regulatory principles and functions a telecommunications regulatory regime should address: price regulation, spectrum management, licensing, universal services and access, cybersecurity, competition and consumer protection, noting that all these should be included in a national ICT policy. The chapter then focuses on the economic situation facing Pacific Island countries and its impact on telecommunications markets; it argues that an independent telecommunications regulator may not be affordable by all PICs and that a regional solution might assist smaller PICs to develop the robust telecommunications market that is needed to support e-government and economic growth. To some extent this chapter addresses the questions posed by Heeks in relation to his model of ICT4D 2.0: whether the policies and procedures are in place to make this revolutionary technology available to all, while Chap. 5 addresses the question of uptake, and how to ensure it makes the most impact for social and economic development.

As Chap. 5 indicates, the number of mobile phone subscriptions has increased dramatically since the introduction of liberalized telecommunications markets in most PICs, and the chapter suggests that mobile technologies have been transformational in the lives of people in PICs in a way that earlier computer-based technologies were not, due to the greater coverage of mobile networks, and the affordability of the technology. The use of mobile technology for communication between government and citizens, business, and other government agencies is identified as a subset of e-government, m-government; the authors argue that mobile technology offers pertinent solutions to e-government challenges that PICs face. Examples of m-government, m-service and m-participation are provided, and their success in delivering e-government solutions assessed. Applying some of the principles articulated in Heeks' ICT4D 2.0 manifesto (Heeks 2009), the authors argue that the considerable potential of m-government in the Pacific will be reached only if applications are developed through a true Web 2.0 participatory process. As noted above, the chapter includes some valuable research on examples of m-government applications in the region and calls for more research, both around ownership structures, pricing and regulatory frameworks, and to ensure that m-government addresses community needs and contributes to desirable and sustainable outcomes in PICs.

In Part 2, individual chapters examine the use of ICTs in different government sectors, using examples from various Pacific Island countries. Chapter 6 starts by examining the role of good governance and robust financial management in building stable democratic states in the Pacific Islands, and countering corruption. It places the implementation of e-government in public financial management in PICs at the heart of the public finance reforms mandated by the World Bank and the International Monetary Fund (IMF), which were adopted by the Pacific Islands Forum and the Forum Economic Minister's Meeting in 2010. The chapter shows that in addition to bringing transparency and accountability to government budgeting, well-functioning financial management information systems also enable PICs to link policy and planning with budgeting for more effective government. However, the very significant ICT investment and capacity building that such systems require, and effective implementation of the public finance reforms is itself dependent on appropriate legal and policy frameworks and capacity building. The chapter notes that development agencies, and in particular the IMF's regional office, the Pacific Financial Technical Assistance Centre (PFTAC), should ensure that external advice and support takes into account local contexts and needs, and that reform is led from within each country. The chapter argues that each government has specific information needs, and its own 'information ecology' and that there is no 'one size fits all' framework for public finance reforms.

In Chap. 7 the potential of e-government applications and basic ICT to contribute to the justice and parliamentary sectors is considered, noting that good functioning in both sectors is essential to the rule of law. The chapter outlines how with the support of regional organizations and development partners PICs are using ICTs to improve their electoral and parliamentary processes and provide citizens with information about parliament and the legislative process. It is suggested that while full implementation of the technologies used by their industrialized Pacific neighbors is beyond their resources, in most PICs the benefits of new ICT applications, and better access to information about parliament, legislation, the electoral process and the justice system is evident. However, this development is still piecemeal, and ongoing progress is very dependent on considerable regional collaboration and knowledge sharing, and on budgets being stretched to include the appropriate maintenance of ICT systems and ICT support. The chapter reiterates the contention made in Chap. 1 that whereas there is considerable interest from global agencies in policy development and e-government implementation in core e-government areas such as public finance management and border control, there is a great deal less international e-government investment is some other areas, and suggests that the motivations of some aid agencies in supporting financial reform may relate to broader agendas as well e-government for development.

In Chap. 8 the considerable role played by regional collaboration in assisting PICs to apply ICTs to the collection, management and utilization of statistics is highlighted. This is a formidable task for most Pacific Islands countries due to lack of resources, capacity and their dispersed rural populations. The chapter notes the critical role that statistics play in development and in monitoring the former MDGs and current SDGs and the potential of ICT solutions to significantly improve the collection of data and the dissemination and analysis of statistics in forms that will ensure its value is maximized. It also notes that while PICs face an increased demand for statistics from the United Nations, its agencies and the World Health Organization, regional collaboration is empowering PICs to focus on statistics that reflect the needs of individual countries, and the value of non-market activity. The *Ten Year Pacific Statistics, 2011–2017* (TYPSS) adopted by PICs in 2009 is seen as an exemplar of regional policy that reflects local ownership and contextualized e-government policy offering a sustainable solution to the challenges faced by PICs in the domain of official statistics.

Chapter 9, on the primary sector in PICs, also highlights the value of regional cooperation in sustainable development. This is to some extent due to the nature of the sector; most issues are common to all countries in the region, but more importantly, the tropical Pacific Ocean is a large ecosystem across which species and pathogens migrate. While at the country level there are few major initiatives led by government agencies (with the exception of land title databases), the inter-government regional agencies with responsibility for research and development in the sector, primarily the Pacific Community (SPC) which is responsible for agriculture, fisheries, forestry and the geosciences, and the Secretariat of the Pacific Regional Environment Programme (SPREP), gather and disseminate crucial information across all parts of the sector, and offer training and online resources and support across the region. In addition, the SPC helps link PICs into global programs (such as the REDD emissions trading scheme) and both agencies have a strong focus on policy development, driving policy processes and sharing policies amongst PICs. The chapter concludes with an account of the world-class PNA fisheries information management system (FIMS) which uses the Forum Fisheries Agency's Vessel Day Scheme for fisheries management. Since all PICs are members of these institutions, it can be said that regional e-government, driven by urgent local needs although funded by global development partners, is thriving in the sector.

Chapter 10 highlights the fact that PICs are among the countries most vulnerable to climate change and to disaster risks, and notes that the emerging concept of 'risk governance' has positioned the management of climate change and disaster risk at the heart of development. It focuses on two regional policy drivers for climate change mitigation: the *2005–2015 Pacific Islands Framework for Climate Change Adaptation* (PIFACC), and the *2005–2015 Pacific Disaster Risk Management Framework for Action* and anticipates transition to an integrated framework for climate change and disaster risk management in the Pacific Islands region: the *Strategy for Resilient Development in the Pacific: An Integrated Approach to Climate Change and Disaster Risk Management* (SRDP) to commence in 2017. The chapter outlines the way in which these programs, which are related to PIC commitments to the United Nations Framework Convention on Climate Change (UNFCCC), use ICTs in early warning systems, adaptation, mitigation, and disaster response and recovery initiatives. At a national level the challenge for individual PICs has been to establish policies and programs that make effective use of the ICT capabilities available to them. The chapter discusses lessons learned from selected PICs on the role of e-government in implementing climate adaptation and disaster risk management.

Chapter 11 notes that the use of ICT in the delivery of healthcare, or e-health has seen limited investment by international development partners, and that the focus of individual countries' health strategies is usually on improving the quality of health data through better Health Information Systems (HIS) to inform policy in the sector. The primary e-government applications used in the sector therefore are central financial and HR systems to manage the large work force, and HIS systems which are required to collect data on 200 indicators for the WHO. Training and tools for analysis of health data are provided by the WHO's Western Pacific Regional Office and the SPC's Statistics Division, to promote the use of health data for policy-making. The SPC's Public Health Division leads the way with regional systems for the collection and dissemination of critical syndromic-surveillance data (used to predict outbreaks of infectious or vector-borne diseases), and supports individual countries to manage their own hospital-based syndromic surveillance systems. Limited use is made of health IT applications routinely used in industrialized countries, such as patient management systems and clinical decision support, due to the lack of financial resources, skilled personnel, and technical infrastructure (only three countries currently use any kind of country-wide patient management system). The chapter notes that some specific (institution-to-institution) telehealth systems have been put in place on an ad hoc basis by committed individuals, and some innovations make use of mobile technology, (some of which have been evaluated—a rare phenomenon in the region), but these are not driven by government. The chapter recommends that e-health applications such as this will need more regional support if they are to be as effective as public health systems are in the region.

In Chap. 12 the education sector is characterized as another where there is little direct e-government investment by development partners and the same pattern of ad hoc initiatives across PICs is identified. As in the health sector, the primary e-government investment in education is in centralized financial administration and HR systems, with consequent improvements in staff management and remuneration in remote areas. However, the chapter reports that many PICs include education in their national ICT plan, highlighting the need to develop ICT and include it in the curriculum in order to support ICT for development. Fewer PICs focus on ICT systems to support the delivery of education, although the regional university, the University of the South Pacific, uses ICT extensively to deliver its programs across the region. Most countries are developing Education Management Information Systems (EMIS), which include data on schools, staffing levels, and school rolls and in some cases attendance data and student achievement. The work of the SPC on a generic Pacific EMIS is noted. The chapter notes that some countries are starting to make use of online learning but this is not systematized, and not well funded. Regional support and more research into the effectiveness of such initiatives would be of value. The chapter concludes that although education is critical to development this is not an area where e-government is reaching its potential to transform practice and contribute to development.

Chapter 13 focuses on the role of civil society in relation to the 'transformative' power of e-government and ICT for development models which anticipate the use of ICT to foster e-participation and e-democracy. The chapter outlines the importance of civil society in emerging democracies and the value of access to government information to allow citizens to hold government to account. It discusses the increasing use of social media to give citizens a 'voice' in PICs but argues that any changes in state-civil society discourse will not occur easily in the Pacific Islands context, where notions of citizen engagement call into question prevailing state-society power relations and where ideas of open government, freedom of information and public discussion of government activities, on or offline, tend to be viewed as challenges to government. The chapter argues that civil society organizations will have to work hard to position themselves as collaborators in public governance through educating the public and leaders alike as to the potential benefits of new patterns of state/civil society discourse that take advantage of new models of e-communication in PICs.

14.3 Identifying Factors Leading to Success in E-Government in PICs

In the examples of e-government innovations above, and in preceding chapters, it is possible to identify some common factors that contribute to their success. Some are related to policy and some relate to the more pragmatic criteria for sustainability that we identified in Chap. 1: resources, capacity (and capacity building), critical mass, leadership and the clear contribution or value to each PIC of the project.

14.3.1 Policy Processes Contributing to E-Government Success

In the domain of policy, two factors stand out. The first is that a key success factor that has been identified is the participation of PICs in the regional and international policy process, and the recognition of their standing as partners in development and in making policy for SIDs in the region, rather than as passive recipients. This emerges strongly from the analysis in Chap. 2, and has considerable impact on the regional policy goals and frameworks that Chap. 3 reports, and on the confidence of leaders in pursuing these goals. Leadership (and ownership) of policy thus emerges as a key criterion within the policy domain, as it does at the level of program implementation.

The second issue related to policy is the need for e-government policy and external funding for innovations to be translated into ongoing budget allocation (a key element in the criteria for sustainability outlined in Sect. 14.4).

Budget allocation is one area where many e-government projects falter in PICs; examples across the region (and in preceding chapters) show that as funding for a development project starts to tail off at the end of its life, governments are not yet anticipating and budgeting for the demands of an ongoing ICT implementation—the renewal of software licenses, maintenance and timely replacement of hardware, and capacity building for system maintenance and use. Provision for ongoing funding for an e-government innovation should be considered at the start of a project by both partners—government and the donor. But this, again, requires a level of political and policy discipline rarely seen in larger wealthier countries. It is even harder to achieve in the shorter timeframes of policy-making that tend to be the norm in Pacific Island countries (Roberts Aiafi 2016) (see also Sect. 14.6 below).

Regional policy, regional intergovernmental organizations (such as PFTAC, and SPC), and regional information systems also contribute to e-government success or failure at national levels. Where support of this kind is strong, each PIC is helped to identify its own needs and implement systems to respond. Strong regional policy 'owned' by all contributing PICs, as in the area of financial management (see Chap. 6), or technical support, as seen in the domains of official statistics, agriculture, disaster management and health data (Chaps. 8–11) can make a significant difference to the success of projects, e.g. the GIS systems put in place by SPC's Geoscience Division, or its Syndromic Surveillance System which has improved the quality of syndromic surveillance in individual countries. The SPC's development of a Pacific EMIS has similar potential. These projects can confidently argue the benefits they bring, and it is such benefits, financial benefits in particular, that fund their maintenance and the capacity building that sustains them, leading to what might be called a 'virtual virtuous circle'. And as examples in the primary industries sector show (Chap. 9), in areas where PICs work together within an information ecology that reflects the biological and environmental ecology they are a part of, they can make very effective use of ICT and e-government solutions.

One final factor also needs to be considered. The example in Chap. 7 of the support given to parliamentary staff in PICs by staff of parliaments in Australia and New Zealand, and in Chap. 8 of staff from Statistics New Zealand sharing their expertise with colleagues in PICs, demonstrate the benefits of development partners working alongside colleagues in government agencies in PICs.[3] Working alongside results in two-way knowledge sharing, and ownership of shared problems can be a key element in sustainability. The level of understanding that grows from this mutual understanding contributes to the essential attributes of sustainability analyzed in Sect. 14.4.

---

[3]Note from the authors: In the fieldwork on which the country reports (Cullen and Hassall 2016a, b, c, d) that have contributed to this volume are based, we encountered some outstanding external consultants, working in partnership and with evident respect for their Pacific colleagues, deeply committed to increasing capacity in the agencies in which they worked.

14.3.2 Where Regional Policy Processes Could Be Strengthened

With the constraints on resources that exist in PICs there are areas where regional collaboration and support could be strengthened and help fill gaps. Chapter 4 pinpoints telecommunications regulation and computer emergency response (to cyber threats) as issues of regional concern to which a stronger regional response would be appropriate, and Chap. 11 highlights e-health and telehealth as areas where regional policy and initiatives would be of value; regional support is especially valuable where there appears to be little development funding within countries.

In addition, while policy dialogue at the regional level identifies needs and assists PICs by coordinating capacity building and development assistance, translating regional policy into national action is highly dependent on local decision-making and leadership. In some of the examples above, this barrier is overcome by strong collaboration at the regional level and possibly the personal commitment of individuals (as in the case with official statistics). But in many cases policy does not translate into national action, calling into question whether the most urgent issues were identified, whether participation in decision-making was inclusive enough, or whether the process for translating policy into national action was clearly enough articulated. Sometimes issues simply fail to make it onto the regional agenda. As Chaps. 5 and 11 note, more benefit could be gained from the use of mobile technology in PICs; more government involvement and leadership, and better consultation with citizens through Web 2.0 might see some real innovations in health education, agriculture and other sectors.

Even in the area of financial reforms, and the implementation of financial management information systems and HR systems, regional and global support needs to takes into account local contexts and needs, (as Chap. 6 argues), and that reform must be led from within each country. In the justice and parliamentary sectors, support from Australia and New Zealand (both members of the Pacific Islands Forum and the SPC) has been significant, but regional support led by PICs could be stronger. Although the leaders of the Pacific Islands Forum (PIF) have made a strong commitment to good governance through the Framework for Pacific Regionalism, and the Forum Secretariat has many initiatives to support good governance throughout the region, including providing independent observers to report on member countries' elections, the Forum could clearly could play a stronger role in these two sectors, which are essential to good governance and the rule of law.

14.4 Sustainability

Another consequence of the scarce resources available in PICs to support e-government initiatives, and their heavy reliance on development partners to finance such initiatives, is the need to avoid waste and the expenditure of scarce resources on projects that are not sustainable. In Chap. 1 we outlined the criteria we expected to emerge as some of the main factors determining the sustainability of e-government initiatives. These were:

- adequate resources to continue innovation in the future including staffing commitments and budgetary allocations for hardware and software (from either development partners or the national government);
- attention to capacity building during implementation of a project and ongoing commitment to staff development in annual budgets;
- a critical mass of developers and users who find sufficient value in a project, and such changes in business process, outputs and outcomes that its continuation is assured (this does not preclude investment in more efficient or updated software);
- strong local leadership, and ownership; and
- a clear contribution to national or local well-being.

These have been clearly demonstrated in the preceding chapters and, furthermore, can be seen to be strongly linked to the success factors noted above. The need for adequate and ongoing resources, for example, is closely related to the link between policy and planning. So, as noted, at the planning stage, development partners and PICs need to consider how ongoing funding for the project will be managed. This means that where e-government projects can demonstrate that they increase income, or make savings through efficiency (either for the agency itself, business or citizens), the increased resource will be directed into ongoing funding for the project. The value of effective ICT-based border control and customs systems is one very clear example (Chap. 6) of this kind of self-sustaining program.

Capacity building is another of the criteria identified, sometimes considered the most important of all. But the need for capacity can be oversimplified and represented simply as IT skills—the ability to maintain equipment, install systems and keep them running. Strong information systems and project management skills are rarer (as they are in most developed countries, where expensive e-government failures are still routinely reported.) But capacity is a multi-faceted criterion, so for effective e-government implementation the need for capacity starts at the very top. It is government ministers who attend Pacific Islands Forums, often prime ministers. So while heads of ministries and senior officials must have a thorough understanding of existing processes, the changes a new e-government system will require, and the benefits it will bring, political leaders must also have a good grasp of these issues. Many PICs, with their frequent changes of government, ministers and ministry staff, find it hard to maintain continuity of policy—and leadership of e-government initiatives suffers as a consequence. At lower organizational levels, some PICs struggle to maintain a base of capable people broad enough to fill roles when someone leaves, or sufficient capacity to ensure an adequate handover period. All these issues contribute to a lack of sustainability of e-government projects.

Two of the sustainability criteria listed above relate to the value of a project, in terms of the improvement it brings to business process, its outcomes, and it contribution to well-being at national or local level. (Several small initiatives in the health sector come within this latter category). Larger projects which clearly demonstrate their value include the FIMS (Parties to the Nauru Agreement) and the border control systems as noted above. When benefits to the community from e-government projects align with increased income (or lower costs) to government and individuals, and also with sustainable development goals, the 'virtual virtuous circle' is evident.

Turning to leadership, it is broadly accepted in the e-government literature, as well as the more extensive information systems literature, that successful ICT projects are dependent on good project management and leadership. In the case of e-government, political as well as organizational leadership is critical; champions must be found as high as possible in the government, and this leadership must be reflected with the same passion and commitment at the top of the organization responsible for the project.

Leadership, along with ongoing budget allocation, is also a key point stressed in Lee's seven 'success factors' (Lee 2014, p. 16), which lead to higher levels of social and economic development (but not sustainable e-government per se). Lee's other factors—the establishment of legal frameworks for ICT and e-government, a coordinating body responsible for e-government, sufficient budget, a national e-government policy, creation of awareness in user groups, and necessary government reform—are clearly desirable and may well be critical in moving PICs further up the e-government rankings, which are based on the indicators from which these factors are derived. But in the early stages of e-government development, which is where PICs find themselves, individual projects are not seen to be dependent on all these. Some PICs, and some regional organizations have made considerable progress on projects which have shown real benefits without having to make the wholesale commitment to e-government that the UN Survey Indicators imply.

14.5 Applying the Theoretical Frameworks Used in This Study

14.5.1 ICT4D 2.0

As noted above, Heeks' ICT4D 2.0 Manifesto and his E-government for Development frameworks have informed some of the analysis in this study. Particularly helpful has been Heeks' rejection of what he describes as the 'techno-centric' and 'invention down' approach of some ICT4D research in favor of a bottom up approach which places emphasis on the technology that is actually used (e.g. mobiles, radio and television), innovative ways of applying technology and innovative business models and scaling up existing applications that work. While this approach aligns well with our focus on the information ecology of e-government innovations, it also reflects the pragmatic approach to technology that is usually adopted in PICs, where the reach of government is smaller, applications that can be sustained on existing technology and operating systems are preferred, and where it is better to depend on hardware and software that can be maintained locally. And although we have not focused very much on radio and television as e-government technologies in this study, we have acknowledged in several places that they are still the most effective way for Pacific Island governments to reach their people (e.g. for disaster warnings and public health messages.)

Moreover, mobile technology, which has seen a dramatic increase in uptake recently, and which has given a tremendous boost to Internet use in PICs in a way it has not in countries which have experienced high levels of Internet use for over a decade, has been a major factor in several successful e-government initiatives (e.g. in disseminating information about imminent cyclones and tsunamis, in disseminating essential information during elections, or in communication with farmers concerning markets, prices and shipping schedules). In Chaps. 8 and 11 we noted that the transformational power of mobile technology in enabling the registration of births in remote areas has helped PICs enact their commitment to the Pacific Civil Registration and Vital Statistics (CRVS) program as well as their commitment to the United Nations Convention on the Rights of the Child. All these examples are ways of exploiting the burgeoning use in PICs of mobile technology that is happening independently of government while taking advantage of the competitive telecommunications markets that PICs government have put in place.

Heeks' 'design-reality gap' construct, which applies an information systems perspective, has been similarly helpful. The model, which argues that many e-government (and basic IT) projects fail because the design does not reflect the reality of the context of implementation on any one of the seven dimensions that Heeks identifies: information; technology; processes; objectives and values; staffing and skills; management systems and structures; and other resources. As Heeks suggests, 'one size fits all' approach to e-government projects can result in a business process and application that is built for a different context and may contain assumptions about how the organization functions and what its resources (i.e. financial and human capacity) are that are simply inaccurate. Heeks also suggests that for individual projects, further investigation can be made into the drivers/enablers (such as *strong external pressure* and *internal political desire*); these factors relate to those we have identified in several of the preceding chapters, in particular in Chaps. 2 and 3, which throw light on global and regional policy processes; the same factors can also be identified in the demands of powerful agencies such as the World Bank (Chap. 6). Heeks' model emphasizes the need to align these with national goals and priorities, and the importance of political championship to ensure success. The constraints Heeks identifies (*poor change management, lack of overall vision and strategy, dominance of self-interest*) can be seen in examples throughout the volume.

14.5.2 Information Ecology

These factors are part of what Heeks describes as an ICT4D 2.0 ecosystem, which he defines as the combination of organizations, policies, culture, technologies and resources which must all come together for ICT to have a transformational effect. As we have seen in many examples in preceding chapters, this does not necessarily mean that the organization which is the agent of change must be a government agency—it can also be an inter-government agency, an NGO or even a civil society organization (CSO). In the Pacific Islands region, intergovernmental organizations such as the Pacific Community (SPC) have played the role of change agent many times and can take credit for some significant e-government applications, which have been of real value in government decision-making. A good example is the Pacific Risk Information System (PacRIS) developed by SOPAC (now the SPC Geoscience Division), the SPC's Pacific Regional Data Repository for Sustainable Energy for All (http://prdrse4all.spc.int) or the Pacific Regional Environment Programme's knowledge bases and online networks (such as the Regional Technical Support Mechanism (RTSM) at: http://rtsm.pacificclimatechange.net). These may look a little different from what are normally thought of as e-government applications in industrialized countries but they provide data and expert knowledge for policy and planning in all PICs and are no less transformative.

In Chap. 1 we identified Heeks 'eco-sytem' as being closely related to the information ecology construct (Bekkers and Homberg 2005; Davenport and Prusack 1997). Applying this construct to e-government in the Pacific islands has been very helpful. It allows us to unpack and take into account complex patterns of governance (not just traditional Pacific forms working alongside a more recently adopted parliamentary system and civil service as outlined in Chap. 1), but also the public governance model (and reform agenda) underpinning interventions of the United Nations and its agencies, as well as the World Bank and International Monetary Fund. These institutions have only in the past decade recognized that financial reforms, for example, can proceed only at a pace that a small developing country can handle (see Chap. 6); and they still persist in recommending decentralization despite the clear benefits for most PICs of a centralized financial and HR system.

Understanding the different agendas, values, and preferred management styles of a large number of stakeholders, all of whom have legitimate interests in policy or project development, or both, helps avoid a focus on ICT applications, and especially on Internet-based applications separately from the business process they are intended to implement and the policy framework that the business process is intended to address. Redesigning business process, as Bekkers and Homberg note, is a political process that changes roles and relationships. "The introduction of ICT in public administration is a social intervention in a policy network, which influences the position, interests, values and (information) domains of the actors involved… it is not a neutral intervention but a political one" (Bekkers and Homberg 2005, p. 9). The impact of e-government on each country's and each agency's information ecology is evidenced in many chapters in this volume, and nowhere more than in Chap. 7 which examines ICT and parliamentary process.

14.5.3 Complexity Theory

These two related sets of theory have helped legitimize the cultural, policy and technology context of e-government in Pacific SIDs, and enabled us to view the constraints that PICs face when introducing e-government innovations as no different from constraints faced in the implementation of any information system. They are simply part of what is a multi-faceted and complex process. This a view shared by Everest-Phillips who argues that the issues faced by the public services of SIDs due to their 'smallness' or 'islandness' are "no different, or different but no less complex than in other contexts" (Everest-Phillips 2014, p4). Moreover, he argues, the solutions proposed must reflect complex adaptive thinking that works "by making small changes, observing the results, and then adjusting" (Ramalingam 2013, cited by Everest-Phillips 2014). While this might appear contrary to the traditional approach to large development projects of planning and tracking milestones, he claims it enables development partners to avoid normative solutions in the face of accelerating change and complexity. More radically, it is suggested, it might even be necessary to abandon the neo-Weberian civil service principles of de-politicization, neutrality, professionalism, continuity and anonymity, which may actually be impossible in the small interconnected societies of SIDs where "formal institutions co-exist with informal networks and… public and private realms of administrative behavior remain intertwined" (p. 17).

This approach to thinking in the public administration/development sector recognizes that the challenges PICs and other SIDS face in terms of the 'diseconomies of scale', 'patronage politics', top heavy political systems, capacity constraints and blurred lines between politics and policy might be better served by a more agile, responsive, individualistic policy response and a unique form of Pacific leadership. We conclude by summarizing what has been learned in the preceding chapters about the policy processes in PICs and whether they can rise to this challenge.

14.6 Applying a Policy Framework Analysis

Contemporary thinking about effective policy planning and implementation focuses on a "policy cycle" approach which, while not intended to be prescriptive, does expect a degree of sequencing through the stages of agenda setting, policy formulation, policy implementation, and policy evaluation. Setting the agenda for e-government policy requires extensive consultation with stake-holders within government and beyond it, on such matters as choice of Internet "backbone" (such as cable, satellite, and Wi-Fi), computer hardware and software, and to identify the information needs for which these technologies are the proposed solution. The government then develops policy options that determine, for example, whether ICT investments will be owned by government, by the private sector, or by a "public-private partnership". Necessary legal reforms will also be scoped, drafted, and forwarded to the legislature. The third policy phase, implementation, the most visible of all and possibly the most challenging, involves the installation of infrastructure and the deployment of software. The effectiveness of this implementation is subsequently evaluated periodically to assess the benefits and provide a basis for necessary adjustments.

Ideally, these policy processes should apply across government departments and institutions. Following an agreed model of policy development creates a sense of direction, purpose and shared timeframes, and allows for training, development and deployment, as well as ongoing budgetary allocation. In practice, however, not all ICT policy process in PICs has followed such a pathway. Draft policies may be shelved while specific projects offered by development partners are installed; hardware may be procured before stakeholder needs have been assessed; and individual agencies may cooperate with project roll-out with varying levels of intensity and commitment. The final phase in the policy cycle, monitoring and evaluation is sometimes omitted, or else undertaken with little public awareness.

Moreover, as noted above, the timeframes of policy-making tend to be shorter in Pacific Island countries (Roberts Aiafi 2016) and, as we have noted throughout the volume, are often driven by international agendas that do not take local needs and context sufficiently into account. In addition, the processes for aligning policy with budgetary allocation is weak. The findings in this study confirm the findings of Roberts Aiafi (pp. 1–11) that policy processes in PICs have tended to be top down, shaped by political and external interests, lack connectedness to society and social needs, and may be driven by ideology or incompatible social constructions. This approach to development is not uncommon. However, there are encouraging signs in many examples presented in the volume that this is changing, and evidence that when e-government solutions are built from the bottom up, meet needs, have clear leadership and value, they are financially and culturally sustainable. Policy processes in PICs, and all stakeholders be they Pacific governments or development partners, therefore, need to ensure that policy reflects "participation, partnership, ownership, understanding and learning" (Roberts Aiafi 2016, p. ii). This might indeed look more like the agile, responsive, individualistic policy process and unique Pacific leadership advocated above, adjusting what works for development partners to what works in the Pacific.

14.7 Conclusion

The major prerequisites for e-government in Pacific island countries have been put in place in recent decades: connection to the Internet and mobile technologies; essential legislation; regulatory institutions; and networked information systems for such core activities as payroll, tax, customs, and civil registry. The challenge in the years immediately ahead will be to extend these capabilities to other government departments, to extend the range of services that these departments deliver to the public and to establish the sustainability of this new level of government operations. Associated challenges include: the creation of a more agile and 'indigenous' policy process (Roberts Aiafi 2016) that helps to balance public policy process and traditional forms of governance; setting priorities based on short and long term needs; identifying solutions that maximize the use of available and affordable technologies, especially those that enhance capacity and generate economic growth to ensure their ongoing viability; selecting innovations that are both sustainable and bring maximum value to the local economy and social well-being; making more government information "open" to the public and establishing ways of receiving and dealing with feedback in order to develop policy discourse as part of local culture and local politics; and finally, learning from past successes and failures, to ensure scarce resources are directed in the most effective way.

Meeting these challenges requires strong political and technical leadership and long-term commitment at national and regional level. Pacific island countries need to take a government-wide approach, with central political leadership (such as through the office of the prime minister) and a centralized approach to e-government implementation. We see a National ICT/e-government committee as essential to provide the necessary policy expertise and technical knowledge to ensure the involvement of all departments and agencies and successful implementation of e-government in all sectors. This is also essential for knowledge sharing, both nationally and regionally, building on successful initiatives and learning from those which are less successful. Regional organizations (e.g. PIF, PFTAC, SPC and others) need to focus attention on how best to translate well-conceived regional policies into national action plans that reflect local realities, and which can in turn feed implementation experience back into regional policies. Regional organizations and international development partners need to continue evolving their recent shift towards in-country program leadership, and participate with PICs in developing a more agile and pragmatic policy process that reflects identified needs, behaviors and values, and mandates the link between policy, planning and budgeting, to ensure new e-government initiatives have ongoing funding to support hardware, software and the necessary skills and training.

Many examples in this volume help identify these factors; there are many others where they can be shown to be lacking, and where the outcomes are less positive. Reflecting local context and local values, ensuring sustainability, and seeking out and nurturing strong leadership will result in e-government becoming a powerful force for economic and social development in Pacific island countries. The challenges they face in maximizing the benefits of e-government may seem different, largely because of the scarcity of resources and capacity, but they are no less complex than in most industrialised countries, and the benefits may be even more transformational. Heeding the lessons from this study may not only help other countries in their development journey, but also development partners seeking to maximise the use of ICTs to make government in the Pacific Island states more efficient, inclusive and accountable and enhance good governance.

References

Bekkers V, Homberg V (2005) The information ecology of e-government: e-government as institutional and technological innovation in public administration. IOS Press, Amsterdam

Bridgman P, Davis G (1998) Aust Policy Handb. Allen & Unwin, Sydney

Bridgman P, Davis G (2003) What use is a policy cycle? Plenty, if the aim is clear. Aust J Public Adm 62(3):98–102

Cullen R, Hassall G (2016a) E-Government in the Pacific Island states: ICT policy and implementation in small island developing states: Cook Islands country report. Victoria University of Wellington, Wellington. http://pippr.victoria.ac.nz/handle/123456789/33

Cullen R, Chan Mow I, Hassall G (2106b) E-Government in the Pacific Island states: ICT policy and implementation in small island developing states: Samoa country report. Victoria University of Wellington, Wellington. http://pippr.victoria.ac.nz/handle/123456789/31

Cullen R, Hassall G (2106c) E-Government in the Pacific Island states: ICT policy and implementation in small island developing states: Vanuatu country report. Victoria University of Wellington, Wellington. http://pippr.victoria.ac.nz/handle/123456789/30

Cullen R, Hassall G (2106d) E-Government in the Pacific Island states: ICT policy and implementation in small island developing states: Solomon Islands country report. Victoria University of Wellington, Wellington. http://pippr.victoria.ac.nz/handle/123456789/35. Accessed 17 Mar 2016

Davenport TH, Prusak L (1997) Information ecology. Oxford University Press, Oxford

E-government benefits study (2003) Canberra: national office for the information economy. Available at: https://www.finance.gov.au/agimo-archive/__data/assets/file/0012/16032/benefits.pdf. Accessed 1 Oct 2016

Everest-Phillips M (2014) Small, so simple? Complexity in small island developing states. Singapore: UNDP Global Centre for Public Service Excellence. Available at: http://www.undp.org/content/undp/en/home/librarypage/capacity-building/global-centre-for-public-service-excellence/complexity-small-islands.html. Downloaded 1 Oct 2016

Gupta MP, Jana D (2003) E-government evaluation: a framework and case study. Gov Inf Q 20 (4):365–387

Heeks R (2009) The ICT4D 2.0 manifesto: where next for ICTs and international development? Development Informatics Working paper series. Manchester: Development Informatics Group, Institute for Development Policy and Management

Kertesz S (2003) Cost-benefit analysis of e-government investments. Cambridge, Mass: Kennedy School of Government Harvard University. http://www.edemocratie.ro/publicatii/Cost-Benefit.pdf. Accessed 1 Oct 2016

Lee YB (2014) E-government for Promoting Sustainable Development in Small Island Developing States (SIDS). Background Paper for the Special Event of the 69th Session of the United Nations General Assembly Second Committee. New York: United Nations. Available at: http://workspace.unpan.org/sites/Internet/Documents/UNPAN93784.pdf. Accessed 1 Oct 2016

Roberts Aiafi P (2016) Public policy processes in the Pacific islands: A study of policy initiation, formulation and implementation in Vanuatu, the Solomon Islands, Samoa and regional inter-governmental organisations. PhD thesis, Victoria University of Wellington, 2016

Correction to: Mobile Technology in Pacific Island Countries: the Potential for M-Government

Amanda H. A. Watson, Joseph Kim Suwamaru, Ioana Chan Mow and Sarah Logan

Correction to: Chapter 5 in: R. Cullen and G. Hassall (eds.), *Achieving Sustainable E-Government in Pacific Island States*, Public Administration and Information Technology 27, https://doi.org/10.1007/978-3-319-50972-3_5

The following corrections were incorporated in the chapter "**Mobile Technology in Pacific Island Countries: the Potential for M-Government**":

The authors' affiliation "**Amanda Watson, Department of Government and International Relations, University of Sydney, Sydney, Australia** and **Sarah Logan, Australian National University, Canberra, Australia**" has been corrected to "***Amanda H. A. Watson, Australian National University, Canberra, Australia*** and ***Sarah Logan, University of New South Wales, Sydney, Australia***" in this chapter.

The erratum book had been updated with these changes.

The updated original online version for this chapter can be found at https://doi.org/10.1007/978-3-319-50972-3_5

R. Cullen and G. Hassall (eds.), *Achieving Sustainable E-Government in Pacific Island States*, Public Administration and Information Technology 27,
https://doi.org/10.1007/978-3-319-50972-3_15

Index

R. Cullen and G. Hassall (eds.), *Achieving Sustainable E-Government in Pacific Island States*, Public Administration and Information Technology 27,
DOI 10.1007/978-3-319-50972-3

D

FSC
www.fsc.org
MIX
Papier | Fördert gute Waldnutzung
FSC® C083411

Kolibri 360 GmbH
Ferdinand-Jühlke-Straße 7
99095 Erfurt, Deutschland
produktsicherheit@kolibri360.de